Lecture Notes in Computer Science 5730

Commenced Publication in 1973
Founding and Former Series Editors:
Gerhard Goos, Juris Hartmanis, and Jan van Leeuwen

Marina L. Gavrilova   C.J. Kenneth Tan (Eds.)

Transactions on Computational Science VI

 Springer

Editors-in-Chief

Marina L. Gavrilova
University of Calgary, Department of Computer Science
2500 University Drive N.W., Calgary, AB, T2N 1N4, Canada
E-mail: mgavrilo@ucalgary.ca

C.J. Kenneth Tan
Exascala Ltd.
Unit 9, 97 Rickman Drive, Birmingham B15 2AL, UK
E-mail: cjtan@exascala.com

Library of Congress Control Number: 2009940672

CR Subject Classification (1998): H.2.8, J.2, G.1.2, H.3.3, I.5.3, H.5, I.2.1

ISSN   0302-9743 (Lecture Notes in Computer Science)
ISSN   1866-4733 (Transaction on Computational Science)
ISBN-10   3-642-10648-X Springer Berlin Heidelberg New York
ISBN-13   978-3-642-10648-4 Springer Berlin Heidelberg New York

springer.com

© Springer-Verlag Berlin Heidelberg 2009
Printed in Germany

Typesetting: Camera-ready by author, data conversion by Scientific Publishing Services, Chennai, India
Printed on acid-free paper   SPIN: 12803156   06/3180   5 4 3 2 1 0

LNCS Transactions on Computational Science

Computational science, an emerging and increasingly vital field, is now widely recognized as an integral part of scientific and technical investigations, affecting researchers and practitioners in areas ranging from aerospace and automotive research to biochemistry, electronics, geosciences, mathematics, and physics. Computer systems research and the exploitation of applied research naturally complement each other. The increased complexity of many challenges in computational science demands the use of supercomputing, parallel processing, sophisticated algorithms, and advanced system software and architecture. It is therefore invaluable to have input by systems research experts in applied computational science research.

Transactions on Computational Science focuses on original high-quality research in the realm of computational science in parallel and distributed environments, also encompassing the underlying theoretical foundations and the applications of large-scale computation. The journal offers practitioners and researchers the opportunity to share computational techniques and solutions in this area, to identify new issues, and to shape future directions for research, and it enables industrial users to apply leading-edge, large-scale, high-performance computational methods.

In addition to addressing various research and application issues, the journal aims to present material that is validated – crucial to the application and advancement of the research conducted in academic and industrial settings. In this spirit, the journal focuses on publications that present results and computational techniques that are verifiable.

Scope

The scope of the journal includes, but is not limited to, the following computational methods and applications:

- Aeronautics and Aerospace
- Astrophysics
- Bioinformatics
- Climate and Weather Modeling
- Communication and Data Networks
- Compilers and Operating Systems
- Computer Graphics
- Computational Biology
- Computational Chemistry
- Computational Finance and Econometrics
- Computational Fluid Dynamics
- Computational Geometry

- Computational Number Theory
- Computational Physics
- Data Storage and Information Retrieval
- Data Mining and Data Warehousing
- Grid Computing
- Hardware/Software Co-design
- High-Energy Physics
- High-Performance Computing
- Numerical and Scientific Computing
- Parallel and Distributed Computing
- Reconfigurable Hardware
- Scientific Visualization
- Supercomputing
- System-on-Chip Design and Engineering

Editorial

The Transactions on Computational Science journal is part of the Springer series *Lecture Notes in Computer Science*, and is devoted to the gamut of computational science issues, from theoretical aspects to application-dependent studies and the validation of emerging technologies.

The journal focuses on original high-quality research in the realm of computational science in parallel and distributed environments, encompassing the facilitating theoretical foundations and the applications of large-scale computations and massive data processing. Practitioners and researchers share computational techniques and solutions in the area, identify new issues, and shape future directions for research, as well as enable industrial users to apply the techniques presented.

The current issue is devoted to selected best papers from the workshops comprising the International Conference on Computational Science and its Applications 2008, which took place in Perugia, Italy, from June 30th to July 3rd, 2008. The fully revised and extensively refereed versions of these papers comprise the current issue.

The 21 selected contributions originate from eight workshops: Virtual Reality in Scientific Applications and Learning (VRSAL 2008), Computational GeoInformatics (CompGeo 2008), Mobile Communications 2008 (MC 2008), Information Systems and Information Technologies (ISIT 2008), Logical, Scientific and Computational Aspects of Pulse Phenomena in Transitions (PULSES 2008), Internet Communication Security (WICS 2008), Geographical Analysis, Urban Modeling, Spatial Statistics (GEOG-AN-MOD 2008) and Computational Geometry and Applications (CGA 2008).

The contents are presented in two parts. Part 1 is entitled Information Systems and Communications and Part 2 is entitled Geographical Analysis and Geometric Modeling.

Part 1 is devoted to state-of-the-art research utilizing advanced virtual reality paradigms, computational geoinformatics methods, mobile communications, information technologies and data transmission mechanisms.

Part 2 takes an in-depth look at the selected computational science research in the areas of geographical analysis, spatial statistics and geometrical modeling. Each paper provides a detailed experimentation or a case study to amplify the impact of the contribution.

In conclusion, we would like to extend our sincere appreciation to all workshop chairs and authors for submitting their papers to this issue, and to all Associate Editors and referees for their meticulous and valuable reviews. We would also like to express our gratitude to the LNCS editorial staff of Springer, in particular Alfred Hofmann, Ursula Barth and Anna Kramer, who supported us at every stage of the project.

It is our hope that the collection of outstanding papers presented in this issue will be a valuable resource for Transactions on Computational Science readers and will stimulate further research into the vibrant area of computational science.

September 2009 Marina L. Gavrilova
C.J. Kenneth Tan

LNCS Transactions on Computational Science – Editorial Board

Table of Contents

Part 1: Information Systems and Communications

Part 2: Geographical Analysis and Geometric Modeling

Volterra – Lax-Wendroff Algorithm for Modelling Sea Surface Flow Pattern from Jason-1 Satellite Altimeter Data

Maged Marghany

Natural Tropical Resources Information & Mapping Research Group
(NATRIM Research Group), Faculty of Geoinformation Science and Engineering
Universiti Teknologi Malaysia
81310 UTM, Skudai, Johore Bahru, Malaysia
maged@utm.my, magedupm@hotmail.com

Abstract. This paper introduces a modified formula for geostrophic current. The method is based on utilization of the Volterra series expansion in the geostrophic current equation. The purpose of this method is to transform the time series JASON-1 satellite altimeter data into a real ocean surface current. Then, the Volterra kernel inversion used to acquire the sea surface current velocity. In doing so, the finite element model of Lax-Wendorff scheme used to determine the spatial variation of current flow. The results show that the new formula of geostrophic current is able to avoid the impact of Coriolis and geoid parameters. The second-order Volterra kernel illustrates an error standard deviation of 0.03, thus performing a better estimation of flow pattern as compared to first-order Volterra kernel. We conclude that modeling of sea surface current by using JASON-1 satellite altimeter data can be operationalized by using the new formula for geostrophic current.

Keywords: JASON-1 satellite altimetry data, Volterra model, Lax-Wendorff schemes, Finite element model, Sea surface current.

1 Introduction

The radar altimeter has been recognized as a powerful tool for coastal hydrodynamics modeling [15]. In this context, the radar altimeter is arguably the most useful of all the satellite for measuring ocean currents. Although satellite altimetry records are still quite short compared to the tide gauge data sets, this technique appears quite promising for sea level change problem because it provides sea level measurement with large coverage. A precision of about 1 mm/year of measurement global change can be obtained [2]. Consequently, the algorithms have been used to retrieve ocean current from radar altimeter data are based on the mechanism of radar pulse interaction with the sea surface. According to Robinson [20], a satellite altimeter is a nadir-viewing radar which emits regular pulse and records the travel time, and the shape of each return signal after reflection from the sea surface. In this circumstance, the travel time can be converted to a distance when the path delays can be estimated. Martins

M.L. Gavrilova and C.J.K. Tan (Eds.): Trans. on Comput. Sci. VI, LNCS 5730, pp. 1–18, 2009.

et al. [16] stated that the height of the sea surface can be determined by combining this distance with the position of the satellite that determined by precision orbit determination and correcting for earth and ocean tides and atmospheric loading. In this circumstance, the actual surface flow, however, cannot be represented by measuring ocean current from altimeter data. Thus, the along-track-only measurements of sea surface heights estimates only the cross-track component of the surface geosgtrophic velocity in which the sea level variations driving the flow are balanced predominantly by the Coriolis force [12]. In this context, the current at the sea surface and the horizontal pressure gradient are proportional to the sea-surface slope measured relative to the equipotential surface (that is, the geoid) [5]. In this case, Balaha and Lunda [4] concluded that a major challenge with altimeter measurements is that sea surface height must be relative to geoid, that is defined as the shape of ocean surface in the absence of all external forcing and internal motion. The sea surface height, however, is dominated by the (time-invariant-at least on oceanographic time scales) earth's geoid [10]. According to Hu et al. [11], the surface currents could be determined if the geoid was accurately known. Therefore, Glenn et al., (1991) introduced synthetic geoid method to compute accurate geoid. In this method, the surface dynamic height is measured by in situ observation in which correlated by regression model with one is measured from radar altimeter. Conversely, the limitation of this method is that the unknown barotropic currents can induce errors in the dynamic height field in which are incorporated in the geoid. Thus, Blaha and Lunde (1992) have used the logical reverse of the synthetic geoid method in which to use the altimeter height measurements to estimate the depth of thermocline; for instance for every 10 cm rise in altimeter height.

One of the techniques that is increasingly used for the ignition of the ocean models for prediction of ocean current is the assimilation of satellite altimeter data into complex ocean models. Researchers have agreed that data assimilation into a numerical model is a useful tool to utilize the altimeter data further [4, 15]. According to Gil et al. [5], the model screens the data and extracts only the components that are explained with the physics retained in the model. Researchers are divided assimilation techniques into two categories: a sequential technique and a variational method [5]. A sequential technique is dominated by potential problem of the low cross-track resolution [6]. The low cross-track resolution has been solved by using Nyquist wavelength, which is twice the cross-track interval [22]. In this context, White et al. [22] concluded that Nyquist wavelength is improved the low horizontal resolution in which is appropriate for measuring ocean current from altimeter Geosat data with the space and time decorrelation scales of ~ 400 km and $40\text{-}135°$. This method nevertheless, is not able to reconstruct mesoscale eddies because the cross-track interval is large relative to eddy size.

By contrast, the variational technique may be more appropriate for data assimilation. In fact, the less well-known components of the model are usually chosen as control variables. According to Ikeda [5], the space of control variables is often composed of initial and boundary conditions. In fact, the variational principle and optimal control have been used to derive an adjoint equation system. In this context, Moore [18] used the variational method, in which a feature model with prescribed vertical profiles was required to generate the first guess of control variables for initializing feature structures. This method, however, has a critical problem is how to acquire the minimum in the cost function. It might be the estimation of gradients of the cost function is an efficient way to solve the adjoint equations. Ikeda [5], therefore, has suggested that solving adjoint equations can be obtained by setting partial derivatives of Lagrange function to zero.

Further, Kelly and Gille [10] developed a simple based assimilation data tool that recovered the mean surface height profile from the residual height data. In their method, the instantaneous velocity profile of boundary jet is assumed to be a Gaussian function. Therefore, they used the least-squares fit to minimize the difference between synthetic height profiles and synthetic mean height profiles. Moreover, Ghil and Malanotte-Rizzoli [4] have implemented dynamical interpolation tool such as Klaman filter to acquire accurate knowledge of ocean currents from radar altimeter satellite data. At this stage, however, data assimilation of satellite altimeter is still primarily a research tool. This is because of data assimilation techniques vary from relatively simple nudging technique to more sophisticated techniques like Volttera filter and adjoint models i.e., finite difference model [15].

At present, few studies have been utilized different altimeter sensors for the sea surface investigations in the South China Sea [12, 15, 17, and 19]. Hu et al. [11] investigated the sea surface height with a period of 3-6 months using six years TOPEX/POSEIDON altimeter data. They reported that the sea surface height variations are associated strongly with current and eddy features. Further, Imawaki et al. [9] introduced a new technique for acquiring high-resolution mean surface velocity. In doing so, they combined the uses of TOPEX/POSEIDON and ERS-1/2 altimeter data and drifter data acquired from 1992 through 2001 to detect the undulation of mean-surface dynamic topography down to the oceanic current. Indeed, the ocean current is always allied with geoid by the amount of sea surface dynamic topography. Maged et al. [15] suggested the modification of the hydrodynamic equation solutions of the mean sea-surface to invert accurate pattern of sea surface current. Therefore, the altimeter data that are both more accurate and more widely distributed in time are required. Furthermore, Ikeda [5] stated that geoid errors can be substantially smaller than the corresponding values of the sea surface topography if knowledge of the long wavelength geoid is improved. The main contributions of this work is to design scheme to reduce the impacts of geoid and Coriolis parameters in continuity equation. In doing so, this study extends the previous theory of geostrophic current by implementing Volterra model and Lax-Wendrof scheme. Indeed, Malaysia is dominated by weak Coriolis parameter due to its location in low latitude zone between 2° N to 7° N. In this context, a weak geostrophic current might be occurred [15].

The main hypothesize in this work is the fact of modeling the altimeter data as series of nonlinear filter. In this context, there is significant difference between the ocean current velocities can be acquired by the inverse filter of linear and nonlinear Volttera kernels. In addition, utilization of Lax-Wendroff scheme can suppress the numerical solution of current velocity is acquired by the inverse filter of Volterra kernels.

2 Data Set

2.1 Altimeter Data and Study Area

Altimeter satellite data acquired in this study are derived from the JASON-1 satellite which is the successor to TOPEX satellite. According to Robinson [2], JASON-1 has the same orbit and ground track as the original TOPEX orbit within 60 s or 500 km behind TOPEX. Although JASON-1 is similar in design to TOPEX, because of advances in electronic miniaturization, its mass is only 500 kg compared with 2400 kg for TOPEX. Unlike TOPEX, JASON-1 carries only the POSEIDON-2 altimeter with dual-frequency k_u-and C-band (5.3 and 13.6 GHZ). In addition, JASON-1 carries the JASON Microwave Radiometer (JMR) for atmospheric correction purpose in which operates at 18.7, 23.8 and 34.0 GHz [2]. The JASON-1 altimeter data of Merged Sea Level Anomaly (MSLA) are acquired on October 2003, April 2003 and March 2005, respectively between 102° E to 114° E and 1 ° N to 10° N in the South China Sea (SCS). Fig. 1 shows the location of the South China Sea (SCS) which it is considered as an equatorial, semi-enclosed sea with a complex topography that includes large shallow regions [1]. The SCS is located between the Asian continent, Borneo, the Philippines, and Taiwan (Fig. 1). The northeastern part adjoins a deep sea basin, while the southern part is a shelf sea with depths less than 200 m. The SCS has two features that have very important and interesting effects on the general circulation: (1) the Coriolis force decreases to zero at the equator where both nonlinear and frictional forces become very important; and (2) the monsoon regime exerts a strong effect on the SCS circulation [1]. Neither of the above phenomena is studied in coastal regions of the world ocean at the present time [9].

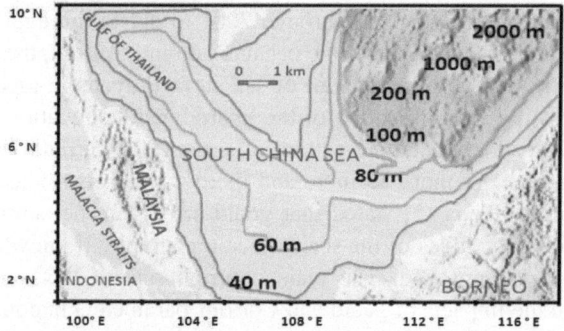

Fig. 1. Location of the South China Sea

2.2 In situ Data Collection

In situ sea surface current measurements are collected between 102° 5'E to 105° 10'E and 2° 5'N to 6° 10'N (Fig. 2). In situ measurements are conducted for three years between 2002 to 2005 along the east coast of Peninsular Malaysia. In doing so, 105 sampling locations are chosen to study of coastal oceanography of the South China Sea (Fig. 2). The field cruises are conducted on separately area by area along east coast of Peninsular Malaysia. In fact, it is difficult to cover a large scale area over than 700 km^2 in short period. In September 2002 and April 2003, in situ measurements are conducted on Terengganu coastal waters. Further, the oceanography cruises are conducted on Phang coastal waters during September 2003 and April 2004. In addition, in situ measurements are acquired in Johor coastal waters in October 2004 and January of 2005. In doing so, Valeport electro-magnetic current meter which is lowered down from the sea surface to water depth of 50 m and recorded the vertical current profile within 5 m water depth interval. The sea surface current data only have considered for this study.

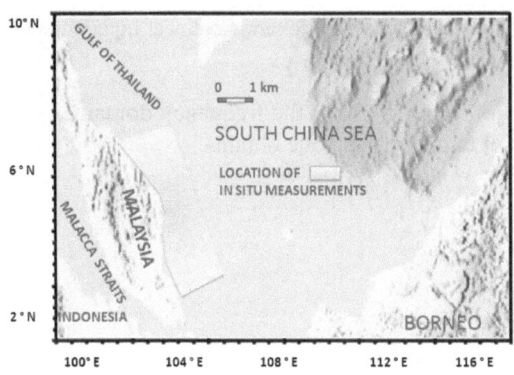

Fig. 2. Location of in situ measurements along east coast of Malaysia

3   Model

3.1   Volterra Model

The Volterra model is based on a Volterra series and also called a "power series with memory," is a function extension of linear (first-order) system to bilinear (second-order), triliner (third-order), and higher-order systems [8]. According to Liska and Wendroff [13], these extensions require knowledge of multidimensional frequency response functions $H(f,g)$ to describe the bilinear system and $H(f,g,h)$ to describe the triliner system in place of the simple one-dimensional frequency response function $H(f)$ for the linear system [21]. Let the output $Y(t)$ due to input $X(t)$ is given by a sum of Volterra functions as follows:

$$Y(t) = y_0 + y_1(t) + y_2(t) + y_3(t) + \dots\dots\dots \qquad (1)$$

where y_0 is constant, $y_1(t)$ is linear output, $y_2(t)$ is bilinear output and $y_3(t)$ is trilinear output (Fig. 3). Therefore, the mean value $\bar{Y} = E[Y(t)]$ satisfies

$$\bar{Y}(t) - \bar{Y} = [y_1(t) - \bar{y}_1] + [y_2(t) - \bar{y}_2] + [\bar{y}_3(t) + \bar{y}_3] + \dots\dots\dots \qquad (2)$$

where, $\bar{Y}(t) = y_0 + \bar{y}_1 + \bar{y}_2 + \bar{y}_3 + \dots\dots\dots$

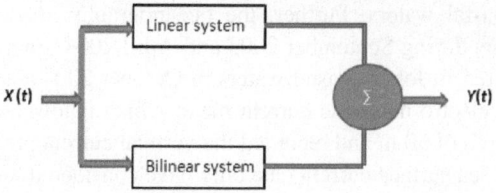

Fig. 3. Block diagram of Volterra nonlinear input/output model

Following Inglada and Garello [8], in the frequency domain, by taking Fourier transforms of both sides of equations (2), one obtains

$$Y(f) = Y_1(f) + Y_2(f) + Y_3(f) + ...$$ (3)

Where,

$$Y(f) = F[Y_1(t) - \overline{Y}] = \int_{-\infty}^{\infty} [Y(t) - \overline{Y}] e^{-j2\pi f t} dt$$ (3.1)

with

$$Y_1(f) = F[Y_1(t) - \overline{Y}_1]$$
$$Y_2(f) = F[Y_2(t) - \overline{Y}_2]$$ (4)

Maged et al. [15] expressed the geostrophic current velocity as a series of nonlinear filters by using Volterra model. This means that the Volterra model can be used to study the geostrophic current energy variations as a function of parameters such as the current direction, or the current waveform [14]. A generalized, nonparametric framework to describe the input–output u and v geostrophic current components relation of a time-invariant nonlinear system. In discrete form, the Volterra series for input, gestrophic current $\vec{u}(t)$, and output, $Y(t)$ as given by Maged and Mazlan [14] can be expressed as:

$$Y(t) = h_0 + \sum_{\tau_1=1}^{\infty} h_1(\tau_1)\vec{u}(t-\tau_1) + \sum_{\tau_1=1}^{\infty}\sum_{\tau_2=1}^{\infty} h_2(\tau_1,\tau_2)\vec{u}(t-\tau_1)\vec{u}(t-\tau_2) +$$

$$\sum_{\tau_1=1}^{\infty}\sum_{\tau_2=1}^{\infty}\sum_{\tau_3=1}^{\infty} h_3(\tau_1,\tau_2,\tau_3)\vec{u}(t-\tau_1)\vec{u}(t-\tau_2)\vec{u}(t-\tau_3) + ...+$$ (5)

$$\sum_{\tau_1=1}^{\infty}\sum_{\tau_2=1}^{\infty}...........\sum_{\tau_k=1}^{\infty} h_k(\tau_1,\tau_2,........\tau_k)\vec{u}(t-\tau_1)\vec{u}(t-\tau_2)...........\vec{u}(t-\tau_k)$$

where, t, τ_1, τ_2 ,...,τ_k, are discrete time lags. The function $h_k(\tau_1 ,\tau_2 ,...,\tau_k)$ is the kth-order Volterra kernel characterizing the system. The h_1 is the kernel of the first order Volterra functional, which performs a linear operation on the input and h_2, h_3 ,...,h_k capture the nonlinear interactions between input and output sea level variations have acquired from JASON-1 satellite data. Consequently, the order of the non-linearity is the highest effective order of the multiple summations in the functional series [21].

General ways to implement equation 5 in real sea level data are acquired from JA-SON-1 satellite data, it is necessary to separate quantities in time domain [$h_k(\tau_1 ,\tau_2 ,...,\tau_k)$] and their corresponding in the frequency domain [$H_2(f_1, f_2,....., f_k)$]. In this context, Fourier transform of first (h_1) and second (h_2) kernels is considered and denoted by

$$H_1(f) = \int h_1(\tau)e^{-j2\pi f\tau}d\tau \tag{6}$$

$$H_2(f_1, f_2) = \iint h_2(\tau_1,\tau_2)e^{-j2\pi(f_1\tau_1+f_2\tau_2)}d\tau_1 d\tau_2 \tag{7}$$

These frequency-domain quantities $H_1(f)$ and $H_2(f_1, f_2)$ are called the first-, and second order frequency–domain kernels, respectively. In addition, equations 6 and 7 describe the output term in general first-order linear and second-order nonlinear/output model of equation 5. A sufficient condition for $H_1(f)$ to exist is that the linear system is stable. $H_2(f_1, f_2)$, however, is occurred when the bilinear system be stable. In addition, $H_2(f_1, f_2)$ is symmetric with respect to its two frequency variables be-cause the quantity $h_2(\tau_1,\tau_2)$ is unique and satisfies $h_2(\tau_1,\tau_2) = h_2(\tau_1,\tau_2)$. In words, $h_2(\tau_1,\tau_2)$ is symmetric with respect to its two variables [13]. Equations 6 and 7, therefore, can yield the formulas for the first two Fourier transform functional in equation (3). In this circumstance, the modification of equation 3 in which based on equations 6 and 7 is given by:

$$Y(f) = H_1(f)X(f) + \int H_2(\alpha, f - \alpha)x(f - \alpha)d\alpha - \overline{y}_2 \partial_1(f) \tag{8}$$

According to Rugh [21], the quantity $\partial_1(f)$ is an approximation to the usual theo-retical delta function and occurs in digital computation of Fourier transforms with sub-records of the finite delta function and $\overline{y}_2$ is given by:

$$\overline{y}_2 = \iint h_2(f_1,f_2)C_{xx}(f_1,f_2)df_1 df_2 \tag{8.1}$$

where, C_{xx} is the autocorrelation function of $x(t) = e^{j2\pi f_1 t} + e^{j2\pi f_2 t}$.

3.2 Volterra Model for Geostrophic Current Equation

The instantaneous surface geostrophic velocity U_g can be obtained from temporal mean velocity $\bar{U}(x,t)$ and anomaly of sea-surface geostrophic velocity $\bar{U}(x,t) = (u',v')$ by following formula [8]:

$$U_g(x,t) = \bar{U}(x) + \bar{U}(u',v') \tag{9}$$

where (u',v') are the sea surface geostrophic components in x and y directions which are estimated from the anomaly ξ of sea surface dynamic topography which practically equivalent to the sea-surface height anomaly ζ_s [9]. It might be the following Volterra model can be used to express the instantaneous surface geostrophic velocity U_g as follows;

$$U_g(x,t) = N(\xi,\zeta_s) + \sum_{i=1}^{\infty} \int_R h_i(\bar{\tau},\xi) \prod_{j=1}^{i} \bar{U}_a(u'-\tau_j)(v'-\tau_i)d\tau \tag{10}$$

where $N(\xi,\zeta_s)$ is the geoid height which is estimated from sea surface dynamic topography ξ and the sea-surface height anomaly ζ_s as described by Imawaki et al. [9]. Following Maged et al. [15], the mathematical expressions for first-order $(D_{1x}$ and $D_{1y})$ and second–order Volterra kernels $(D_{2xx}$ and $D_{2yy})$ of instantaneous surface geostrophic velocity U_g are as follows:

$$D_{1x}(f_x,f_y) = \left[\xi_x \bar{f} + \frac{\partial N(\bar{\xi})}{\partial \bar{\xi}} . (\bar{U}_{t_0} - \beta(\frac{\partial \zeta_s}{\partial y} - \frac{\partial \xi}{\partial y}) \right] \tag{11}$$

$$D_{1y}(f_x,f_y) = \left[\xi_y \bar{f} + \frac{\partial N(\bar{\xi})}{\partial \bar{\xi}} . (\bar{U}_{t_0} - \beta(\frac{\partial \zeta_s}{\partial x} - \frac{\partial \xi}{\partial x}) \right] \tag{12}$$

$$D_{2xx}(f_{1x}, f_{1y}, f_x - f_{1x}, f_y - f_{1y}) = \left[\xi_x \frac{\partial D_{1x}(f_{1x}, f_{1y})}{\partial \zeta}(f - f_1) - \xi_x \bar{f} + \frac{\partial N(\bar{\xi})}{\partial \bar{\xi}}.(\bar{U}_{t_0} - \beta(\frac{\partial \zeta_s}{\partial y} - \frac{\partial \xi}{\partial y}) \right] \times$$

$$\beta \frac{\partial \xi}{\partial y} \bar{f} \tag{13}$$

$$D_{2yy}(f_{1x}, f_{1y}, f_x - f_{1x}, f_y - f_{1y}) = \left[\xi_y \frac{\partial D_{1y}(f_{1x}, f_{1y})}{\partial \zeta}(f - f_1) - \left[\xi_y \bar{f} + \frac{\partial N(\bar{\xi})}{\partial \bar{\xi}}.(\bar{U}_{t_0} - \beta(\frac{\partial \zeta_s}{\partial x} - \frac{\partial \xi}{\partial x}) \right] \right] \times$$

$$\beta \frac{\partial \xi}{\partial x} \bar{f} \tag{14}$$

where $f_x, f_y, \bar{f}$ are the frequency domain in x and y dimensions while $\bar{f}$ is average frequency domain. Furthermore, β is the ratio of gravity acceleration to the Coriolis parameter expressing the effect rotation of the Earth [15]. In order to estimate the sea surface current from the altimeter data, we assume that there is a non-linear relationship between 2-D Fourier transform of mean surface slope variations obtained from JASON-1 satellite altimetry data ($F(f_x, f_y)$) and the first-order (D_{1x}, D_{1y}) second –order Volterra kernels (D_{2xx}, D_{2yy}) as follows:

$$F(f_x, f_y) = U_x(f_x, f_y).D_{1x}(f_x, f_y) \tag{15}$$

$$F(f_x, f_y) = U_y(f_x, f_y).D_{1y}(f_x, f_y) \tag{16}$$

$$F(f_{xx}, f_{yy}) = U_x(f_{xx}, f_{yy}).D_{2xx}(f_{xx}, f_{yy}) \tag{17}$$

$$F(f_{xx}, f_{yy}) = U_y(f_{xx}, f_{yy}).D_{2yy}(f_{xx}, f_{yy}) \tag{18}$$

According to Maged and Mazlan [14], the resultant of current velocity $\vec{U}(x,t) = (u', v')$ can be obtained by the inverse of Volterra model as

$$\vec{U}(f_x, f_y) = \frac{FT\left[\prod_{j=1}^{i} x(t) \right]}{D(f_x, f_y)} \tag{19}$$

where $FT\left[\prod_{j=1}^{i} x(t)\right]$ is the linearity of the Fourier transform [7]. According to

Maged and Mazlan [14], the inverse filter $P(f_x, f_y)$ is used since $D\,(f_x, f_y)$ has

not a zero for (f_x, f_y) which indicates that the mean current velocity should have

variation values in x and y directions. The inverse filter $P(v_x, v_y)$ can be given as

$$P(f_x, f_y) = \begin{cases} [D_x(f_x, f_y)]^{-1} & if\,(f_x, f_y) \neq 0, \\ [D_y(f_x, f_y)]^{-1} & otherwise. \end{cases} \tag{20}$$

Therefore, the resultant current velocity can be calculated by using equation 20 into equation 19 as

$$\vec{U}(f_x, f_y) = FT\left[\prod_{j=1}^{i} x(t)\right].P(f_x, f_y) \tag{21}$$

Following Maged et al. [15], the current direction can be obtained by using this formula,

$$\theta = \tan^{-1}(\frac{U_y}{U_x}) \tag{22}$$

According to Maged and Mazlan [14], equation 21 is required standard mathematical method to satisfy demand of the accurate ocean current circulation in large scale basin such as the South China Sea. In words, the discrete finite difference method such as Lax- Wendrof scheme is used to determine pattern of current movement.

3.2 Lax-Wendrof Scheme

The second-order accurate dispersive Lax-Wendrof scheme is used for sea surface current flow in JASON-1 altimeter data which can be written in predictor–corrector form in a staggered grid [13]. The staggered grid consists of a primary grid where points are labeled with (i, j, k) and a dual grid where points are labeled with (i + 12, j + 12, k + 12) (Fig.4). In the two-dimensional (2-D), the centers of the cell edges in the primary grid are given by (i + 12, j), etc. Following Maged et al. [15], 2-D, numerical flow for the two-step forms of Lax-Wendrof scheme is given by:

$$U_{i,j}^{n+1} = U_{i,j}^{n} + \frac{\Delta t}{2\Delta x}(v(U^{n+1/2}_{i+1/2,j+1/2}) + v(U^{n+1/2}_{i+1/2,j-1/2}) - v(U^{n+1/2}_{i-1/2,j+1/2}) -$$

$$v(U^{n+1/2}_{i-1/2,j-1/2})) + \frac{\Delta t}{\Delta y}(G(U^{n+1/2}_{i+1/2,j+1/2}) + G(U^{n+1/2}_{i-1/2,j+1/2}) - G(U^{n+1/2}_{i+1/2,j-1/2}) - \qquad (23)$$

$$G(U^{n+1/2}_{i-1/2,j-1/2}))$$

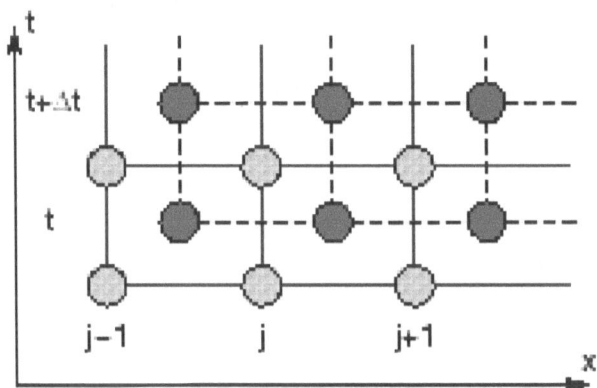

Fig. 4. Staggered grid

where v and G are smooth functions as described by Liska and Wendroff [13]. The values at the center of all faces of the primary cell on time level $n + 1/4$ are computed using the analog of 2-D predictor by

$$U_{i,j+1/2,k+1/2}^{n+1/4} = \frac{1}{4}(U^{n}_{i,j,k} + U^{n}_{i,j+1,k} + U^{n}_{i,j,k+1} + U^{n}_{i,j+1,k+1}) + \frac{\Delta t}{4\Delta y}(GU^{n+1/6}_{i,j,k+1/2})) \qquad (24)$$

The finite difference scheme model is applied with traditionally rectangular grids due to most their simplicity. A model numerical solution is enhanced by using the curvilinear nearly orthogonal, coastline-following grids. This technique is implemented to overcome the problems raised because of complicated SCS topography, and boundary conditions.

3.3 Boundary Data Requirements for Solution

The scheme is giving in equation 24 is the simplest possible one as regards the requirements for the introduction of the boundary condition properly that is required to compute a solution using the the staggered grid. Four points boundary data of U must be specified at each boundary in order to acquire a unique solution. At the left-hand boundary, only U is specified in time at $n+1/4$ but both are required to find v and G at this point. According to Liska and Wendroff [13], some special boundary method is then needed to provide U at this boundary point. Let assume $U(0,t)$ is specific as the left-hand boundary data. As indicated in Fig. 4, some interpolation usually requiring iteration because of the starting location on the G-characteristic striking the boundary at $(n + \frac{1}{4})(\frac{\Delta t}{4})$ which is unknown. Consequently, let assume an initial guess of $\phi = \frac{(\Delta x / 4)}{4}$, $U(\phi)$ and $G-(\phi)$ are computed from

$$G- = \frac{-\phi}{\Delta t / 4} \qquad (25)$$

$$U(\phi) = \frac{\phi U_4^n + (\frac{\Delta x}{4} - \phi)U_1^n}{\frac{\Delta x}{4}} \qquad (26)$$

Equation 26 is then used to compute the actual distance ϕ and the iteration continues according to some convergence criterion. In general, equation 26 provides the second boundary variable directly as part of the solution procedures and in particular can accommodate changes from one current pattern to another [13].

4   Results and Discussion

The proposed method to simulate the sea surface current movements in the South China Sea by using first-order Volterra kernel is showed in Fig.5 whereas the sea surface current patterns derived by second order Volterra kernel is presented in Fig.6.

The difference between first-order and second order Volterra kernels can be shown in sea surface pattern flows. It is interesting to find that the first-order Volterra kernel produces rough and fuzzy computer graphic pattern flow as compared to the second-order kernel. This evident is clear in Fig.5a and Fig.5c as result of the increase of sea surface velocity with maximum value of 1.3 m/s. This could be attributed to impact of the first-order Volttera model's linearity. In words, the first order kernel cannot capture the nonlinearity between altimeter return pulse and sea surface geophysical parameters. According to Ikeda [6], geoid and mean dynamic topography associated with pulse time delay which can produce errors in measuring altimeter data of Merged Sea Level Anomaly (MSLA). Thus, the nonlinearity feature becomes important in determining the actual flow pattern. In this context, the second-order kernel can produce a smooth computer graphic of current flow pattern in which the physical morphology of computer graphic flow pattern is distinguished clearly (Fig.6). This can be explained the ability of the second-order for fuzzy pattern reduction in computer graphic. Indeed, the nonlinearity can be solved with implementing second-order Volterra kernel as compared to first-order kernel. This study confirms finding of

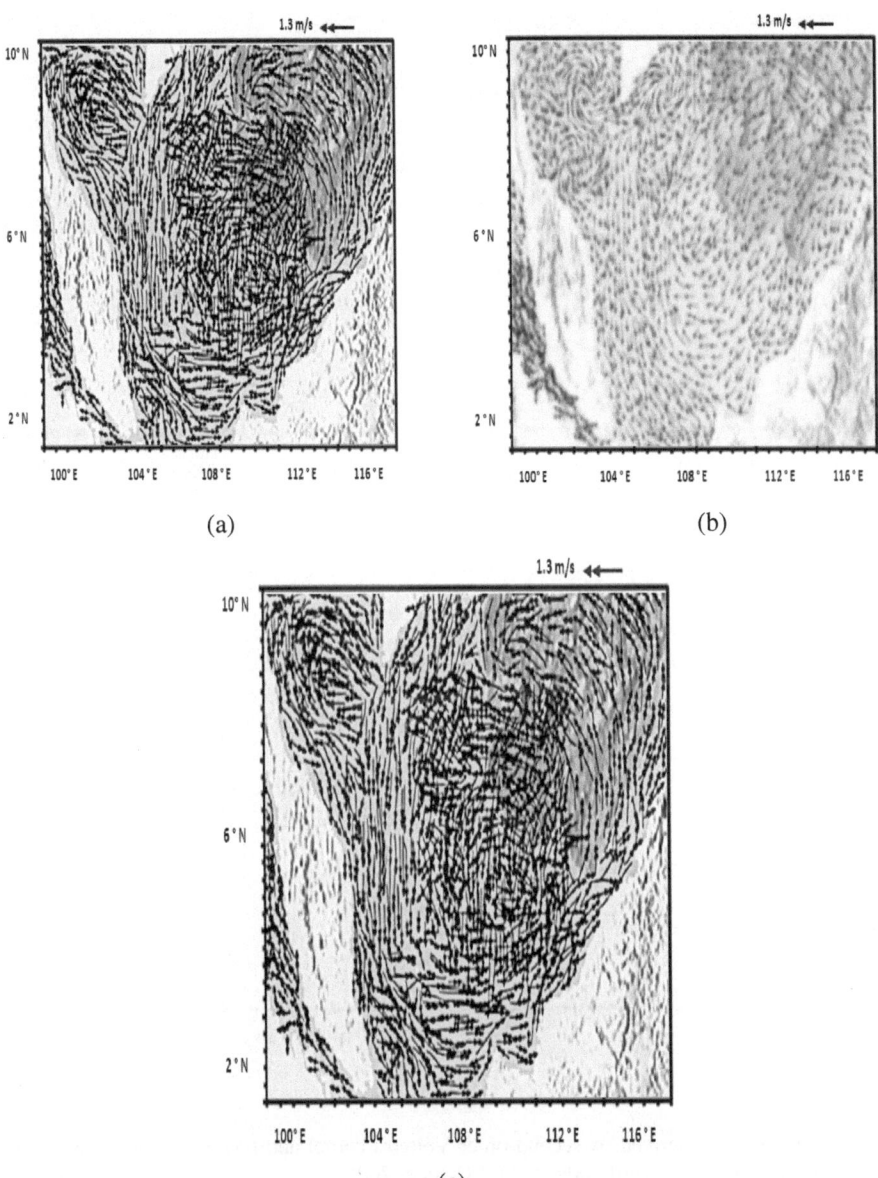

(a)   (b)

(c)

Fig. 5. Flow patterns derived by first-order Volterra kernel and Lax-Wendroff scheme during (a) October 2003, (b) April 2004, and (c) October 2005

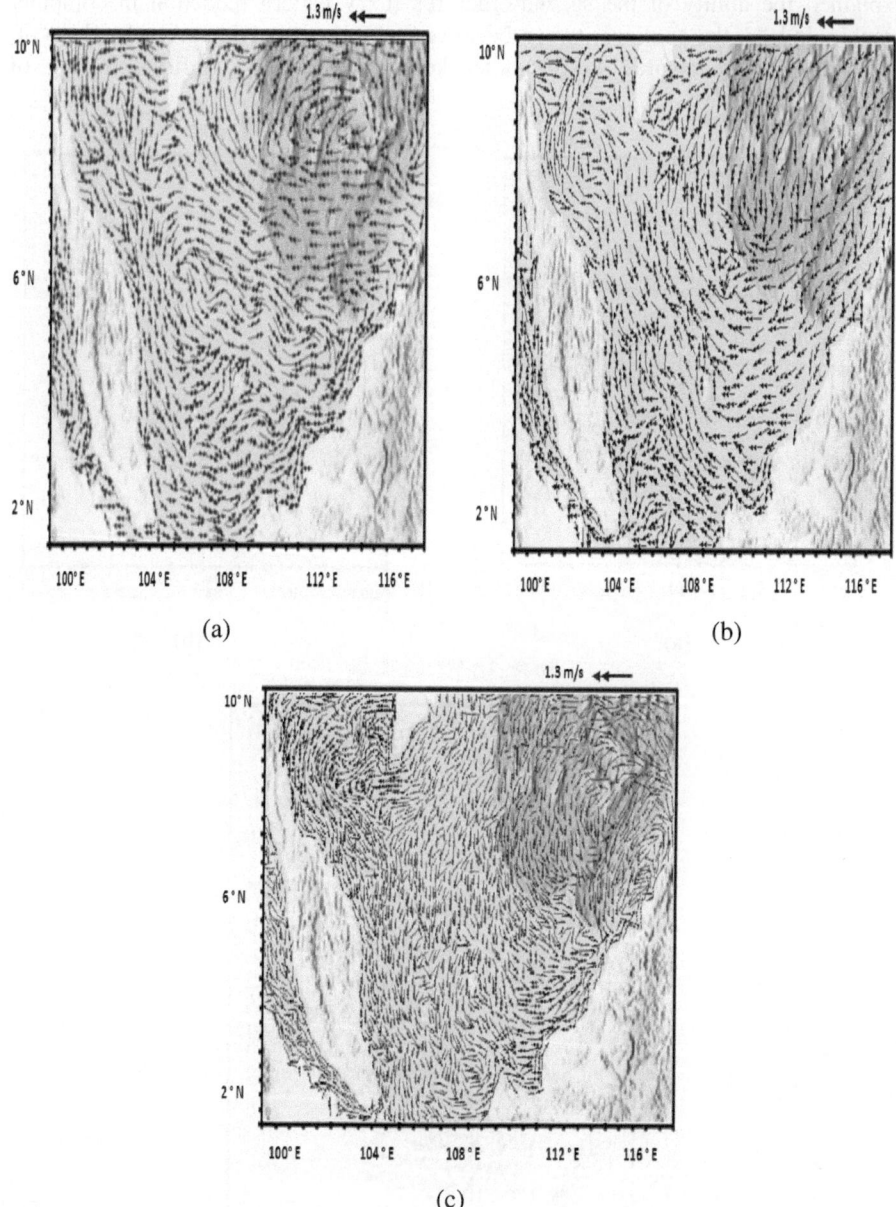

Fig. 6. Flow patterns derived by second-order Volterra kernel and Lax-Wendroff scheme during (a) October 2003, (b) April 2004, and (c) October 2005

Liska and Wendroff [13]. Both kernels, however, are agreed that the maximum current magnitude is 1.0 m/s in the northern and eastern coast of Malaysia whereas; the minim current velocity is 0.4 m/s that is observed in the Gulf of Thailand. Further, the east coast of Malaysia is dominated by southwards current flow in which is moving parallel to coastline. Part of these currents was turned southeast and moving parallel to Borneo coastline with the current velocity of 1.0 m/s. These findings are agreed with Wryki [9], Alejandro [1] and Mohd et al. [17].

The receiver-operator–characteristics (ROC) curve in Fig.7 indicates significant differences in modeling sea surface current by using first and second orders Volterra kernels. For the ROC curve, this evidence is provided by an area difference of 30% for first-order kernel and 6% for second-order Volterra kernels with in situ measurements. This is proved with a p value below 0.05. Further, Fig. 8 shows an exponential relationship between current velocity and the standard deviation of the estimation error for the sea surface current velocity. The maximum error standard deviation is 0.7, corresponding to the current speed value of 1.3 m/s which is estimated by first-order Volterra kernel whereas   for second-order Volterra kernel, the minimum error standard deviation of 0.03 occurs in a region of the current speed of 1.3 m/s.

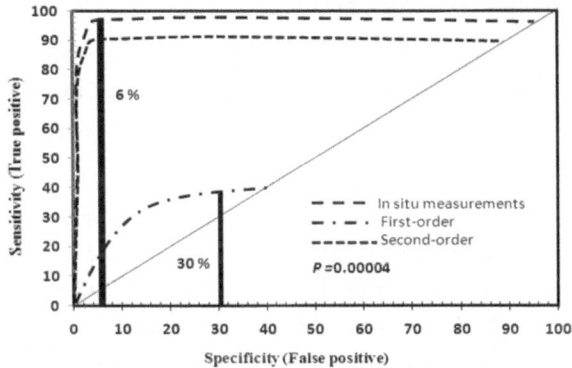

Fig. 7. ROC curve for first and second Volterra kernels

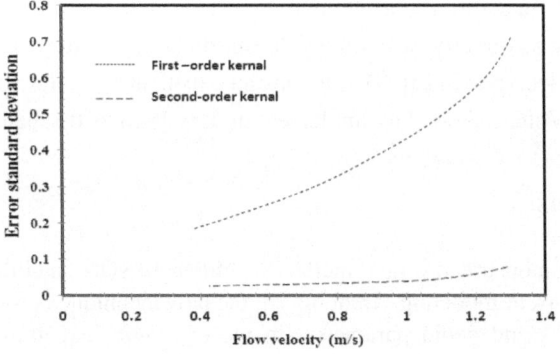

Fig. 8. Accuracy assessment of first and second-order Volterra kernels

The validation of the new approach of using Volterra-Lax-Wendroff scheme can be discussed by ROC and error standard deviation results. ROC suggests non-significant difference between second-order Volterra kernel and in situ measurements as compared to first-order Volterra kernel which confirms the study of Maged et al., [15]. The reason is that the bilinear frequency response function i.e. the second-order frequency-domain kernel is symmetric with respect to its two frequency variables. Hence, the bilinear output will contain mixing frequencies and this input/output pair can be used to determine $H_2(f_1, f_2)$ in a nonlinear differential equation [13]. $H_1(f_1)$ however, does not exist due to weak linear relationship between the geostrophic current, altimeter data of Merged Sea Level Anomaly (MSLA) and in situ measurements. This leads that the linear system be unstable in first-order Volterra kernel. Moreover, second-order Volterra kernel produces new equation which allows studying the nonlinearities of sea surface current modeling from JASON-1 altimeter data by using a fast Fourier transform. Therefore, the inversion of the second-order kernel avoids the use of iterative data assimilation techniques and is only based on the physical knowledge of estimating geostrophic current from altimeter data [8].

The lax-Wendroff scheme with the second-order kernel leads to produce a good discrimination for different flow patterns in computer graphic. Indeed, Lax-Wendroff scheme employs a numerical filtering device that has been called a "dissipative interface" which leads to a local smoothing (dissipation) of the data within each time step (interface) to which it is applied [13]. In addition, the dissipative interface can be useful to provide some dissipation to non-dissipative schemes and to suppress nonlinearity instabilities. In these circumstances, the combination between second-order and Lax-Wendroff scheme provides smoother and accurately flow pattern as compared to first-order kernel. In addition, Lax-Wendrof scheme can handle mixed sub and supercritical current flows directly, with no regard for directional nature of the computation (left to right or vice versa) [13]. This is thus an excellent indicator for the validation of second-order Volterra kernel by implementing Lax-Wendroff scheme.

5   Conclusions

This work has demonstrated a new method to utilize JASON satellite altimeter data for sea surface current modeling. In doing so, the new technique is used to reduce the impact of Coriolis and geoid parameters in the continuity equation of geostrophic current. Two procedures are involved: first and second order Volterra Kernels and Lax-Wendroff Scheme. It is interested to name this algorithm as Volterra-Lax-Wendroff scheme. The modification of geostrophic current formula by using second-order Volterra kernel and Lax-Wendroff scheme produces smoother computer graphic of current flow as compared with first-order. The second-order Volterra- Lax-wendroff scheme has a lower standard deviation error of 0.03 for maximum current velocity of 1.3 m/s than first-order Volterra kernel due to its weak linearity/nonlinearity. In conclusion, the combination of second-order Volterra kernel and Lax-Wendroff has improved modeling of sea surface current from JASON-1 satellite data. This new approach can be used to solve the nonlinearity in altimeter satellite data due to Coriolis and geoid parameters.

References

1. Alejandro, C., Nasir, M.S.: Dynamic behavior of the upper layers of the South China Sea. In: Proceeding of the national conference on climate change, Universiti Pertanian Malaysia, Serdang, Selangor, Malaysia, August 12-13, pp. 135–140 (1996)
2. AVISO/Altimetry, AVISO user handbook for merged TOPEX/POSEIDON products. AVI-NT-02-101, ed. 3.0. AVISO (1996)
3. Balaha, J.B.L.: Calibrating altimetry to geopotential anomaly and isotherm depths in the western North Atlantic. Journal geophysical research 97, 7465–7477 (1992)
4. Gil, M., Malanotte-Rizzoli, P.: Data assimilation in meteorology and oceanography. Advance geophysics 33, 141–266 (1991)
5. Glenn, S.M., Poorter, D.L., Robinson, A.R.: A synthetic geoid validation of Geosat mesoscale dynamic topography in the Gulf Stream region. Journal geophysical research 96, 7145–7166 (1991)
6. Ikeda, M.: Mesoscale Variability revealed with sea surface topography measured by altimeters. In: Ikeda, M., Dobson, F.W. (eds.) Oceanographic applications of remote sensing, pp. 15–25. CRC, New York (1991)
7. Inglada, J., Garello, R. (1999). Depth estimation and 3D topography reconstruction from SAR images showing underwater bottom topography signatures. In: Proceedings of IGARSS 1999 (1999)
8. Inglada, J., Garello, R.: On Rewriting the Imaging Mechanism of Underwater Bottom Topography by Synthetic Aperture Radar as a Volterra Series Expansion. IEEE Journal of oceanic engineering 27, 665–674 (2002)
9. Imawaki, S., Unchida, H., Ichikawa, K., Ambe, D.: Estimating the High-Resolution Mean sea-Surface Velocity Field by Combined Use of Altimeter and Drifter Data for Geoid Model Improvement. Space science reviews 108, 195–204 (2003)
10. Kelly, K., Gille, S.: Gulf Stream surface transport and statistics at 69° W from the Geosat altimeter. Journal geophysical research 59, 3149–3161 (1990)
11. Hu, J., Kawamura, H., Hong, H., Kobashi, F., Wang, D.: 3-6 Months Variation of Sea surface Height in The South China Sea and Its Adjacent Ocean. Journal oceanography 57, 69–78 (2001)
12. Hawang, C., Chen, S.A.: Fourier and Wavelet analyses of TOPEX/Poseidon derived Sea Level anomaly over the South China Sea: A contribution to the South China Sea Monsoon Experiment. Journal of geophysical research 105, 28785–28804 (2000)
13. Liska, L., Wendroff, B.: Composite schemes for conservation laws. SIAM Journal numerical analysis 35(6), 2250–2271 (1998)
14. Maged, M., Mazlan, H.: Three–Dimensional Reconstruction of bathymetry Using C-Band TOPSAR. Data. Photogrammetri-Fernerkundung Geoinformation, S469–S480 (June 2006)

15. Maged, M., Mazlan, H., Cracknell, A.: Volterra algorithm for modeling dea surface current circulation from satellite altimetry data. In: Gervasi, O., Murgante, B., Laganà, A., Taniar, D., Mun, Y., Gavrilova, M.L. (eds.) ICCSA 2008, Part II. LNCS, vol. 5073, pp. 119–128. Springer, Heidelberg (2008)

16. Martins, C.S., Hamann, M., Fiuza, A.F.G.: Surface Circulation in the Eastern North Atlantic from Drifters and Altimetry. Journal of geophysical research 107, 10.1–10.22 (2002)

17. Mohd, I., Ahemd, S., Wah, F., Chin, K.: Water Circulation Pattern from Sea Surface Current and Chlorophyll-A Derived Using Satellite Data in the South China Sea. Paper Presented at the 2nd International Hydrographic and Oceanographic Conference and exhibition 2005 (IHOCE 2005), Kuala Lumpur, Malaysia, July 5-7 (2005)

18. Moore, A.M.: Data assimilation in quasigeostrophic open ocean model of the Gulf Stream region using adjoint method. Journal physical oceanography 21, 398–427 (1991)

19. Morimote, A., Matsuda, T.: The application of Altimetry data to the Asian Marginal Seas. In: Proceedings of IEEE International IGARSS 2005, vol. 8, pp. 5432–5435 (2005)

20. Robinson, I.S.: Measuring the oceans from Space. In: Principles and Methods of Satellite Oceanography. Springer and Praxis Publishing, Chichester (2004)

21. Rugh, W.: Nonlinear system theory: The Volterra/Wiener Approach. Johns Hopkins University Press, Baltimore (1981)

22. White, W., Ttai, C., Holland, W.R.: Continuous assimilation of simulated Geosat altimetry sea level into an eddy-resolving numerical model. I. Sea level differences, Journal geophysical research 95, 3219–3233 (1990)

23. Wyrtki, K.: Physical Oceanography of the South-East Asian Waters. In: NAGA Report, vol. 2, Univ. Calif. Scripps Inst. Ocean, La Jolla (1962)

Guidelines for Web Usability and Accessibility on the Nintendo Wii

Valentina Franzoni and Osvaldo Gervasi*

Department of Mathematics and Computer Science, University of Perugia,
via Vanvitelli, 1, I-06123 Perugia, Italy
osvaldo@unipg.it

Abstract. The aim of the present study is to propose a set of guidelines for designing Internet web sites usable and accessible with the Nintendo Wii console. After an accurate analysis of usability issues and of the typical Wii Internet users, twelve usability guidelines will be proposed. These guidelines are focused on visibility, understandability, clickability and compatibility. We then restructured a sample web site according to the guidelines. To prove their effectiveness, we performed the usability tests on a sample of forty individuals, selected among the various categories of potential users of the Nintendo Wii console, after having visited the restructured and the original web sites. The analysis of the resulting information confirmed that the restructured web site is more usable than the original and the improvement is more pronounced for weak categories (elderly and individuals with no experience with web browsing). Furthermore the adoption of the guidelines reduces the difficulties experienced by users with different expertise, in visiting a web site with the Wii console.

1 Introduction

The development of computer graphics engines made several commodity consoles produced by the major operators in the entertainment market available. These consoles are usually enhanced with Internet access, web surfing and other communication facilities that open new perspectives to facilitate collaboration among people and to experience the Internet for seniors and people with disabilities. Access to Internet through a console has some limitations compared to a personal computer, in particular the screen resolution, which impacts the access to information through the Web. Web designers have to take the issues related to this new class of users into account and adopt a layout for the web site that is suitable to be visited this way.

In the present paper we introduce twelve guidelines to adapt the content of web sites to the Nintendo Wii console in order to guarantee the compliance of the web content to the rules of usability and accessibility. The proposed guidelines will allow people to get information and use web services really usable and

---

* Corresponding author.

M.L. Gavrilova and C.J.K. Tan (Eds.): Trans. on Comput. Sci. VI, LNCS 5730, pp. 19–40, 2009.

accessible, both using PC and the Nintendo Wii interface, offered as an alternative for PC web browsing. Although a detailed study was been carried out in this work only for Wii, we expect the same guidelines to be valid even using other consoles and mobile devices.

The usability is defined by the ISO standard 9241-11 and in ISO/IEC standard 9126[1]. During the last 10 years some usability guidelines were defined by J. Nielsen, D. Norman and other usability experts, concerning visibility, affordance, conceptual models, feedback, user protection and flexibility[2,3,4]. The usability of a web site is also a key factor in increasing the user trust of a web site and the degree of web site loyalty[5].

To adapt a product to the needs and expectations of the final user, one has to understand the characteristics of the users, their activities and the organizational and social contexts in which they are living and operating. Concerning the understanding and using friendliness, then, one can take advantage of the available consolidated usability principles and guidelines, based on the acquaintances of behavioral and cognitive aspects in human-computer interaction[6,7,8,9]. The aim of ISO norms and usability guidelines is to make sure that the design of the interface and the contents planning of web sites are user-centered. The users know their competences, culture, needs, limits and attitudes and it is necessary their involvement in planning, realizing and managing a web site through usability tests.

Accessibility is defined and analyzed in several notes of the World Wide Web Consortium (W3C)[10]. The Italian Parliament has promulgated a law to define the rules to benefit disabled users in accessing computers[11]. We want to guarantee accessibility to individuals belonging to categories experiencing difficulties in using web technologies: the elderly, sick people, ipoliterates, users who operate in a low illuminated atmosphere, and/or using old machinery, users who have few competences and abilities to use Web and computers.

The paper is organized as follows: in section 2 a brief description of the Nintendo Wii console is given; in section 3 the users who performed the usability tests are presented; in section 4 our twelve usability guidelines for the Nintendo Wii console are described; section 5 discusses the usability tests made on a set of forty users and the results obtained; in section 6 some conclusions are illustrated.

2   The Nintendo Wii Console

The Nintendo Wii console makes it possible to surf the Internet in an easy way. The browser Opera for Wii, based on Opera 9.10, has been adapted for the innovative motion-sensitive controller, called WiiRemote: the user's actions are caught by the integrated motion sensors and sent via a bluetooth link to the SensorBar, that is positioned in front of the television screen. The WiiRemote is designed to be ergonomic and easy to use. A few buttons, pushable with one hand, allow one to use advanced controls. A symmetrical design and the possibility to adjust the relative settings, allow full access to left-handed people. The WiiRemote includes also a built-in speaker and a rumble feature, to make browsing interaction more realistic. An external socket is integrated to connect the Nunchuck, a second motion-sensitive controller with more buttons and a control stick, that expands the control system for two-hand actions. Movements are computed by a CPU Power PC, a 512 Mbyte flash memory and an ATI graphic accelerator. It is possible to connect up to 4 WiiRemotes, making it possible to surf the Web in a very innovative way: together with family and friends. The users can participate together in the browsing sessions and are allowed to point on the web site elements, to increase communicative interaction. During the web browsing sessions it is possible to zoom in and out (automatic or progressive zoom): the Outline Font feature focus the enlarged text elements. It is also possible to scroll the visible area, to set the single-column mode (that reduces the layout, viewing the components in the same order in which they are written in the markup code and resizes the text areas), and other settings.

3   The User Experience

According to the studies on web usability made by the Nielsen Norman Group in the last years[2], a percentage of 66% of web sites are not usable (users can not complete the requested tasks during usability tests) for generic PC web browsing. In most cases users leave a web page after 35 seconds on average, if it is not immediately useful or interesting, This behavior and the subsequent research of new web pages to find the requested information, is known as "information foraging": the more new informative resources are easily available, the fewest users spend time on any one of them. So users, like opportunist predators, prefer to make quick use of immediately available portions of information rather than dwell on them to choose the best ones. Concerning Wii web browsing, we argue that the percentage of web sites that present the discussed problems is rising, depending both on the lower precision of the Wii pointing device respect to the mouse and on its extreme easiness of use. Therefore the range of users increases to a potentially less expert public, from professionals and enthusiasts of technologies to pure newbies, who approach the Web exclusively through the Nintendo Wii.

The users of the Wii console are particularly diverse. We decided to refer novice users of new technology, usually cut off by the number of potential users of web browsing instruments in the assessment of usability. On the other hand, we took into account the pleasantness of the web site, exposed in its fundamentals in the ISO/IEC 9126 standard, because we don't want to penalize advanced users. However we focused our attention on users who approach the Web, even for the first time, using the Wii console.

3.1   Teenagers

We may focus our attention on teenagers, still the main users of game consoles and guarantors of the success of the Web. The ability of new generations to handle interactive mechanisms better than adults, in particular if they include the use of fingers like the mouse, joystick and joypad, is due – whenever real – to a precocious formation and habit in using such instruments, not to a pseudo-darwinian evolution from the "Homo sapiens sapiens" to the "Homo technologicus", as mass media like to refer. In spite of this, according to the studies and usability tests by the Nielsen Norman Group[2], teenagers are inclined to worry about Internet and its dangers, like viruses, spam, malware and phishing and they avoid clicking on unknown elements, downloading and freely surfing the Web. We will focus for them on affordance and on simplification of contents and interfaces.

Moreover, Hoa Loranger of the Nielsen Norman group refers in the same tests that, in presence of technical obstacles, teenagers tend to give up rather than try to solve the problem and in general have less success than adults with web browsing, because they are less patient and thus they abandon pages easily. It is important for them to guarantee light layout and contents, to avoid also the problem of the slowness of page loading. Regarding the frequency of use, we still don't have statistical information about the Wii platform, but we can refer to studies on the use of the Web and of videogames by teenagers, in fact the Wii is first of all a gaming console. According to a study conducted by the Italian "Minors and Media Centre" in the year 2006 on 2000 students of secondary and high schools in 18 Italian cities, 89% of them use videogames one or more times in a day. Then according to a research conducted by the Italian "Observatory on Minors" in the same year on 400 Italian teenagers, only 72% of them make use of the Internet, even occasionally. Although a comparison can not be precise, we know approximately 17% of Italian teenagers in the year 2006 did not browse the Web, but played videogames. Considering that both browsing and console-gaming phenomena are increasing, we can assume that in the next years also the percentage of teenagers using Internet through gaming consolesr will encrease.

3.2   Elderly People

Other interesting typical users could be elderly people (over 65 years old), who represent a big percentage of the population. In Italy seniors are over 20% of the population, second in the world after Japan. Nintendo gave a lot of publicity to the suitability of Wii for elderly users and in general for whom has not experience with web browsing and with the use of new technologies. Obviously only a small number of them could try to use the Wii console, but the Wii could be right for them, because of the extreme simplicity of its interface and it could be used for physical and mental training of healthy seniors. First of all, elders living at home with their family will be more likely to use the Wii, than couples living alone, people attending associations, who receive voluntary or sanitary services, living in residential facilities or in retirement homes that grant recreationals and lastly singles, if self-sufficient or adequately supported. We consider it really important, to design a web site, to reduce for them any interaction, guaranteeing all the functionalities.

3.3   Sick and Disabled People

Users with special needs represent a very large part of the world population. According to recent studies by the World Health Organization (WHO), almost one hundred and fifty million people are affected by a personal mobility, sensorial or cognitive problems. This kind of research does not include all hidden cases of people with problems for cancer, surgery, early stages of muscolar dystrophy, chronic complications of diabetes, and in general who suffer fatigue, widespread pain, fatigue vision, impaired concentration et cetera. It is very important for us to consider all these cases during the planning of designing or restoring web pages for Wii. In fact, thanks to its simple interface, its low price and some game packages like "Wii Sports" or "Wii Fit" (or user developable software), the Wii console offers an amusing way to keep moving and simply control individual critical parameters, if necessary with the support of a professional staff. Wii can be used for training and rehabilitation for single persons or groups in public and private health structures. We consider the Wii console right for rehabilitation of both orthopedic and neurological patients, such as handicapped people. In fact, thanks also to its rumble and sound features, that are available also for the web browsing, the Wiiremote provides a sensory experience that could be useful to stimulate and exercise patients who are afflicted by problems of perception, memory and attention. Figure 1 shows a great testimony of Wii capabilities: a young patient, affected by cerebral palsy from birth, unable to walk and talk, is playing videogames using the Wii console. Moreover, group gaming could help socialization of individuals with relationship problems, such as autistic people or who have mood disorders, like depression. In Padova University, Prof. Luciano Gamberini[12], of the Human Technology Lab of the faculty of Psychology, started in September 2007 to develop systems for research and clinical intervention based on the Wii console. For this aim we invite web designers to reduce interaction down to the bare to essential and to respect all the guidelines about accessibility.

4   Web Usability Guidelines on the Wii Interface

In the most economically advanced countries, Internet is replacing television programs. Several technologies have been utilised to surf the Internet using television as an output device. The access of multimedial information through TV needs interactive capabilities for which the TV interface is not specialized enough and several adjustments and integrations need to be introduced in order to make the interface efficient and comfortable. Research in this area is going in the direction of "integration through simplification": the automatisms which are being developed and the corresponding services that gradually appear in our life go in the direction of a reduction of interactivity, because on TV, for example, it is difficult to scroll pages or browse hypertexts. So, while on one hand the proposed medium loses the entertainment capability of TV, on the other it renounces the interactive abundance of PCs. Wii inverts this trend, proposing a suitable interface for both group entertainment and interactive web browsing. Nevertheless,

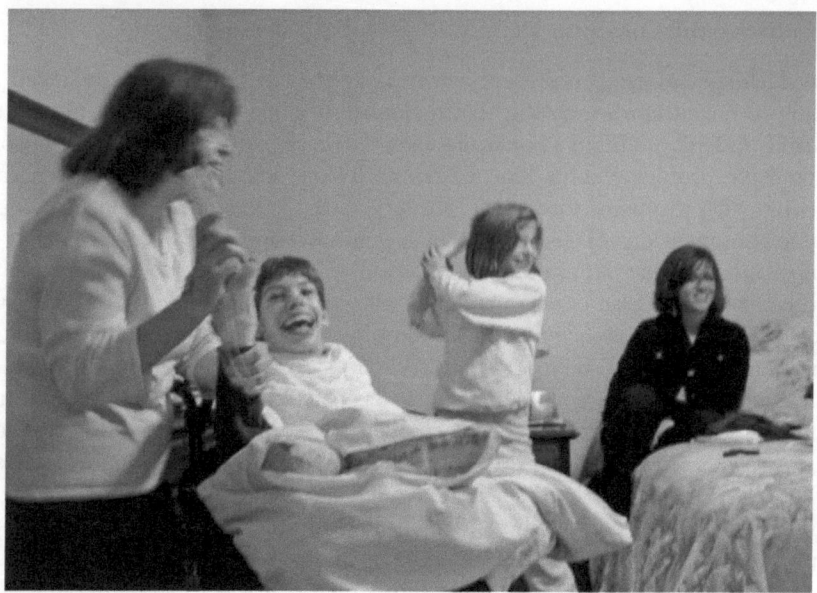

Fig. 1. A great testimony from a special Wii user: "*My little brother Stephen, who was born with cerebal palsy and is unable to walk or talk, is able to play video game baseball against his little sister with the Wii*" (Matt Clark, used with permission)

the majority of web sites are not designed to be surfed on Wii or any other device except computers. Furthermore a lot of them do not comply with accessibility and usability guidelines for the Web. So even on Wii Internet Channel they can not be properly visualized and their interactive features are not well operating.

The usability guidelines for the Wii interface are presented in table 1 and discussed in the following sections.

4.1   The Restoration of Visibility

The first and most important problem that we found on the Wii Internet Channel is related to TV resolution and image quality. To improve reading intelligibility of text elements, Wii developers have reduced the screen resolution so that the width of the page appears in 800 pixels. In spite of this, the visibility of the text is very low and the most used sizes of the text body - 10,12,14 - are clearly inadequate for a comfortable displaying, without making extreme use of the zooming and scrolling features on the WiiRemote. Because of the flicker effect of the screen, given by the low updating frequency of the TV image, text elements are not easily readable, inducing eye fatigue, inversely proportional to the dimension of the font: the edges of the characters are not defined and they result hard and difficult to read. According to the first guideline we suggest adopting a font size of at least 20 for text (26 recommended, 30 for titles) and a line spacing of at least 1.5 lines. For the other components the same rules apply: all of them

Table 1. The 12 accessibility guidelines for browsing the Web with the Nintendo Wii console

Guideline	What to do	How to test it
First	Balance the size and spacing of the fonts and all other components	Check for dimensions (at least 20 for the font, 26 recommended) and line spacing (at least 1.5 lines)
Second	Space out any component from the edges of the page	Check for a margin of 15 pixels in each direction
Third	Test color matching and contrast using specific software	Use specific software
Fourth	Simplify the design	Dispose elements in a maximum of 4 columns (2 for text blocks)
Fifth	Simplify the content	Adjust text to a cognitive level of secondary school: both language and syntax, both length. Edit images with a level of detail understandable with a maximum of 2 zooming operations
Sixth	Reduce interaction to essentials, preserving all functionalities	Do not include in the same page different components with the same interactive function, except for very distant points of the page
Seventh	Space clickable components, keeping them in the same visual area	Check for a line spacing of 1.5 lines for text links and at least 10 pixels for buttons and similar elements
Eighth	Don't use components that don't allow spacing	Check for combo-boxes, drop-down and pop-up menus
Ninth	Use only the supported formats	Check for the specification of the Wii Internet Channel
Tenth	Use CSS for the design	
Eleventh	Guarantee light content	
Twelfth	Adopt the guidelines for web accessibility and usability	

must be large and spaced enough. From our tests on resolution, it is evident that most TVs (especially old models) lose some pixels on the edges and, because of their rounded shape, the first visible pixels get deformed. It is necessary, then, to space out any component from the edges of the page. A margin of 15 pixels in every direction is an appropriate check point. We note that, because of low resolution, colors with a low contrast become illegible and cause tearing. Very intense colors cause discomfort to the eyes, because of the flicker effect, and combined with any other color they tend to merge with them, causing bruising and tearing even to those who do not have particular problems with eyes. The third guideline recommend to test color matching and contrast using specific software, like the "Accessibility Color Wheel" by Giacomo Mazzocato[13]. Using these utilities, we can test colors and combinations and see how they are perceived by the person presenting different forms of colorblindness. In that way we can choose restful colors for all.

4.2   The Restoration of Understandability

The use of large fonts and components allows one to see every element correctly, but does not guarantee a simple interpretation of them. In this case preserving the design and the same content is unfeasible, but even keeping the same elements disposed in a different design is not a good idea. In fact, if the number of elements is high, enlarging them we will obtain a too dense page or a too long one. In the first case, the web page becomes a complex and confusing scheme to understand. In the second, we force the user to a long search using the scroll feature, that increases the number of users abandoning the page before reaching the desired content, or unable to identify its position.

We define the guidelines (fourth and fifth) to simplify the design and to simplify the content. The fourth guideline will make the design clearer and more pleasant and the fifth one will reduce the length of the web page and make it more readable. As a point of control for the fourth one we propose to verify that images, buttons, links, blocks of text are not flooding the page and that they are disposed in a maximum of 4 columns, better two in case of text blocks. The point of control for the fifth guideline is to readjust the text to a cognitive level of secondary school (that is right for teenagers, elders and people with higher levels of instruction but low capability of attention) and edit images so that they present levels of detail understandable with a maximum of two zooming operations. Bringing the text at this level means both to simplify the language and syntax, and to reduce the length. A good way to do so is to provide images that can communicate the proposed concepts. Furthermore, we want to remember that interactivity is another point that may create confusion using web pages. From our tests we see the seriousness of this issue: it is often difficult, even for general users, to identify and interpret the functionality of the various components. We propose the sixth guideline to reduce interaction to essentials, preserving all functionalities. A point of control for it can be to avoid the use in the same page of different components with the same interactive function, except when it is necessary to access them from very distant points of the page. An exception can be the logo in the header, that can be clickable as a link to the home page.

4.3   The Restoration of Clickability

In most web sites, the design includes lists and menus or other components with no vertical spacing. Due to the nature of the controller (sensible to very small movements) and considering the shape and size of the associated pointer (a little hand of 50x70 pixels with a slightly active area) it is rather difficult to stop and click on small or close components and the palm of the little hand may also cover adjacent links. Enlarging the components, as we proposed above, can be only a partial solution, because adjacent components will have an uncertain

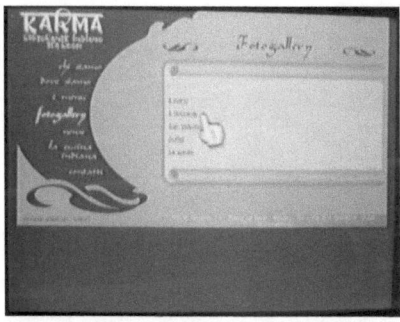

Fig. 2. Example of a menu not usable with Wii (left hand side) and a usable one (right hand side)

clickability. To this end, we propose the seventh guideline, advising to space clickable components, keeping them in the same visual area.

As point of control, a line spacing of 1.5 lines for text links and at least 10 pixels for buttons and similar elements, such as picture thumbnails, should be considered. Because especially combo boxes, drop-down and pop-up menus cause clickability problems, we specify to avoid to use of components that do not permit spacing as shown in Figure 2.

4.4 The Restoration of Compatibility

We will now highlight problems related to the content designed for the Web but not compatible with the browser Opera for Wii.

First of all, we can see from the specification of the console that not all available formats for the web components are supported by the Wii Internet Channel, so we recommend using only the supported formats.

The last free version of the Wii browser, as at the time of writing, can support Ajax with JavaScript 1.5, Flash up to 7, Cookies, SSL, MP3. The Wii browser does not support ActiveX, PDF and any format of video codec and it is not possible to download any file from the Internet.

To access video clips, it will be necessary to use Flash instead of video formats like AVI or DivX. To access text and other components, we have to put them into the page and not into a PDF file, thus avoiding Flash games that require keyboard and adapt the related actions to the Wii controllers. In the case of keyboard controls, several USB keyboards are compatible with the Wii, but they are not a default device, so if you consider the use of the keyboard, one should express this requirement.

The compatibility with Ajax includes the full support of Cascading Style Sheets (CSS). Using CSS for the design is fundamental to the usability of our web site, because the browser Opera for Wii provides a single-column mode that uses CSS for a correct view: it reduces the layout to a single column, visualizing the components in the same order in which they are written in the markup code

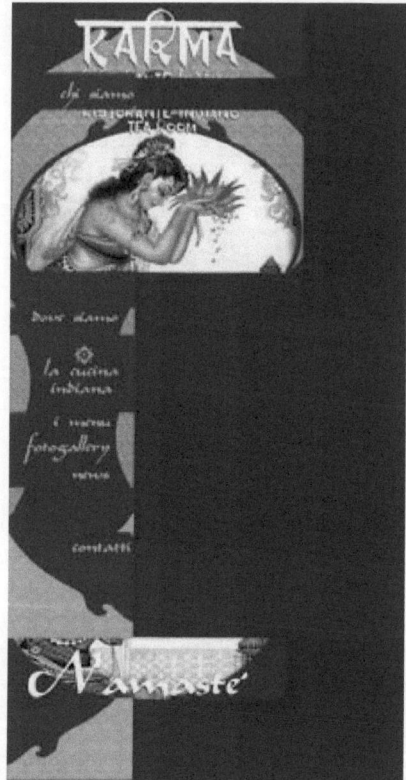

Fig. 3. Example of a page designed with tables (wrong, left hand side) and how it appears in the Wii-single column mode (right hand side)

and adapts the text size to optimize the readibility. The main features of CSS are disabled, except the background color, while the text color is adapted to enhance the contrast. The use of CSS allows the tool to control what visual information is important (like content) and what is unnecessary (like background images). If we use the old method of tables to design HTML pages, in the single-column mode we will obtain a sequential visualization of the content of the cells, that leads to wrong page in most cases (see Figure 3).

The use of CSS is also a requirement for accessibility and allows us to use functionalities like hovers, that otherwise could be designed only through the use of JavaScript, making the page heavier. Another problem that afflicts the Wii Internet Channel is the slowness of page loading, that induces the user to abandon the site and that very much depends on the proposed content. It is necessary, then, to guarantee light content, regulating the use, among others, of decorations and animations. Finally, we reaffirm the importance of maintaining the compatibility with the guidelines for web accessibility and usability.

5   Usability Tests

Our tests were conducted on a sample of 40 users, according to suggestions given by J. Nielsen and M. Visciola.[14] selected among people who don't work in the fields of new technologies, marketing, web design or usability, because these people are not significant user models. According to the suggestions of the Nielsen Norman Group[2], we assessed that the insiders have too many specific expertise and are not able to use a web site as anyone would do. We included in our study, on the other side, users without previous experience of the Internet. In web usability tests this class of users is not generally considered significant, but in our case the neophyte user is one of the most important user models to study, because the Wii console is proposed as an instrument to approach the Web even for the first time.

Our 40 users were selected also on the basis of their age, dividing them in 4 classes, according to the definitions established by the Italian Institute for Statistic (ISTAT): teenagers (12-19 years old), juveniles (20-39), adults (40-64), elders (over 65). The users were selected among the four categories in the same percentage.

For our tests we chose an exhisting web site, named site B, related to a restaurant, because we considered it a suitable target for our users. It was developed in 2003 using old methodologies, like tables for the graphic planning and the use of images for the text link presentation. It included a home page with very complex elements and some internal pages with one template. We redesigned it applying the twelve usability guidelines for the Wii interface. We used XHTML and CSS standards and some simple PHP techniques to optimize maintenance and to manage the contact form, and named it site A.

To keep usability both for Wii and PCs, we used JavaScript code for the platform identification and we created different CSS for the visualization respectively on Opera for Wii and on all other web browsers. In figure 4 is shown the Javascript code used to select the right CSS when the connection is made using the Wii console.

We carried out the usability tests on both websites. We made the test on a cathode ray tube TV, with a 22" display 4:3, to which we connected the Wii console and an ADSL line. We put our users at a distance of 2 meters, which we consider appropriate for this type of TV and compatible with the furniture provision in a ordinary room; the selected distance is also the recommended one

```
<script type="text/javascript">
if (navigator.platform == "Nintendo Wii") {
document.write('<link rel="stylesheet" type="text/css" href="stile-wii.css">');
}
else {
document.write('<link rel="stylesheet" type="text/css" href="stile.css">');
}
</script>
```

Fig. 4. The javascript code used to intercept a connection made through the Wii console in order to use the most appropriate CSS

Fig. 5. The environment we used for tests is shown: notice how the user is using the console for visiting the web site and one of the two video-cameras we used for video-recording the session

in Wii specifications. We made our tests in an adequately lit environment, shown in Figure 5, with the light source put behind the user, in a position that avoids reflections on the screen.

Users were introduced to a small room for tests, separated from the waiting room so they did not perceive any disruption and to avoid anyone to listen to our tests from outside. In the room two other people were present to record data, at a distance of about 1 meter and outside the view field, to avoid distracting the user.

We carried out the tests asking people to perform some tasks on both web sites: the redesigned web site and the original one. Our test users had not been previously trained for it. The purpose of the test was not explained to users before its conclusion, to avoid any influence during the evaluation of both test web sites. Once seated in our workstation, we introduced them to the Internet Channel fundamentals. Before beginning, we proposed a minute to try the different functions of the WiiRemote on the homepage of the site proposed, except for clicking, to avoid the beginning of navigation. We finally explained that they had to surf two different web sites, with similar content, in each of which they were asked to repeat the same 3 tasks, with different levels of difficulty.

We based our tests on the *thinking aloud* method, in which we ask users to report aloud each thought while operating with our interface. Hearing the thoughts of users helps to understand why they are doing what they are doing, and to record any difficulty. One facilitator talked to the user, while the other only recorded times and operation numbers. The role of the first facilitator was to explain the features of the remote control and to give any clarification on it during the test, solving any problem in real time depending on the WiiRemote's functions and on the nature of the assigned tasks.

We did not suggest how to complete tasks and did not provide information about the structure and the contents of both web sites. To count operations, we considered only clicks on clickable elements, both successful (such as transitions from one page to another) and unsuccessful (such as inaccurate clicks and repeated ones). We also registered each task if it was successful or not.

In figure 6 a screenshot of a sample video showing a woman performing the test related to the usability of Nintendo Wii is shown. The video is composed of 2 separate windows. The main window shows the user performing the test. The small window on top (right hand side) shows the operations performed on the page by the user. A similar video was produced for each user (40) performing the test.

At the end of the practice tests, we briefly interviewed the users, recording data relating to their age, sex, experience with the Internet, and their evaluation of both sites, first A then B, with the following questions:

Fig. 6. A screenshot of a sample video showing a woman performing the test related to the usability of Nintendo Wii

Q.1: Did you like to browse this web site?
Q.2: Did you like anything particular?
Q.3: Didn't you like anything particular?
Q.4: How would you define this experience of web browsing: funny, relaxing, difficult, boring, or other?
Q.5: Would you use again this web site on Wii if possible?

We also made two video recordings for each test: one camera was pointed to the screen, the other on the user, both with audio track, in order to view the user movement on the interface and on the WiiRemote and face expressions. The tasks have been assigned in sequence on site A and then repeated on site B. Both sites present roughly the same structure: we decided to test first the redesigned site, then the original, otherwise the user could be conditioned, knowing how to complete the tasks, altering the measured times. The three tasks assigned were at different levels of difficulty: find the address of the restaurant, find the price of defined dishes, find a way to book a table.

Levels were established on the basis of the depth of the links and the number of ways in which it was possible to end the task. The first task was easy, the second was of high difficulty and the last of medium difficulty. In this way the hardest test should not be influenced by tiredness.

5.1   Results

From our study we obtained three types of data:

1. Objective data, relating to times and number of operations for each assigned task.
2. Subjective data, obtained with the feedback in the final interview.
3. Subjective data recorded by our facilitator, relating to any difficulty that users experienced during the practice of the tests, on the basis of the thinking aloud method.

All data were recorded for sites A and B separately and registered on a database. During the transcription, data have been classified by type (2 or 3).

The results showed that users consider the redesigned site better than the original and in particular more easy and pleasant to use, so that our guidelines are valid. In particular the improvement is more evident approaching weak groups, namely elders and people without experience in the Internet. The use of our guidelines reduces the gap among the various ranges of people in the approach to the technology of the Wii web browsing.

5.2   Subjective Data from Feedback

Data recorded with interviews (type 2) revealed that the redesigned site had a higher approval rating. Both the general indicators (the answer to the questions Q.1 and Q.5 gave a much higher value in the redesigned site than in the original.

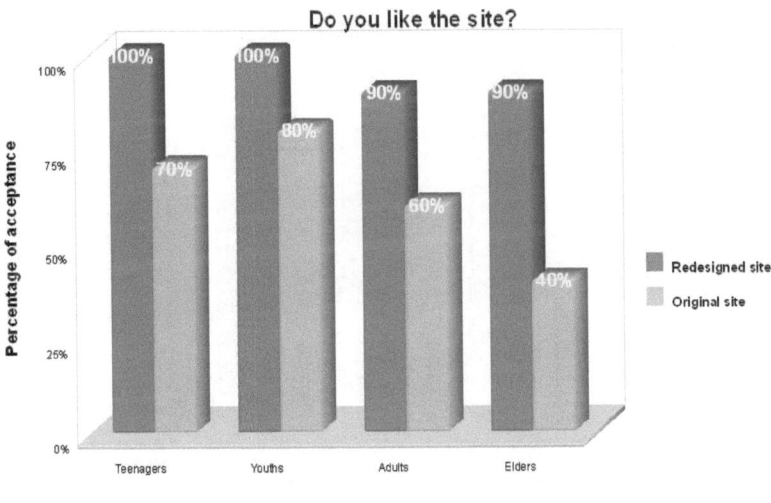

Fig. 7. Percentage of acceptance by the users of the redesigned and the original sites, as a function of the user's age class

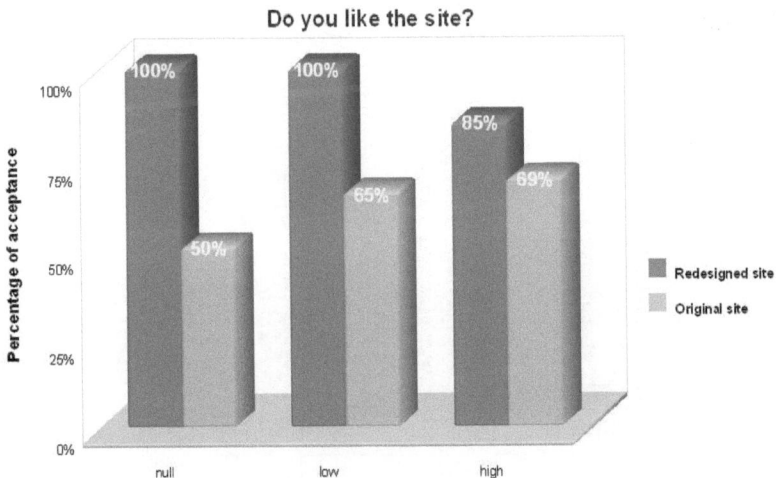

Fig. 8. Percentage of acceptance by the users of the redesigned and the original sites, as a function of the user's experience of the Internet

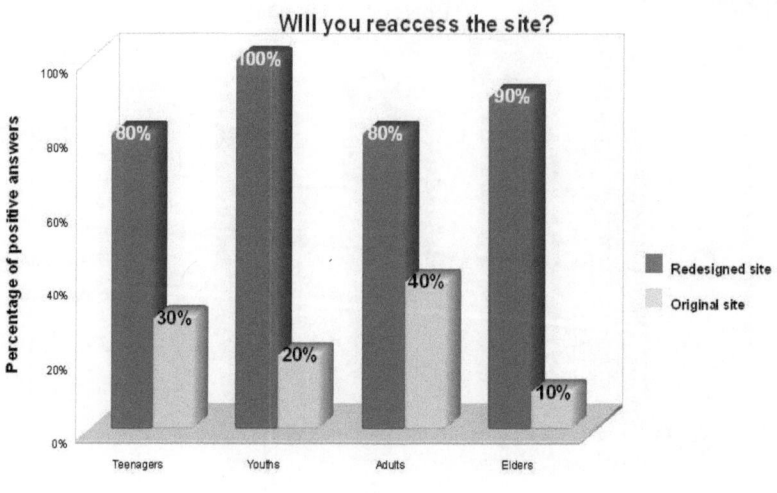

Fig. 9. Degree of intention of users to revisit the redesigned and the original sites, as a function of the user's age class

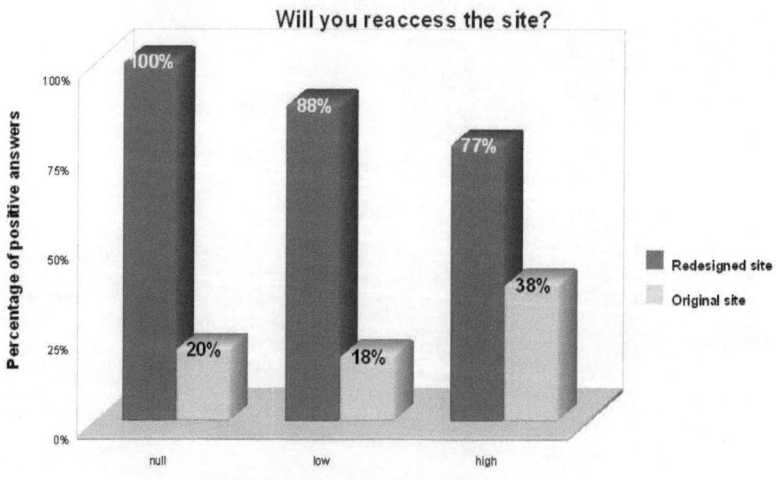

Fig. 10. Degree of intention of users to revisit the redesigned and the original sites, as a function of the user's experience with the Internet

The degree of acceptance of the two sites is shown as a function both of the user's age class (figure 7) and of the user's experience with the Internet (figure 8).

In addition, the intention of users to visit again the two sites is shown as a function both of the user's age class (figure 9) and of the user's experience with the Internet (figure 10). The figures show that the approval of the redesigned site is more evident when we get closer to the weak groups and the gap among the different groups declines considerably.

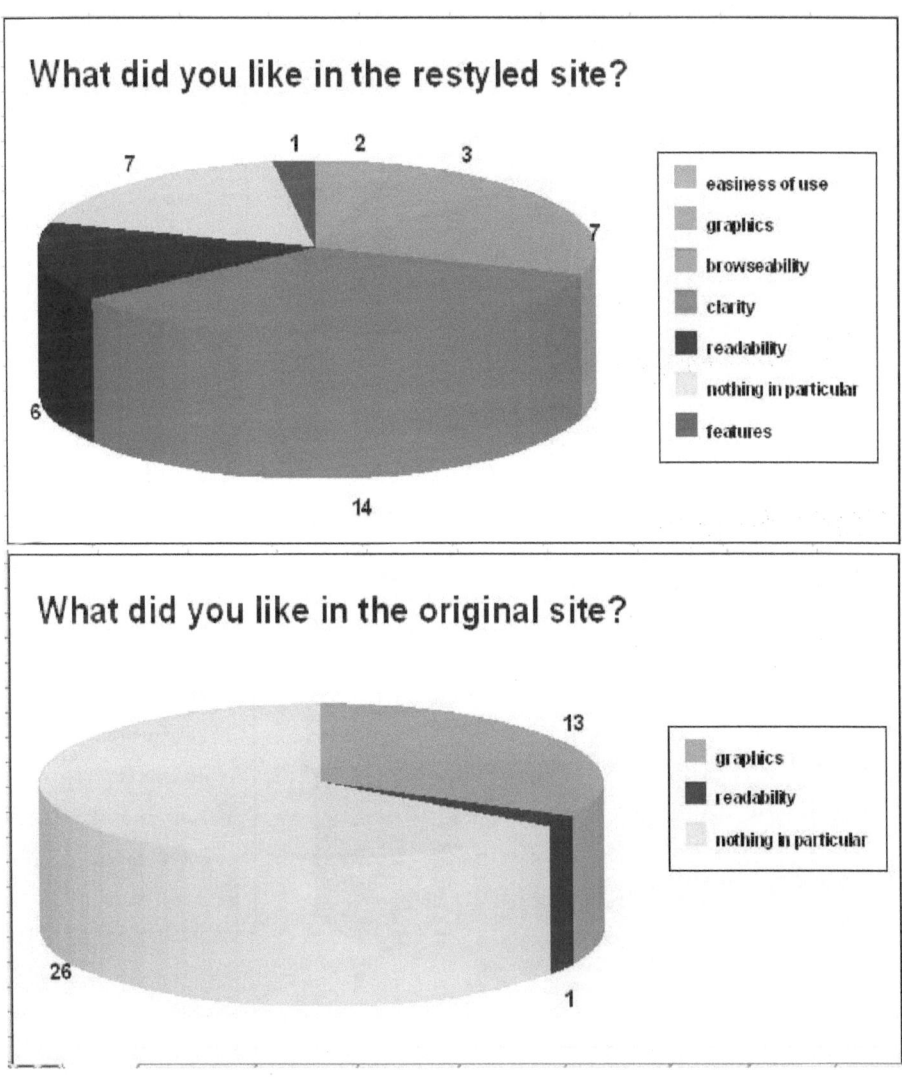

Fig. 11. What users did like in the redesigned (top) and the original (bottom) websites, as recorded by the facilitator in the final interview, with their frequency expressed by numbers

5.3   Subjective Data Recorded by the Facilitator

Each user reported a number of difficulties variable up to 2 (if none, we registered 'none'), for a total of 80 responses. The difficulties experienced by the users were: length of text, contents organization, frequent zooming, fear of clicking and visibility. Fear to click was registered when user kept pointer on a link

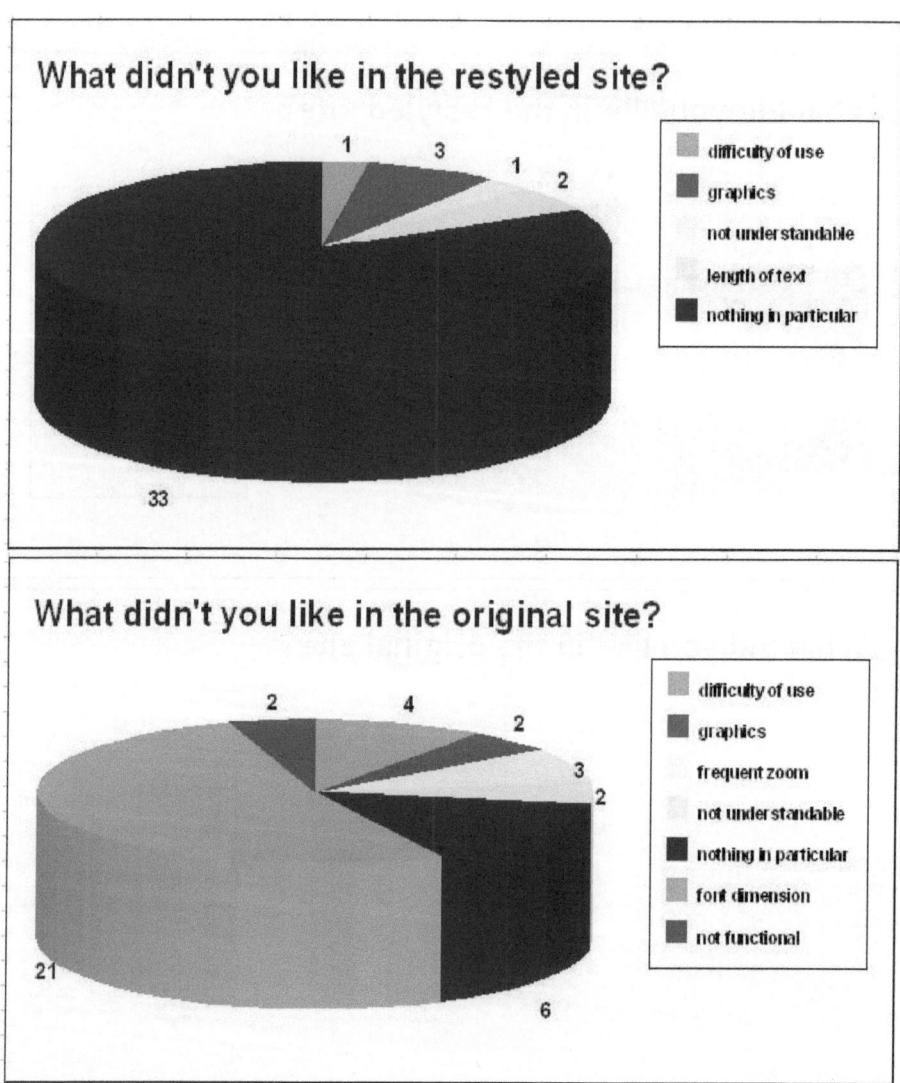

Fig. 12. What users did not like in the redesigned (top) and the original (bottom) websites, as recorded by the facilitator in the final interview, with their frequency expressed by numbers

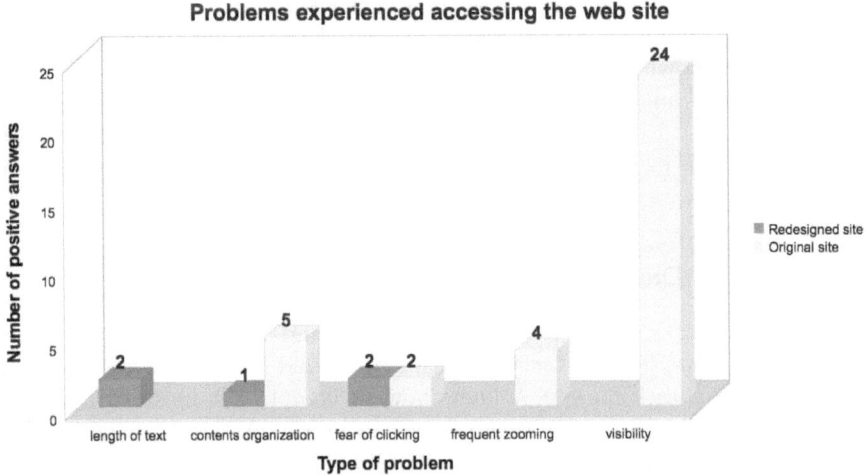

Fig. 13. Problems experienced by users accessing the redesigned and the original web-sites

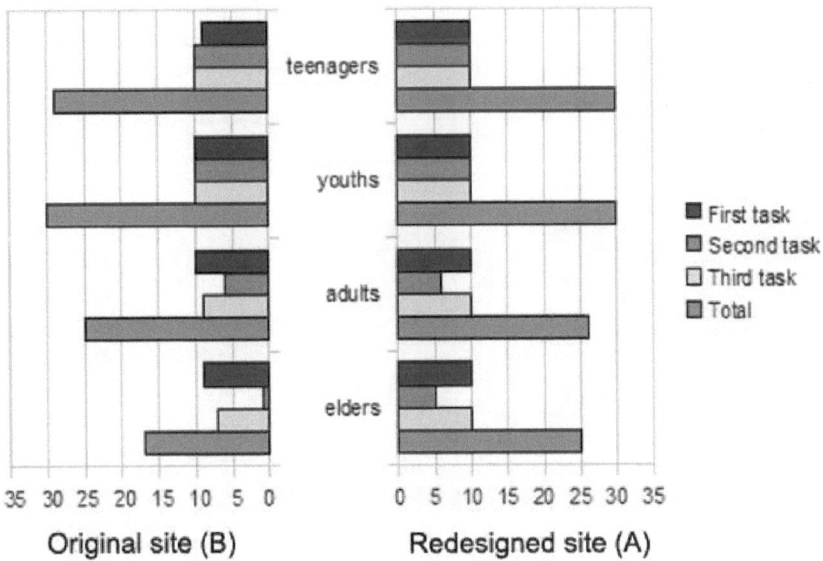

Fig. 14. Number of completed tasks for the redesigned site (lhs) and the original one (rhs) plotted as a function of the user's age class

without clicking on for over 5 seconds and reported aloud observations about it: this difficulty had the same attendance on both sites.

In figures 11 and 12 the final statistics related to the data collected by the facilitators are shown, as a functions of the characteistics of the two web sites.

The validity of our guidelines is further confirmed by the collected responses, as shown in figure 13, where the problems experienced accessing the two sites are shown and it is evident that the problems with the redesigned site are almost solved.

5.4   Objective Data

The number of tasks completed is 111 for site A and 101 for site B. We notice that the solving method for tasks in both sites A and B was the same, so users who completed a task on site A knew how to complete it in site B. In spite of it, we have a higher percentage of success in site A than in site B. Even analyzing these data by age (as shown in figure 14) and experience in the Internet (as shown in figure 15), we can notice that the redesigned site allows users to complete a higher number of tasks, particularly evident in elders. It is also evident that the redesigned site tends to reduce the intergenerational gap and the different experience with the Internet, confirming the subjective data obtained from feedback.

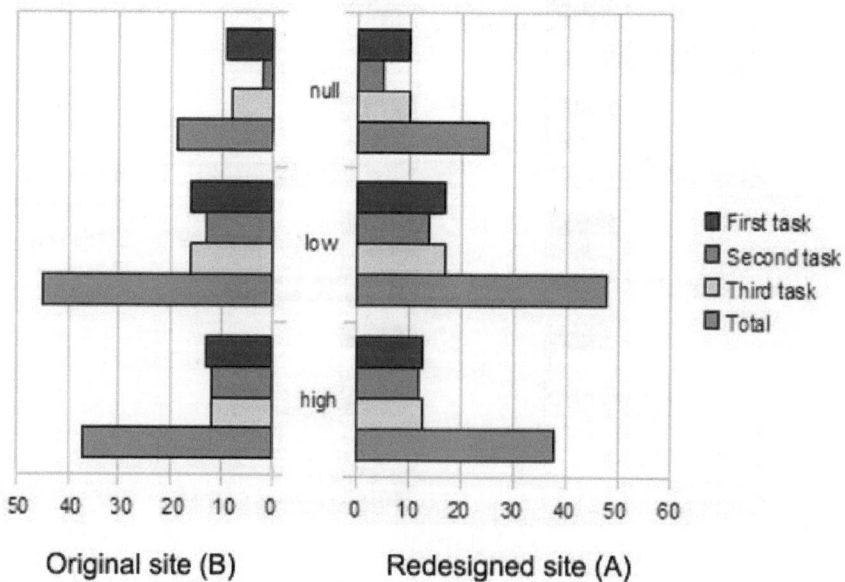

Fig. 15. Number of completed tasks for the redesigned site (lhs) and the original one (rhs) plotted as a function of the user's experience with the Internet

6   Conclusions and Future Developments

The proposed twelve guidelines for the web designers in order to maximize the usability of the Nintendo Wii console have been used to redesign an existing web site. The results of the tests made on a set of 40 users grouped by 4 classes of age, related to the execution of three tasks on both web sites were recorded. The data analysis clearly shows that the site following the guidelines is more usable and the improvement is more pronounced for weak categories (elderly and individuals with no experience with Internet browsing), so the use of the proposed guidelines reduces the difference between the diverse users approaches to this technology. Indeed, all graphs related to the collected data show that the redesigned site has received a larger consensus, particularly in weaker categories. It was also clear that the issues studied are actually really relevant for usability, and that the use of the guidelines solve the problems. We can conclude that the proposed guidelines maximize the usability of a website surfed with the Nintendo Wii console.

In recent years we have seen an important increase in worldwide use of web browsing-enabled mobile devices, such as PDAs, smartphones, portable gaming consoles. These devices have similar problems of visibility, clickability, understandability and compatibiliy of the Nintendo Wii platform. Furthermore, it is also more feasible to connect these devices to TVs, to use them as external displays. With few appropriate adjustments, in particular about compatibility and points of control, our guidelines can be considered valid also for web usability on mobile devices. This is particularly relevant for future developments, considering that the Wiiremote, that can be connected to a PC or to any other device capable of processiong the bluetooth signal, can be used to perform simulations and education. With the Wii balance board we could surf the Internet or move on virtual paths with a satellite vision (like Google Earth). With the three controllers (Wiiremote, Nunchuck and the balance board) installed, an advanced control of the surfing experience is possible, even for simulate different environments, like multitouch. Nintendo itself started to exploit these possibilities, adapting the game "Trauma center" to the Wii console, born for the touch screen of Nintendo DS: the game offers a simulating environment for surgery, so precise that it has been used by doctors for their first tests.

The twelve guidelines proposed for optimizing the web browsing using the stimulating environment of the Wii console can contribute to the development of web sites accessible to weak categories and of applications in several disciplines where the characteristics of the pointing devices will make the surfing experience useful and fruitful.

References

1. Abran, A., Khelifi, A., Suryn, W., Seffah, A.: Usability Meanings and Interpretations in ISO Standards. Software Quality Journal 11(4), 325–338 (2003)
2. Nielsen, J., Loranger, H.: Prioritizing Web Usability. New Riders Press, Berkeley (2006)

3. Lazar, J. (ed.): Universal Usability: Designing Computer Interfaces for Diverse User Populations. John Wiley & Sons, Chichester (2007)
4. Bias, R.G., Mayhew, D.J. (eds.): Cost-Justifying Usability. Academic Press, Boston (1994)
5. Flavian, C., Guinaliu, M., Gurrea, R.: The role played by perceived usability, satisfaction and consumer trust on website loyalty. Information and Management 43(1), 1–14 (2006)
6. Stewart, T.: Ergonomics standards concerning human-system interaction: Visual displays, controls and environmental requirements. Applied Ergonomics 26(4), 271–274 (1995)
7. Helander, M.: A Guide to Human Factors and Ergonomics, 2nd edn. CRC Press, Boca Raton (2005)
8. Sears, A., Jacko, J.A. (eds.): The Human-Computer Interaction Handbook, 2nd edn. CRC Press, Boca Raton (2007)
9. Sharp, H., Rogers, Y., Preece, J.: Interaction Design: Beyond Human-Computer Interaction, West Sussex, England. John Wiley & Sons, Inc., Chichester (2007)
10. Web content accessibility Guidelines 1.0 (1999), http://www.w3.org/TR/WAI-WEBCONTENT/; Techniques for Web Content Accessibility Guidelines 1.0 (2000), http://www.w3.org/TR/WAI-WEBCONTENT-TECHS/; Techniques for Web Content Accessibility Guidelines 2.0 (2007), http://www.w3.org/TR/WCAG20/
11. Legge 9 gennaio 2004, n. 4, by Italian Parliament about the dispositions to benefit disabled users in accessing computers instruments (Disposizioni per favorire l'accesso dei soggetti disabili agli strumenti informatici)
12. Human Technology Laboratories, Università degli Studi di Padova: http://psicologia.unipd.it/htlab
13. Mazzocato, G.: The accessible colorwheel: http://gmazzocato.altervista.org/colorwheel/wheel.php
14. Visciola, M., Vanderbeeken, M.: Co-creation and partecipatory democracy: how web usability and strategic communication can facilitate them. User Experience Magazine 54, Usabilty Professionals' Association Pubblication (2006); Visciola, M.: Usabilità dei siti web, curare l'esperienza d'uso in Internet, Apogeo, Milano, Italy (2006)

Improving Urban Land Cover Classification Using Fuzzy Image Segmentation

Ivan Lizarazo[1] and Paul Elsner[2]

[1] Cadastral Engineering and Geodesy Department
Universidad Distrital Francisco Jose de Caldas, Bogota, Colombia
ilizarazo@udistrital.edu.co
[2] School of Geography, Birkbeck College, University of London, London, UK
p.elsner@bbk.ac.uk

Abstract. The increasing availability of high spatial resolution images provides detailed and up-to-date representations of cities. However, ana-lysis of such digital imagery data using traditional pixel-wise approaches remains a challenge due to the spectral complexity of urban areas. Object-Based Image Analysis (OBIA) is emerging as an alternative method to produce landcover information. Standard OBIA approaches rely on ima-ge segmentation which partitions the image into a set of 'crisp' non-overlapping image-objects. This step regularly requires significant user-interaction to parameterise a functional segmentation model. This paper proposes fuzzy image segmentation which produces fully overlapping image-regions with indeterminate boundaries that serves as alternative framework for the subsequent image classification. The new method uses three stages: (i) fuzzy image segmentation, (ii) feature analysis, and (iii) defuzzification, that were implemented applying Support Vector Machine (SVM) techniques and using open source software. The new method was tested against a benchmark land-cover classification that applied standard crisp image segmentation. Results show that fuzzy image segmentation can produce good thematic accuracy with little user input. It therefore provides a new and automated technique for producing accurate urban land cover data from high spatial resolution imagery.

1 Introduction

Urban land cover classification from remotely sensed images is important for a variety of applications ranging from climate models downscaling to hydrological modeling and cities planning [1]. It remains a challenging problem due to the complex composition and irregular sizes and patterns of urban surfaces [2]. This issue is aggravated with high-spatial resolution images that increase the spectral variability within individual land cover units [3]. Traditional pixel-based classification algorithms rely mainly on spectral information and are often not capable to resolve such complex spectral and spatial signals [4].

Recently developed object-based methods for classifying high spatial resolution images focus on image segmentation, a process which produces a set of image-objects - also referred to as image-regions. Carefully derived image objects

M.L. Gavrilova and C.J.K. Tan (Eds.): Trans. on Comput. Sci. VI, LNCS 5730, pp. 41–56, 2009.
© Springer-Verlag Berlin Heidelberg 2009

represent, partly or completely, geographic objects of interest [5]. Once image segmentation has been conducted, attributes of image-objects are measured to enhance the set of attributes used for the subsequent classification. This object-oriented classification method, also known as OBIA, is increasingly being used as alternative to the conventional pixel based methods [5] and has substantially improved the thematic accuracy of image classification [6] [7] [8].

Image segmentation is recognized as a critical step for object oriented image analysis [9]. It has been reported that a good segmentation produces meaningful image-objects, which is achieved when the overall differences between the image-objects and the associated reference objects are as low as possible [10] [11].

1.1   Crisp Image Segmentation: From pixels to Non-overlapping Image-Objects

Current techniques for image segmentation divide an image into a set of non-overlapping regions whose union is the entire image [12]. Attributes used for segmentation may include intensity, colour, edges and texture [13]. The objective of segmentation is to decompose an image into parts that are meaningful with respect to a particular application. Segmentation allocates each pixel to one specific region –that is, it labels a pixel to the segment it lays in [14]. However, segmentation does not assign meaning to the resulting regions. Such task is conducted in a subsequent classification stage.

Segmentation is often viewed as an enabler process that allows the image analyst to change from pixel as unit of observation to working with larger image-objects or image-regions having a degree of geometric and spectral homogeneity [15]. Once homogeneous image-objects have been produced at one or several nested scales, then they may be classified using traditional algorithms such as maximum likelihood or new techniques such as knowledge-based and fuzzy classification [5]. Attributes that can be measured on image-objects may be spectral (variations in tone or colour) or spatial (shape and spatial patterns). The spatial attributes may refer to the structure or texture of the object –understood as tonal variation focused on the object of interest– or to the broader relationship between the object and the remainder of the scene –usually referred to as context– [16].

Image segmentation may be formally described as follows. Let R represent the entire geographic region occupied by an image. Segmentation is a process that partitions R into n image-regions $R_1, R_2, \ldots, R_n$ such that:

$$\bigcup_{i=1}^{n} R_i = R. \tag{1}$$

$$R \quad \text{is a connected set,} \quad i = 1, 2, \ldots, n. \tag{2}$$

$$R_i \cap R_j = \emptyset \text{ for all } i \text{ and } j, \quad i \neq j. \tag{3}$$

$$Pred(R_i) = TRUE \text{ for } \quad i = 1, 2, \ldots, n. \tag{4}$$

$$Pred(R_i \cup R_j) = FALSE \quad \text{for any adjacent image-regions } R_i \text{ and } R_j. \quad (5)$$

In equations 1 to 5, $Pred(R_k)$ is a logical predicate defined over the points in set R_k, and $\emptyset$ is the null set. The symbols $\cup$ and $\cap$ represent set union and intersection, respectively. A image-region is a connected set of pixels. Two regions R_i and R_j are considered adjacent if pixels lying on their boundaries are neighbors. A pixel may have either 4 neighbors (i.e. two *horizontal* and two *vertical*) or 8 neighbors when the *four* diagonal neighbors are also considered. Equation 1 indicates that every pixel is allocated to a image-region. Equation 2 requires that pixels in a region be 4- or 8- connected. Equation 3 indicates that the image-regions must be disjoint (i.e. they have no pixel in common). Equation 4 indicates what properties must be satisfied by the pixels in a image-region –for example, $Pred(R_i) = TRUE$ if all pixels in R_i have their intensity level within certain interval. Finally, equation 5 denotes that two adjacent image-regions R_i and R_j must be different in the sense of predicate $Pred$.

A traditional image segmentation produce crisp image-objects that may represent real-world objects, part of objects or just noise. In many cases, it is often difficult to determine if a pixel belongs to one or another particular image-object because spectral features may not have sharp transitions at geographic object's boundaries. Regularly, it is problematic to identify uniform and homogeneous image-objects from remotely sensed images because of sensor noise, shadows and occlusions.

Existing software implementations of image segmentation are not automated and require that users manually select a suitable set of parameters values for a given data set. Then, users need to establish correspondence between the image-objects and the real-world geographic objects (classes) and to analyze and define the set of spectral, geometric, or textural image-regions characteristics (also referred to as *feature vector*) to be used as basis for the classification process [17]. Particularly in urban landscapes the objects of interest can have a large variety of sizes. This makes it necessary to produce and interactively analyze image-objects at different scales so that an 'optimal' combination is identified. This analysis is a time-consuming process and introduces a substantial element of subjectivity into the process [18].

What is hence needed is an image classification procedure that requires none or very little user-driven input for the parameterization of a robust and accurate segmentation model. This paper aims to propose a method that meet this objective more closely. The central conceptual approach is that of a *fuzzy image segmentation*, as opposed to 'hard' or 'crisp' image segmentation methods that have until now been implemented.

1.2 Fuzzy Image Segmentation: From Pixels to Fully Overlapping Image-Regions

The adoption of fuzzy sets concepts into the segmentation process has been suggested in the computer vision literature [14] since one decade ago but this suggestion appears not to have translated in actual applications. Within the framework of environmental remote sensing, its application is very recent [19,20]. Fuzzy image segmentation aims to include uncertainties due to sensor noise, limited sensor resolution, atmospheric dispersion, mixed pixels and vague class descriptions. Due to these factors, geographic objects usually appear in remotely sensed images as fuzzy objects with no clear boundaries. This is often true for both natural objects and man-made objects appearing in urban scenes. In order to deal with the ambiguity existing in the image and the uncertainty of the image analysis process, this paper proposes a new method for image classification based on fuzzy image segmentation.

The result of a fuzzy segmentation is the transformation of an image, composed by n multispectral channels, into m fully overlapping image-regions R_i where m denotes the number of target land-cover classes. Each image-region R_i has indeterminate boundaries, that is, it is a fuzzy image region characterized by a membership function $\mu(X_j)$ whose values can be anywhere in the range from 0 to 1. This means that equations 2, 3 and 5 that apply to traditional segmentation do not hold for fuzzy segmentation. Instead, each pixel belongs to all m image-regions with varying degrees of membership (Equation 3). There is also no condition of spatial connectedness (Equation 2), dissimilarity between regions (Equation 5), and membership values are not longer constrained to be either 0 or 1.

While crisp-image objects are commonly defined by their outline, fuzzy image-regions lack a sharp boundary. Fuzzy image-regions can be interpreted as a subtype of spatial vague objects and therefore they can be defined by the points belonging to it [21]. Hence, fuzzy image-regions are aggregations of pixels carrying a membership value which indicates the degree to which each location belongs to one or another region, each region representing a particular target land-cover class. The higher a membership grade is for a point, the more a point belongs to the class; a membership grade 1 indicates it belongs fully to the class whereas membership grade 0 indicates the opposite.

2   Case Study: Land-Cover Classification of Washington DC Mall Area

The potential of fuzzy segmentation was tested with a case study based on an urban image of the Washington DC Mall area that was collected by the Hyperspectral Digital Imagery Collection Experiment (HYDICE) sensor on August 23, 1995. HYDICE is a push broom aircraft sensor system which operates in the spectral range of 400 to 2500 nm with 10 nm spectral resolution. The spatial resolution of the image is 3 m. The original data set is 1,280 x 307 pixels and comprises 191 spectral bands. A false colour composition from this image is shown in Figure 1(a). A number of studies have previously been carried out with this data set [22] [23] [24]. This makes it possible to directly evaluate the potential of the proposed approach in reference to those benchmark studies.

The challenge for the classification of the Washington Mall data set arises from the fact that (a) the building roofs of that area consist of a large number

of materials, and (b) a number of the materials used in the roofs are the same
or similar to those used in the streets [22]. An *independent component analysis*
(ICA) technique was applied to summarize the hyper-spectral data as the ICA
bands provide an un-mixed representation of the original data [25]. The first
three ICA components are shown in Figure 1(b), 1(c) and 1(d). Six ICA bands
were used as input for the image segmentation stage as the best nongaussian
and mutually independent components for the hyper-spectral data set.

Seven land-cover classes were defined as target classes following the conven-
tions used in previous experiments using the same dataset [22]: (i) Road, (ii)
Roof, (iii) Shadow, (iv) Grass, (v) Trees, (vi) Water and (vii) Trail. Training
and testing samples were collected from a reference dataset which was supplied
with the original data [22]. For collecting the samples, the ground reference
dataset was first divided into two approximately equal parts. Then, one half was

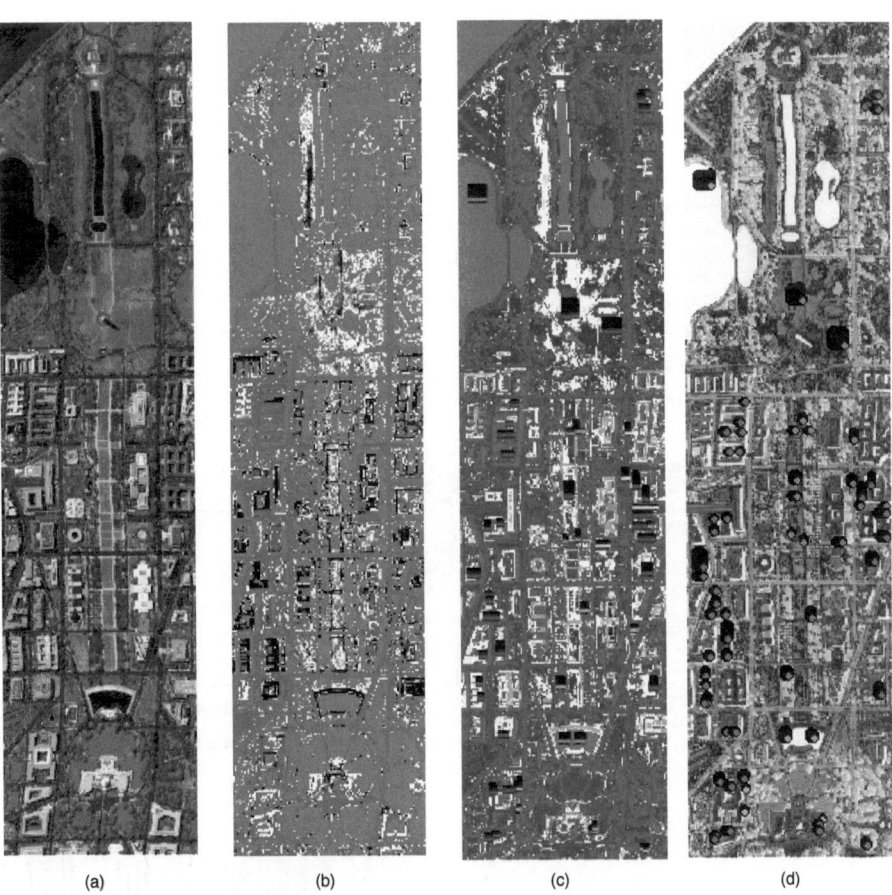

(a)   (b)   (c)   (d)

Fig. 1. (a) False color image composition using R = Band 63 (794 nm), G = Band 52
(676 nm), B = Band 36 (553 nm). (b) ICA band 1. (c) ICA band 2 with training sites
overlaid. (d) ICA band 3 with testing sites overlaid.

used for deriving the SVM models, and the other half to subsequently evaluate classification accuracy. Table 1 provides an overview of training and testing samples for each land-cover class. Training sites are shown in Figure 1(c). Testing sites are shown in Figure 1(d).

This case study can be sub-divided into three stages which are illustrated in Figure 2: (i) Fuzzy Image Segmentation in which the input ICA channels are converted into image-regions, each representing the fuzzy membership for one specific target land-cover class; (ii) Feature Analysis in which properties of the image-region set are analysed to derive additional attributes for the subsequent image classification; and (iii) Defuzzification in which each pixel is assigned to one of the seven target classes, leading to a single image representing the final

Table 1. Number of pixels for training and testing data

Class	Training Data	Testing Data
Road	234	185
Roof	1830	1979
Shadow	69	30
Grass	973	955
Trees	226	179
Water	621	582
Trail	95	80
Total	4048	3990

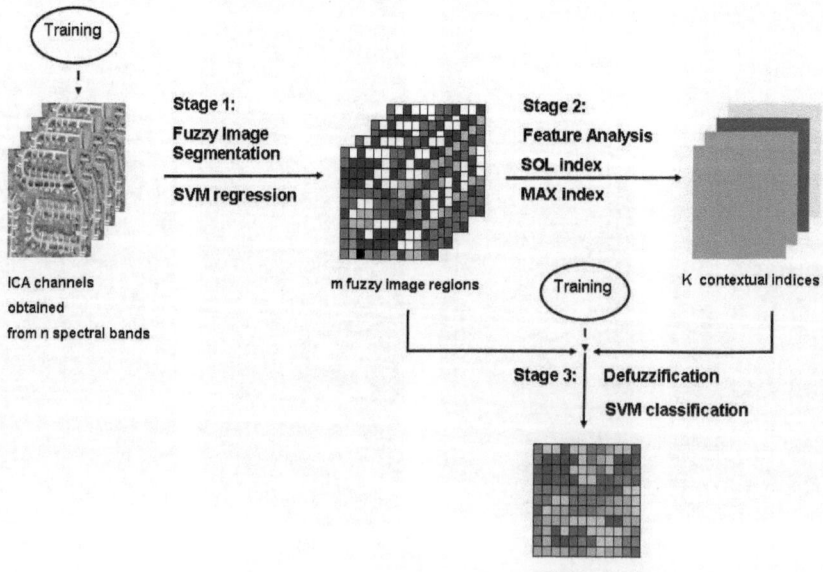

Fig. 2. Main stages of the proposed classification method: (i) Fuzzy Image Segmentation, (ii) Feature Analysis, and (iii) Defuzzification

land-cover classification. Every stage of the method is discussed in the following sections.

2.1 Fuzzy Image Segmentation

Support Vector Machines (SVM) is a machine learning technique which is increasingly popular due to its ability to create accurate classification and regression models using a small number of training samples [26,27,28]. SVM was used to build a binomial regression model where every target land-class is expressed as a quantitative value which is a linear function of n predictor multi-spectral channels. For this regression task, one SVM model is fitted to every class using the *one-against-all* approach. The output of this process was a set of $m = 7$ new image-regions, where m is the number of intended land-cover classes. Each new image-region is composed by pixels expressing degrees of membership to one specific class. The values were normalized to be in the range $[0, 1]$, where 0 indicates no membership, 1 denotes full membership, and any value in-between represents partial membership. At every location, the sum of membership values is constrained to be in the range $[0, 1]$.

For implementing this model, the SVM technique was used to produce a regression with the Gaussian Radial Basis function (RBF)[29]:

$$K(x, x^{'}) = e^{-\frac{|x-x^{'}|^2}{2\sigma^2}} \ . \tag{6}$$

With the RBF kernel, there are two parameters to be determined in the SVM model: C and γ. To get good generalization ability, the following automated validation process was run for tuning SVM parameters:

- Consider a grid space of (C, γ) with $C = 2^{(1:6)}$ and $\gamma = 2^{(-4:4)}$.
- For each hyper-parameter pair (C, γ) in the search space, conduct ten cross fold validation in the training set.
- Choose the parameter pair (C, γ) that leads to the lowest balanced error rate.
- Use the best parameter pair to create a model as the predictor.

Table 2 list parameters of the best SVM regression models tuned from the training sample using an epsilon band equal to 0.1. These models were applied to the whole image dataset to produce fuzzy image-regions representing membership values to every land-cover class.

The results of the fuzzy image segmentation stage are displayed in figures 3 and 4 which show the fuzzy image-regions for each target land-cover class.

2.2 Feature Analysis

Once fuzzy image-regions have been established, a number of operations may be applied to further enhance the set of attributes that can be utilized for the subsequent image classification. A number of operations for both conventional

Table 2. Summary of best SVM regression models to produce fuzzy image-regions

Class	Cost	Gamma	Number of Support Vectors
Roads	16	2	1032
Roof	8	2	1338
Shadow	64	4	1294
Grass	32	0.5	1159
Trees	64	0.25	523
Water	32	16	2196
Trail	16	4	1312

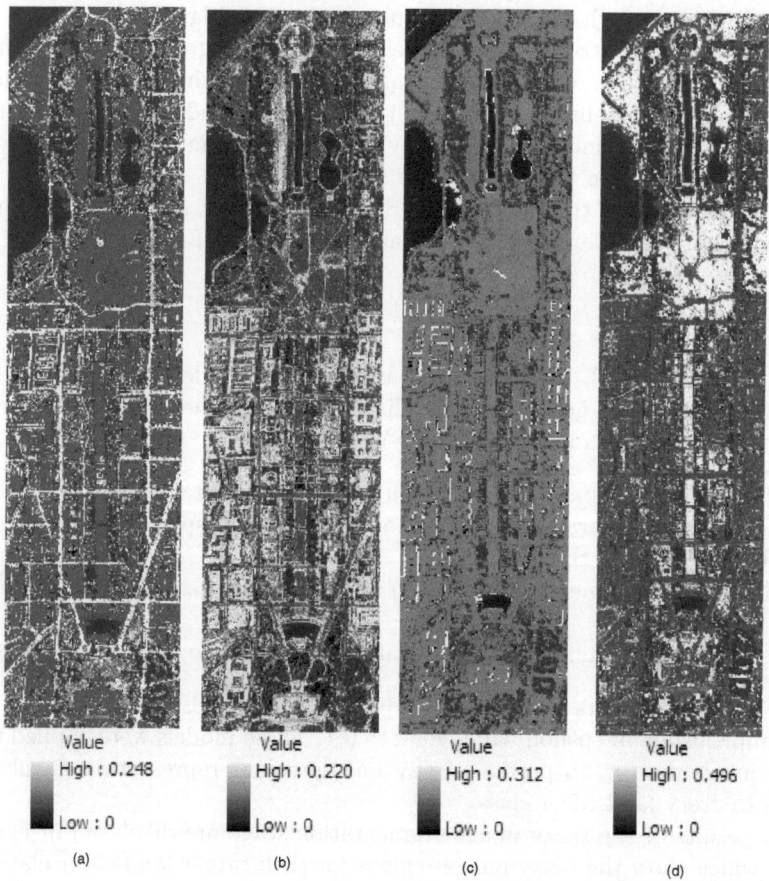

Fig. 3. (a) Fuzzy image-regions for target class *road*. (b) Fuzzy image-regions for target class *roof*. (c)Fuzzy image-regions for target class *shadow*. (d) Fuzzy image-regions for target class *grass*. White corresponds to higher membership values and black correspond to lower membership values. Gray tones represent degrees of membership in between. Contrast stretching was applied to all images for clarity.

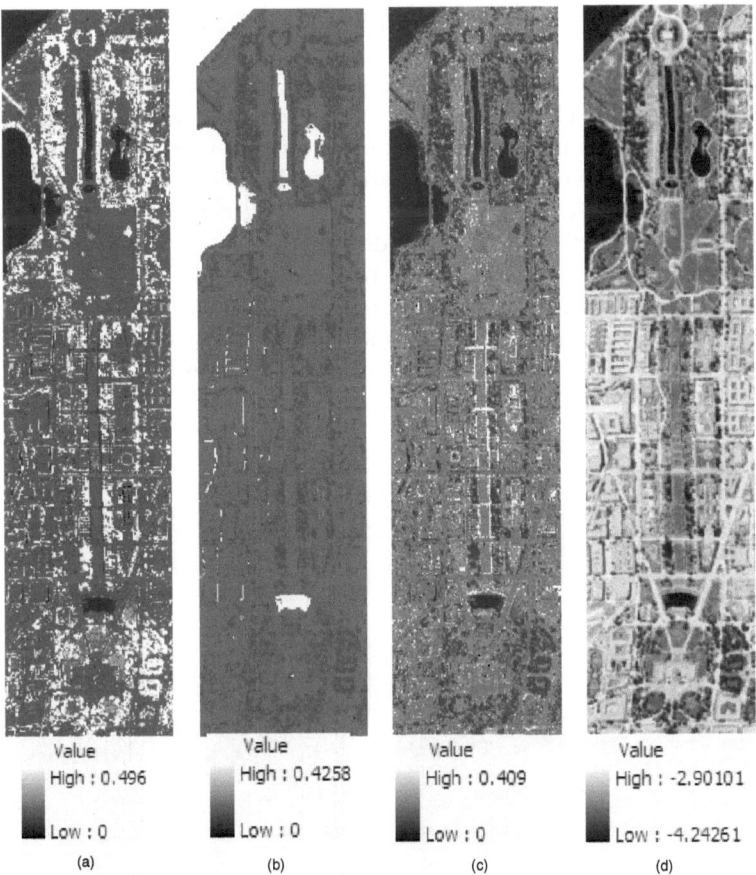

Value
High : 0.496
Low : 0
(a)

Value
High : 0.4258
Low : 0
(b)

Value
High : 0.409
Low : 0
(c)

Value
High : -2.90101
Low : -4.24261
(d)

Fig. 4. (a) Fuzzy image-regions for target class *trees*. (b) Fuzzy image-regions for target class *water*. (c) Fuzzy image-regions for target class *trail*. (d) *SOL* index between *road* and *roof*. Light tones in (a), (b) and (c) represent high degrees of membership and black correspond to low degrees of membership. Gray tones represent degrees of membership in between. In (d) ligh tones represent high degree of overlapping. Contrast stretching was applied to all images for clarity.

fuzzy sets and spatial fuzzy sets are potentially available for this. These include comparisons between fuzzy sets (intersection, union, difference); specific operators to reduce or stretch the elements of fuzzy sets (concentration, dilation, intensification); geometric measurements for spatial fuzzy sets (minimum bounding rectangle, convex hull, area, perimeter, distance); specific operators for defuzzification (alpha cuts); and a number of metrics for description or comparison of spatial fuzzy sets (global confusion, overlapping between fuzzy regions, memberships function as mass distribution) [21][30][31][32].

In this paper, two different indices were used to represent contextual relationships between fuzzy image-region. The first index is the *Sum of Logarithms* Index (*SOL*), defined as:

$$SOL = \ln\left(\mu_{iA} * \mu_{iB}\right) = \ln\left(\mu_{iA}\right) + \ln\left(\mu_{iB}\right) \qquad (7)$$

where ln is the natural logarithm, and μ_{iA} and μ_{iB} are the membership values of the i_{th} location to the classes A and B, evaluated as the average of membership values on a 3x3 pixels window. The SOL index measures the overlapping between two fuzzy image-regions and indicates areas exhibiting high membership values to more than one target class. SOL index values were calculated for the following pairs of classes as they were visually identified as potential source of spectral confusion: road & roof, road & grass, roof & grass, roof & trail and shadow & water. The results are shown in figures 4 (d) and 5. As second index the *Maximum Membership* Index (MAX) was determined by extracting the largest membership value of the ith pixel. The result is shown in Figure 6 (a).

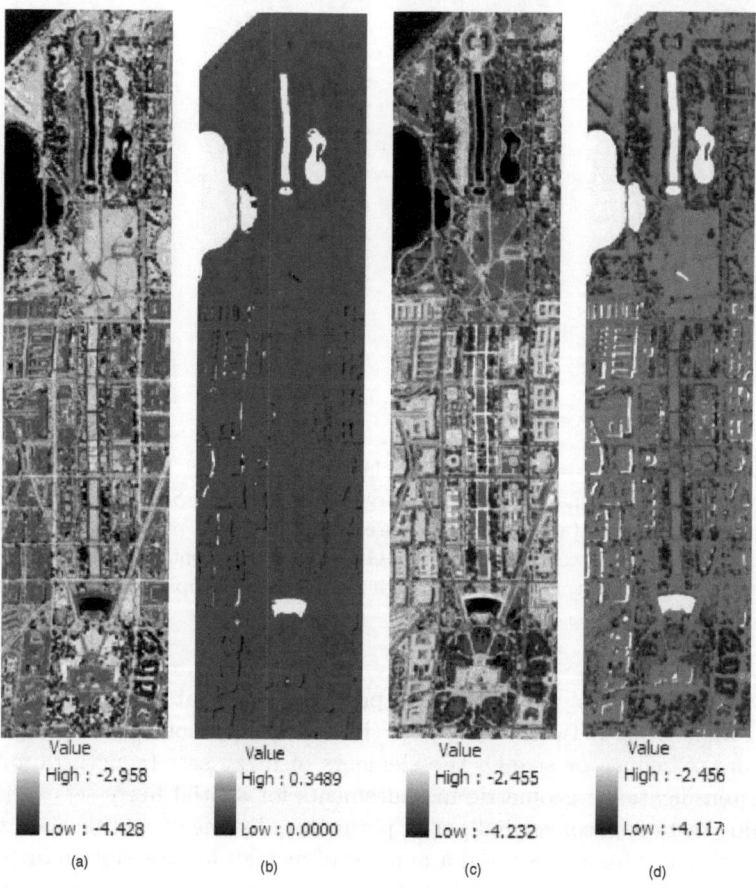

Fig. 5. (a) SOL index between road and grass. (b) SOL index between roof and water. (c) SOL index between roof and trail. (d) SOL index between shadow and water. Light tones represent higher degrees of overlapping. Contrast stretching was applied to all images for clarity.

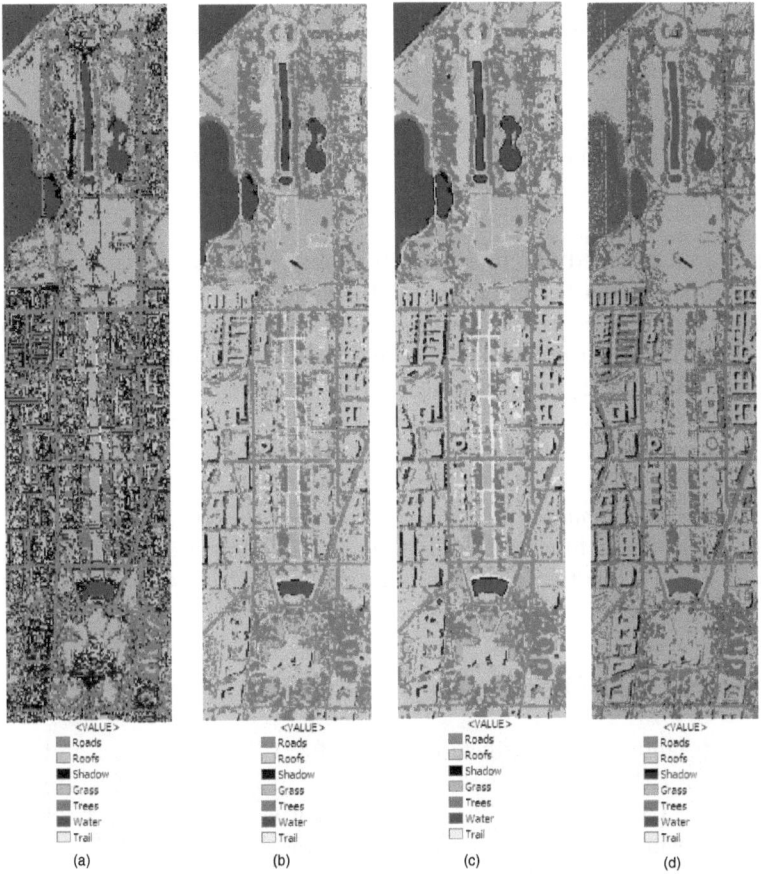

Fig. 6. (a) MAX index representing the class with the highest membership value at every point. (b) Classification using SET-1 predictors combination. (c)SVM classification using SET-2 predictors combination; (d) Benchmark classification using DAFE bands [22]. Classified images show Road in brown, Roof in orange, Shadow in black, Grass in light green, Trees in dark green, Water in blue and Trails in yellow.

2.3 Defuzzification

In the defuzzification stage, the actual classification was performed to assign each pixel to one of the seven target land cover classes. For this, a SVM-based classification technique was applied using two different sets of predictor variables as follows:

- SET-1: 7 fuzzy image-regions plus 5 SOL indices plus 1 MAX index; and
- SET-2: 5 SOL indexes plus 1 MAX index.

For tuning SVM parameters, the same automated procedure outlined in section §2.1 was applied. A SVM-based classification model was built using the *one-against-one* technique. In the kernel transformed space, qualitative (categorical)

Table 3. Summary of best SVM classification models to produce land-cover classes

Method	Class	Cost	Gamma	Number of Support Vectors
SET-1	All classes	16	2	152
SET-2	All classes	16	1	135

responses are expressed as a linear function of the corresponding set of predictor variables. This option fits a multinomial logistic model [29]. The output of the defuzzification stage was one single new image that represents target land-cover units. Table 3 list parameters of the best SVM classification models tuned to produce the final land-cover classification.

All the stages of the new classification method were implemented using R, a free software environment for statistical computing and analysis [33]. In addition to the R base package which provides basic statistic capabilities, the additional packages $rgdal$, $fastICA$, $sp - maptools$, and $e1071$ were used. They provide, respectively, functions for reading/writing images, analysis of independent components, creating and manipulating spatial classes and applying SVM algorithms.

3   Results and Discussion

The output land-cover classification using fuzzy image-regions and methods SET-1 and SET-2 are shown as thematic maps in Figures 6(b) and 6(c). Percentage of Correct Classification (PCC) for SET-1 was 94 +/- 1.5 %. Global Kappa Index of Agreement (KIA) value for this classification was 0.91. Table 5, shows the error matrix for classification using SET-1 predictors. It can be seen that lowest producer's accuracy corresponded to the classes Grass (0.88) and Trail (0.73). Lowest user's accuracy corresponded to classes Road (0.52) and Shadow (0.58).

For SET-2 classification, PCC value was 96 +/- 1.4% and the KIA value was 0.94. Major misclassification problems corresponded to confusion between the classes Road/Grass and Shadow/Water. Overall, both the SET-1 and SET-2 methods proved to be straightforward procedures that produced acceptable thematic accuracies for most practical purposes. Table 3 lists the parameters used to test the statistical significance of the accuracy values obtained. A visual assessment shows that both SET-1 and SET-2 methods produce appealing results whose accuracy seems to be similar.

For comparison purposes, a reference classification obtained using 9 bands extracted from the original dataset using Discriminant Analysis Feature Extraction (DAFE) techniques is shown in Figure 6(d) [22]. The best thematic accuracy (PCC) achieved in a study using nonparametric weighted feature extraction (NWFE), an improved DAFE technique, was 92.3 % [23]. The best thematic accuracy (PCC) achieved in a study using multi-level crisp segmentation and a laborious combination of spectral, textural and shape features for

Table 4. Comparison of thematic accuracy between fuzzy image segmentation and benchmark studies. For fuzzy image segmentation, two different sets of predictor variables. In SET-1, predictors are fuzzy image-regions (FIRs) and contextual indices. In SET-2, predictors are contextual indices only.

No	Predictors	PCC (%)	$\hat{K}$	Var.of. $\hat{K}$	Z
Fuzzy Image Segmentation					
SET-1	FIRs + SOLs + MAX	94 +/- 1.5	0.91	0.0005	38.97
SET-2	SOLs + MAX	96 +/- 1.7	0.94	0.0004	41.60
Benchmark Studies					
NWFE [23]	Five NWFE bands	92.3 +/- 1.6	—	—	—
Crisp Segmentation [24]	Shape, Texture, Spectral	96.8	—	—	—

Table 5. Error matrix corresponding to classification method SET-1

Map	Road	Roof	Shadow	Grass	Trees	Water	Trail	Total	USER
Road	162	32	0	114	0	0	5	313	0.52
Roof	22	1872	0	1	0	0	15	1959	0.98
Shadow	0	4	30	0	0	18	0	52	0.58
Grass	0	11	0	838	0	0	2	851	0.98
Trees	1	2	0	2	179	0	0	18	0.97
Water	0	0	0	0	0	564	0	564	1.00
Trail	0	9	0	0	0	0	58	67	0.87
Total	185	1979	30	955	179	582	80	3990	
PROD.	0.88	0.97	1.00	0.88	1.00	0.97	0.73		

image-objects was 96.8 % [24]. It is hence evident that, while the proposed fuzzy image-segmentation method performs similarly to that of the those benchmark studies, it demands less user skills and labour.

In the experiments reported in this paper, the use of SOL indexes increased the thematic accuracy of the classification compared with earlier results using a much simpler overlapping index known as $ANDI$ [19]. It is further interesting to note that by using only contextual indices as predictor variables (i.e. SET-2) it was possible to achieve similar accuracies than by using a much more complex combination of predictors (i.e. SET-1). These results suggest that the full potential of the new method relies on a proper exploitation of the contextual properties existing between fuzzy image-regions obtained in the segmentation stage. For such purposes, a number of operations from the mathematical morphology realm, such as adjacency and distance between fuzzy spatial objects [34] may be explored in the future to enhance the new method.

Finally, it is worth to note that a further comparison between the obtained results and existing vector data may provide a more rigorous test of the method especially for urban areas where the borders of the objects are more accurately defined. Unfortunately, such accurate vector data was not available to include such a comparison in this paper.

4   Conclusions

This paper introduced a new image segmentation approach based on fuzzy set theory. Its implementation for the Washington DC Mall data set demonstrates that it proved to be a robust and accurate method for supervised image classification of high-resolution data. Fuzzy image segmentation hence constitutes a viable alternative to traditional OBIA procedures based on hard image segmentation. The implemented case study used support vector machine algorithms which require little training data and a basic level of user input. This represents a considerable step forward in the search of an automated segmentation and object-oriented image analysis for operational image classification.

It is important to note that there are a number of additional machine learning algorithms which are potentially useful for conducting the *fuzzification* and *defuzzification* stages of the new method. Although *SVM* techniques produced very accurate results in this paper's tests, their computational cost may be a problem when classifying huge datasets. Further research will be necessary to investigate whether other machine learning techniques like *Random Forests* or *Generalized Additive Models* could be automated to produce similar accuracy while demanding less hardware resources.

Acknowledgments. The authors are very grateful to Dr David A. Landgrebe (Purdue University, USA) who provided them with the DC Mall dataset. The research reported in this paper was partially funded by a Birkbeck International Research Studentship.

References

1. Elvidge, C.D., Sutton, P.C., Wagner, T.W., Ryzner, R., Vogelman, J.E., Goetz, S.F., Smith, A.J., Jantz, C., Seto, K.C., Imhoff, M.L., Wang, Y.Q., Milesi, C., Nemani, R.: Urbanization. In: Gutman, G., et al. (eds.) Land Change Science, Kluwer Academic Publishers, Dordrecht (2004)
2. Mesev, V.: Remotely Sensed Cities. Taylor & Francis, Abington (2003)
3. Thomas, N., Hendrix, C., Congalton, R.G.: A comparison of urban mapping methods using high-resolution digital imagery. Photogrammetric Engineering and Remote Sensing 69(9), 963–972 (2003)
4. Wilkinson, G.G.: Results and implications of fifteen years of satellite image classification experiments. IEEE Transactions on Geoscience and Remote Sensing, Vol 43(3), 433–440 (2005)
5. Jensen, J.R.: Introductory Digital Image Processing - A Remote Sensing Perspective. Prentice-Hall, Englewood Cliffs (2006)
6. Civco, D.L., Hurd, J.D., Wilson, E.H., Song, M., Zhang, Z.: A comparison of land use and land cover change detection methods. In: 2002 ASPRS-ACSM Annual Conference and FIG XXII Congress (2002)
7. Song, M., Civco, D.L., Hurd, J.D.: A competitive pixel-object approach for land cover classification. International Journal of Remote Sensing 26(22), 4981–4997 (2005)

8. Blaschke, T., Burnett, C., Pekkarinen, A.: Image Segmentation Methods for Object-based Analysis and Classification. In: de Jong, S.M., van der Meer, F.D. (eds.) Remote Sensing Image Analysis: Including the Spatial Domain. Springer, Heidelberg (2006)

9. Baatz, M., Schape, A.: Multiresolution Segmentation: An Optimization Approach for High Quality Multi-scale Image Segmentation (2000)

10. Meinel, G., Neubert, M.: A comparison of segmentation programs for high resolution remote sensing data. International Archives of Photogrammetry, Remote Sensing and Spatial Information Sciences, 1097–1102 (2004)

11. Neubert, M., Herold, H., Meinel, G.: Evaluation of Remote Sensing Image Segmentation Quality - Futher Results and Concepts. In: Proceedings of First International Conference on Object-Based Image Analysis (2006)

12. Haralick, R.M., Schapiro, L.G.: Computer and Robot Vision. Addison-Wesley, Reading (1992)

13. Pratt, W.: Digital Image Processing. Wiley, Chichester (2001)

14. Bezdek, J.C., Pal, M.R., Keller, J., Krisnauram, R.: Fuzzy Models and Algorithms for Pattern Recognition and Image Processing. Springer, Heidelberg (1999)

15. Benz, U.: Definiens Imaging GmbH: Object Oriented Classification and Feature Detection. IEEE Geoscience and Remote Sensing Society, Newsletter (2001)

16. Tso, B., Mather, P.M.: Classification Methods for Remotely Sensed Data. Taylor & Francis, Abington (2000)

17. Navulur, K.: Multi-Spectral Image Analysis Using the Object Oriented Paradigm. CRC Press, Boca Raton (2006)

18. Schiewe, J., Ehlers, M.: A novel method for generating 3D city models from high resolution and multisensor remote sensing data. International Journal of Remote Sensing 26(4), 661–681 (2005)

19. Lizarazo, I., Elsner, P.: Fuzzy Regions for Handing Uncertainty in Remote Sensing Image Segmentation. In: Gervasi, O., Murgante, B., Laganà, A., Taniar, D., Mun, Y., Gavrilova, M.L. (eds.) ICCSA 2008, Part I. LNCS, vol. 5072, pp. 724–739. Springer, Heidelberg (2008)

20. Lizarazo, I.: SVM-based segmentation and classification of remotely sensed data. International Journal of Remote Sensing Vol 29(24), 7277–7283 (2008)

21. Verstraete, J., Hallez, A., De Tre, G.: Fuzzy Regions: Theory and Applications. In: Geographic Uncertainty in Environmental Security, pp. 1–17. Springer, Heidelberg (2007)

22. Landgrebe, D.A.: Signal Theory Methods in Multispectral Remote Sensing. Wiley, Chichester (2003)

23. Kuo, B., Landgrebe, D.A.: Improved Statistics Estimation and Feature Extraction for Hyperspectral Data Classification, PhD Thesis and School of Electrical & Computer Engineering Technical Report TR-ECE 01-6 (2001)

24. Aksoy S. and Akcay H.G: Multi-resolution segmentation and Shape Analysis for Remote Sensing Image Classification. In: 2nd International Conference on Recent Advances in Space Technologies, Istanbul, Turkey, June 9-11 (2005)

25. Hyvrinen, A.: Independent Component Analysis. Wiley, Chichester (2001)

26. Cortes, C., Vapnik, V.: Support-vector network. Machine Learning 20, 273–297 (1995)

27. Benett, K.P., Campbell, C.: Support Vector Machines: Hype or hallelujah? SIGKADD Explorations 2(20) (2000)

28. Posdnoukhov, A., Kanevski, M.: Multiscale Support Vector Regression for Hot Spot detection and modelling. Research report No. 006-007. University of Lausanne (2006)

29. Meyer, D.: Support Vector Machines: The interface to libsvm in package e1071. Technische Universitat Wien, Austria (2007)
30. Burrough, P.A., van Gaans, P.F.M., Hoostmans, R.: Continuous classification in soil survey: spatial correlation, confusion and boundaries. Geoderma 77, 115–135 (1997)
31. Cheng, T., Moleenar, M., Lin, H.: Formalizing fuzzy objects from uncertain classification results. International Journal of Geographical Information Science 15(1), 27–42 (2001)
32. Dilo, A., de By, R.A., Stein, A.: Metrics for vague spatial objects based on the concept of mass. In: IEEE International Fuzzy Systems Conference (2007)
33. CRAN, The R Foundation for Statistical Computing (2008), http://www.r-project.org/
34. Bloch, I.: Fuzzy Spatial Relationships for Model-Based Pattern Recognition in Images and Spatial Reasoning under Imprecision. Springer, Heidelberg (2006)

Connecting the Dots: Constructing Spatiotemporal Episodes from Events Schemas

Arie Croitoru

The University of Alberta,
Edmonton, AB T6G-2E3, Canada
croitoru@ualberta.ca

Abstract. This paper introduces a novel framework for deriving and mining high–level spatiotemporal process models in in-situ sensor measurements. The proposed framework is comprised of two complementary components, namely, hierarchical event schemas and spatiotemporal episodes. Event schemas are used in this work as the basic building model of spatiotemporal processes while episodes are used for organizing events in space and time in a consistent manner. The construction of event schemas is carried out using scale-space analysis from which the interval tree, a hierarchical decomposition of the data, is derived. Episodes are constructed from event schemas using by formulating the problem as a constraint network, in which spatial and temporal constraints are imposed. Consistency is achieved using a path–consistency algorithm. Once created, possible episodes can be derived from the network using a shortest–path search.

1 Introduction and Background

While constantly imbedded in a flux of change we manage to perceive our environment and reason about it in space and time by adopting a discrete view, in which change is composed of a set of events – bounded parts with temporal and spatial relations between them. While segmenting change into salient events, another process becomes apparent: a natural tendency to organize events in *hierarchical structures*. Key motivations for employing a hierarchical approach in cognitive processes such as problem solving are order and robustness [1][2]. The mechanisms that govern the creation of hierarchical event structures, or *event schemas*, are the subject of on–going research, and recent work shows that the creation of an event schema can be a *top–down* or a *bottom-up* process [2].

The study of events and processes within the context of Geographic Information Systems (GISs) has also received considerable attention. A primary motivation for this interest stems from the realization that current GISs are still largely driven by the "snapshot paradigm", thus making the representation of events and processes implicit rather than explicit [3][4]. As a result, new paradigms have been suggested in which the modeling of spatiotemporal phenomena is based not only on objects and states, but also on events and processes [5][3]. The development of an event-based GIS paradigm has been recently reinforced

M.L. Gavrilova and C.J.K. Tan (Eds.): Trans. on Comput. Sci. VI, LNCS 5730, pp. 57–76, 2009.

by the introduction of the SPAN/SNAP formal ontology frameworks for representing dynamic phenomena [6][7]. Hierarchical event structures have also been introduced in SNAP/SPAN in the form of *granularity trees*.

Although the theoretical foundation for modeling events and processes in space and time has developed considerably in recent years, the tools available for detecting salient spatiotemporal events and constructing meaningful processes from them is still lagging behind. Detecting events requires the ability to effectively *segment* spatiotemporal data into salient parts. Once events and their corresponding locations and time intervals are derived, it is required to organize them in a consistent manner in space and time into processes. In essence, this organization process amounts to finding a meaningful *order* that does not include any inconsistencies. The ordering of temporal events has been studied extensively due to the fundamental role it has in natural language understanding, scheduling and planning, and expert systems [8][9]. However, the development of tools for event ordering in both *space and time* (for example, events relating to a spatiotemporal phenomenon) has not received comparable attention.

Thus, the primary challenge addressed in this work is twofold. **First**, we seek to develop a method for segmenting spatiotemporal data into salient events. Inspired by recent cognitive research, we wish to develop the ability to discover event schemas from spatiotemporal data using a bottom-up data-driven process. In this context, we are particularly interested in developing a hierarchical event segmentation scheme. In conjunction, we seek to explore how event schemas could be used for top-down spatiotemporal data mining. **Second**, we seek to develop a method for discovering possible orderings of a set of salient spatiotemporal events in space and time. Here, our main interest is in developing an approach that would consider *both space and time* when deriving partial orders of spatiotemporal events. The overarching objective of this research is to enable event and process mining in a set of point low-level observations made over time by a finite array of sparsely distributed *in-situ sensors*, such as a sensor network.

1.1 A Motivating Example

To demonstrate the motivation for this work consider the problem of analyzing the behavior of a storm using data from a network of in-situ sensors, such as the GoMOOS system – an ocean monitoring network that was deployed in the Gulf of Maine [10]. The network is comprised of 12 buoys, namely A01, B01, C02, D01, E01, F01, I01, J02, K01, L01, M01 and N01 (Fig. 1(a)). Each buoy is equipped with various sensors, a processing unit, a controller, a navigation aid, a GPS positioning system and a communication system. The sensors onboard each buoy are capable of making a range of measurements, such as air and water temperature, wind direction and speed, wave height and salinity.

Storms along the cost of Maine, such as the one depicted in Fig. 1(b), can cause significant flooding due to high waves, which may result in significant damage and even loss of life. Consequently, the study of such events and their causes is of great interest to government (e.g., weather prediction and analysis) and private (e.g., insurance) agencies. Following such a storm event, a weather

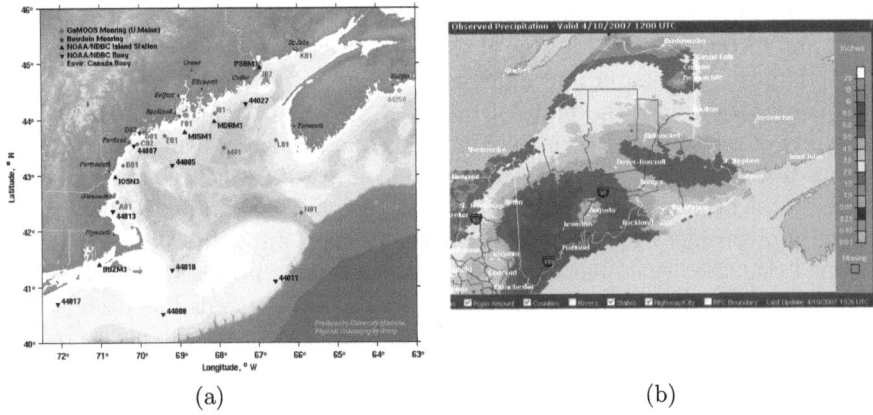

(a)   (b)

Fig. 1. Monitoring a storm in the Golf of Maine. (a) map of the Gulf of Maine showing the GoMOOS buoy locations (adopted from [11]); (b) a precipitation map of a storm event in the Gulf of Maine (adopted from [12]).

monitoring agency might be interested in studying the storm event as captured by the GoMOOS sensor observations, for instance, through the qualitative query Q_1: "given three consecutive days, $\{t_1, t_2, t_3\}$, find all sensors for which wave height increased during the first day t_1, decreased and increased during the second day t_2, and decreased during the third day t_3". Once all matching events were detected, the agency is interested in understanding how the storm evolved as a process by finding possible (or a feasible) sequences of events, for instance through the query "given the time intervals and the corresponding locations of the set of events that were detected by Q_1, find a consistent order of the events". This query may result in one or more feasible orders of events, which may give experts an insight into how storms develop and progress.

2   Related Work

The issue of scale and granularity has been well recognized in various application domains, such as climate change and impact [13] and ecosystem dynamics [14]. The importance of hierarchy, scale, and granularity in spatial and spatiotemporal data modeling, management and analysis has also been long recognized in GIS and GIscience research [15][16] . This has led to the development of conceptual and theoretical frameworks for modeling and representing spatiotemporal dynamic phenomena, such as [17] and [18]. In more recent work [19][4], a representational framework in which the hierarchical structure of spatiotemporal events is based on zones, sequences, and processes (in a hierarchical ascending order) is suggested.

The event mining approach presented here is closely related to recent work on temporal pattern discovery using scale space [20][21]. The work presented in this paper also utilizes scale space, but differs significantly from the work presented in [20][21]. First, the work presented here does not focus on a specific (optimal) scale, and instead utilizes a range of scales to discover the *hierarchical structure* of the data set to discover and mine patters.

In conjunction with the body of literature related to spatiotemporal events, the study of relations between time intervals and reasoning using time intervals was studied extensively, in particular in the context of artificial intelligence [22]. However, the primary focus of this work is on temporal constraints alone, without taking into consideration spatial the spatial domain. In addition, it is generally assumed that the events participating in the reasoning process are well defined. Similarly, the analysis of relations between temporal intervals has been studied in the context of constraint satisfaction [23] (for a recent survey please refer to [24]). Once more, the primary focus in these works is on the temporal domain alone. The work presented in this paper builds on these bodies of literature, but extends them by incorporating spatial and temporal constraints into one constraint satisfaction problem.

finally, the work presented here on event ordering is related to the recent work on event ordering presented by [25], in which the authors suggest to derive linear ordering of events using topological sorting. Once orders are derived in the temporal domain, the spatial context is explored by examining whether event locations follow a linear pattern. However, this would is limited both by the its ability to support different temporal constraints and non–linear motion.

3   Mining Event Schemas[1]

3.1   Granularity in Dynamic Phenomena Ontology

The notion of hierarchical structure and whole-part relations is essential in ontological frameworks. Reistma and Bittner [6] have addressed the issue of scale and granularity in the context of the SNAP/SPAN ontological framework by introducing the concept of *granularity tree* structures that are based on part-whole relationships. In their framework, a granularity tree G is a pair $(P, \subseteq)$ in which P is a set of objects and $\subseteq$ is a binary relation. Granularity trees are governed by a set of seven Axioms [6], which focus on the parent–child relation in a granularity tree rather than child–child relationships.

3.2   Scale-Space as a Hierarchical Event Segmentation Scheme

Scale-space serves in the proposed framework as a hierarchical segmentation mechanism. The approach taken in scale-space analysis is to consider *all* possible scales (or a scale range) instead of focusing on a single scale [27]. The construction of a scale space representation is carried out by imbedding the signal f into a one–parameter family of derived signals, in which the scale is controlled by a scale parameter s. More formally, given a signal $f(x) : \mathbb{R} \to \mathbb{R}$ in which $x \in \mathbb{R}$ and a varying scale parameter $s \in \mathbb{R}^+$, the *scale space representation* $L : \mathbb{R} \times \mathbb{R}^+ \to \mathbb{R}$

---

[1] An extended version of this section has appeared in [26].

is defined as $L(x,s) = g(x,s) * f(x)$, such that $L(x,0) = f(x)$, and $*$ is the convolution operator [27]. The scale space representation is therefore constructed by iteratively convolving f with $g(x,s)$ where the scale factor s is increased continuously. This results in a set of increasingly smoothed signals from which the *deep structure* is derived, as discussed below. For a one-dimensional signal, $g(x,s)$ is taken as the one-dimensional Gaussian kernel [28][27].

The scale-space representation of f can be used to derive the inner structure of the signal. This process is based on tracking salient features across different scales in the scale-space representation of f. A natural choice of features would be extremum points, i.e. points for which the nth derivative is zero and the $n+1$ derivative is non–zero. In the general case, the trajectory of extremum points form arch-like contours in the scale space which reveal the *deep structure* of the signal, that is, the evolution of extremum points across different levels of scale. The apex of a scale space arch (a catastrophe) is a point which satisfies:

$$\frac{\partial^n L}{\partial x^n} = 0, \; \frac{\partial^{n+1} L}{\partial x^{n+1}} = 0, \; \frac{\partial^{n+1} L}{\partial x^{n+1} \partial s} \neq 0. \tag{1}$$

Extremum points in a Gaussian scale space are eliminated in pairs. An example of the deep structure of a sample signal (Fig. 2(a)) is given in Fig. 2(b).

As was indicated by [29], the tracking of extremum points in a Gaussian scale-space ensures the resulting deep structure will have some key characteristics that are essential in the context of this work. In particular, it was shown that scale-space trajectories will not intersect [30], that as the scale parameter s increases new extremum points will not be generated, [31][32], and that the scale-space representation of almost all signals uniquely determines the signal up to a constant multiplier [33]. The scale-space of a signal is therefore a *unique fingerprint* of the signal, and could therefore be used in data mining.

The deep structure can be used for partitioning scale-space. Such a partitioning can be based on the extents of the arches generated by the extremum points: by partitioning the time-scale space using the *bounding box* of each arch, the entire scale space plane is partitioned into salient regions that span over time and scale intervals. In this partitioning the bounding box of an arch is defined as the following: the height of the box is derived from the scale of the arch apex, while the width of the box is defined as the difference between the time values of the two extremum points that comprised the arch at $s = 0$. The width is defined

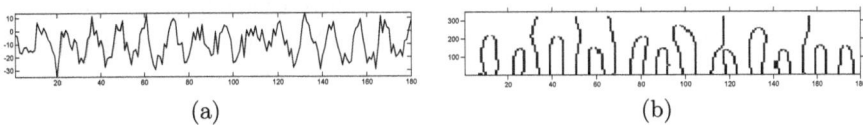

(a)   (b)

Fig. 2. An example of a scale-space signal analysis: (a) temperature measurements (monthly averages over 15 years) of a weather station; (b) The deep structure of (a) which was erived by following extremum points

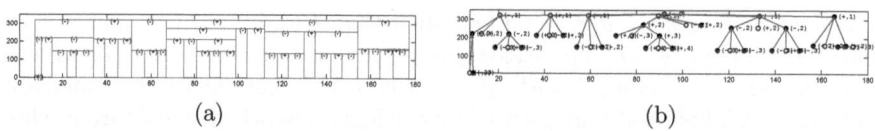

Fig. 3. The interval tree of the signal in Fig. 2(a): (a) The interval tree; (b) The same interval tree in graph representation

this way to overcome the well-known drift effect: as scale increases the location of an extremum point may drift from its true location due to a diffusion process.

The delineated bounding boxes can be used for generating a *complete hierarchical tessellation* of scale space, i.e. the entire scale space is tessellated without gaps. The tessellation is applied repeatedly: for each bounding box that extents from t_{min} to t_{max} in time and from 0 to s_{max} in scale, a horizontal line $s = s_{max}$ is created. The intersection of this line with adjacent bounding boxes (or with the limits of the scale-space) creates three subregions: a central region, c, that corresponds to the bounding box of the arch, and two adjacent (empty) subregions, l and r, on the left and right of c. By repeating this process for every bounding box a complete tessellation is created. This tessellation process results in a hierarchal decomposition of the scale-space domain which can be represented as a *ternary interval tree* [28][34]. An example of an interval tree of the signal in Fig. 2(a), both as a scale-space tessellation and as a tree graph, can be seen in Fig. 3(a) and Fig. 3(b).

3.3   The Ordered Granularity Tree (OGT): From Interval Trees to Spatiotemporal Granularity Trees

The interval tree, as described in Section 3.2, can serve as a mechanism for generating granularity trees from spatiotemporal data. However, the granularity tree does not constrain the temporal relation between parts of the same spatiotemporal granule. More specifically, given a parent node u with several child nodes v_1, v_2, v_3, the granularity tree Axioms described in section 3.1 along with the four constraints do not specify the *order* of the children of u, which may be essential for the understanding or identification of a spatiotemporal event.

In order to resolve the temporal ambiguity additional *temporal order* Axioms were suggested in [26] between the child nodes that belongs to the same parent node in a granularity tree. This is done using utilize Allen's interval algebra [35], as well as the Axioms and conditions described in Section 3.1. Consequently, the ordered granularity tree not only maintains the "part of" relationship, but also the temporal order of the parts.

3.4   Discovering Granularity Patterns

The matching problem is at the heart of the data mining process addressed in this paper: its goal is to identify identical (or similar) events by finding a correspondence mapping between a query granularity tree representing an event schema and a set of granularity trees (forest) that were extracted from sensor observations. Let us begin by exploring the correspondence mapping problem:

Definition 1. *Let G_1 and G_2 be two ordered granularity trees and let $V(G_1)$ and $V(G_2)$ be the set of label nodes nodes (vertices) of G_1 and G_2 respectively with labels l from the alphabet Σ. Let $M \subseteq V(G_1) \times V(G_2)$, and let $v_1, v_2 \in V(G_1)$, $u_1, u_2 \in V(G_2)$. $M : G_1 \rightarrow G_2$ is a correspondence mapping if for the pairs $(v_1, u_1), (v_1, u_1) \in M$ the following conditions are fulfilled:*

1. $v_1 = v_2 \rightarrow w_1 = w_2$
2. $Child(v_2, v_1) \rightarrow Child(u_2, u_1)$
3. $LeftOf(v_1, v_2) \rightarrow LeftOf(u_1, u_2)$
4. $Label(v) = Label(M(u)) \; \forall v \in G_1 \;, \; \forall u \in G_2$

It should be noted that the label of each node can be utilized for encoding important information regarding the hierarchical and temporal order of a granularity tree not explicitly represented by the graph structure. More specifically, in our matching process each node is labeled with a label $\ell = (d, p, o)$, in which d is the node's depth label $(d \in \mathbb{Z}^+)^2$, p is the extremum type label $(p = \{+, -\})$, and o is the temporal order label $(o = \{l, c, r\})$. Our alphabet is therefore $\Sigma = \{d, +, -, l, c, r\}$. Based on this correspondence mapping scheme the data mining problem is defined as:

Definition 2. *Given a query ordered granularity tree G_Q and a forest F of n ordered granularity trees $\{Q_1, Q_2 \ldots Q_n\}$, find all possible correspondence mappings between G_Q and F.*

3.5   Ordered Tree Matching

Based on the proposed framework, the problem of event schema mining is now reduced to an ordered tree matching problem. An overview of the tree matching problem and the various related algorithms and techniques that were developed can be found in [36], [37] and [38]. To solve the tree matching problem in our implementation we utilize GraphGrep, a recently developed exact matching algorithm for graph querying [38][39]. The primary motivation for selecting this tool was twofold. First, unlike other domain–specific tools, GraphGrep is a universal graph querying tool that does not make any assumptions about the type or topology of the graphs. Second, GraphGrep increases the querying efficiency by utilizing an indexing process that reduces the matching search space (Further details about GraphGrep, including its time complexity, can be found in [39]).

4   Discovering Spatiotemporal Episodes

As was shown in Section 3.4, given data from a set of in–situ sensors and a query pattern, a pattern mining process can be applied to the sensor data to detect

---

$^2\mathbb{Z}^+$ denotes the positive integers.

event schemas. Each of these events can then be associated with a *time interval* based on its bounding box as derived from the granularity tree. As the in-situ sensors are imbedded in space, the detected time intervals can also be associated with a set of *locations*. Given a set of events and their corresponding time intervals $\mathcal{I} = \{i_1, i_2, ..., i_n\}$ and a set of corresponding locations $\mathcal{P} = \{p_1, p_2, ..., p_n\}$ our goal is to derive a partial order of intervals in space and time. The discovery of such orderings, or *spatiotemporal episodes*, depends on the ability to successfully generate feasible partial orders.

4.1 Constraining the Episode Generation Process

Clearly, it is possible to generate many different orderings (and episodes) which will result in very different spatiotemporal process models. As $|\mathcal{I}|$ and $|\mathcal{P}|$ become larger the number of possible orderings will increase dramatically and the problem will quickly become unmanageable. In order to overcome this, the proposed approach is based on constraining the solution space, therefore limiting the number of possible orderings that can be found. Since the ordering should be valid both in space and time, two types of constraints are applied:

- **Spatial constraints:** It is assumed that spatiotemporal as process moves in space, its movement is gradual and continuous. For example, as a storm or a cloud of plume passes over a domain, it moves continuously through the domain and does not exhibit an erratic behavior of "jumping" from one location to another. Consequently, if an event schema related to the underlying spatiotemporal process was discovered at a given point p_a in the domain, the next possible locations in which the process can be expected to be detected are points in the neighborhood of p_a. The neighborhood between points in $\mathcal{P}$ can be defined using the Delaunay triangulation, the dual graph of the Voronoi diagram, that ensures that nearby locations will be connected. The Delaunay triangulation is therefore used as a spatial constraint which determines how points are connected in space. It should be noted that the triangulation process defines only connectivity constraints between points in the form of edges between points. However, it does not define the edge directionality. As will be shown later, the edge directionality along the path will be resolved using temporal constraints.
- **Temporal constraints:** Events detected at different locations are associated with their corresponding time intervals $\mathcal{I}$. As the spatiotemporal process moves in space relations between time intervals can be formed. For example, if the process moved from point p_a to p_b then one might expect that the process will be detected at p_i *before* it is detected at p_j. This can be translated to a temporal constraint between the two time intervals of the event schemas, i_a and i_b, that were detected at p_a and p_b, i.e. i_a *before* i_b. In general, given two time intervals, there are seven basic relations that can be defined [35], namely *before, meets, overlap, starts, during, finish* and *equal* (abbreviated by *b, m, o, s, d, f*, and = respectively). With the exception of *equal*, all other relations have an inverse (*bi, mi, oi, si, di*, and *fi*), thus a

total of 13 unique relations can exist between two time intervals. Using these interval relations, it is possible to constrain the temporal relations between the intervals comprising the spatiotemporal episode, i.e. "find a sequence of intervals that are related only by the b relation". In addition to these qualitative constraints, it is possible to impose quantitative constraints on the sequence of intervals, i.e. if a b relation is imposed one can require that the time difference between the end of i_a and the beginning of i_b will be smaller than dt time units.

4.2   Deriving the Spatiotemporal Constraint Network

The combination of the two constraint types results in a spatiotemporal constraint network. The structure of the network is constrained by the Delaunay triangulation, which indicates which nodes in $\mathcal{P}$ (i.e. in–situ sensor locations) are connected by an arc. The temporal relations between nodes are constrained by the interval relations between the corresponding intervals in $\mathcal{I}$. More formally, a constraint network $\mathcal{N} = (\mathcal{P}, \mathcal{C}_\mathcal{S}, \mathcal{C}_\mathcal{T})$, where $\mathcal{P}$ is a set of nodes, $\mathcal{C}_\mathcal{S}$ is a set of spatial constraints such that $\mathcal{C}_\mathcal{S} \subseteq \mathcal{P} \times \mathcal{P}$, $\mathcal{C}_\mathcal{T}$ is a set of temporal constraints associated with edges such that $\mathcal{C}_\mathcal{T} \subseteq \mathcal{I} \times \mathcal{I}$. It should be noted that $\mathcal{C}_\mathcal{T}$ does not include constraints between nodes that are not defined in $\mathcal{C}_\mathcal{S}$.

The constraint network has a central role in deriving possible spatiotemporal episodes: a spatiotemporal episode can be derived by finding a *path* (or a set of paths) in the network, such that for every pair of consecutive nodes along the path all constraints between the nodes are fulfilled. More formally:

Definition 3. *Given a spatiotemporal constrain network $\mathcal{N} = (\mathcal{P}, \mathcal{C}_\mathcal{S}, \mathcal{C}_\mathcal{T})$, a spatiotemporal episode is a sequence of nodes $p_1, p_2, \ldots, p_n$ within $\mathcal{N}$ for which for $i = 1, \ldots, n - 1$ there exist an edge $\mathcal{C}_\mathcal{S}(p_i, p_j)$ between p_i and p_{i+1} and all temporal constraints $\mathcal{C}_\mathcal{T}(p_i, p_j)$ between p_i and p_{i+1} are fulfilled.*

Based on this definition, the problem of finding a spatiotemporal episode is transformed into a search for a path in the constraint network graph. However, this search can not be carried out directly on the "raw" constraint network derived directly from $\mathcal{P}$ and $\mathcal{I}$. To illustrate, consider the example in Fig. 4 (left), including three nodes (i, j and k). Let us assume that following an event schema mining process four possible schemas were found, one for nodes i and j and two for node k (namely k_1 and k_2). The time interval for each event schema is shown in Fig. 4 (right). Based on this, the following temporal relations exist: $\mathcal{C}_\mathcal{T}(i,j) = \{b\}$, $\mathcal{C}_\mathcal{T}(j,k) = \{b, bi\}$, and $\mathcal{C}_\mathcal{T}(i,k) = \{b, bi\}$. A possible spatiotemporal episode that may result from this data is $i \rightarrow j \rightarrow k_2$, which may correspond, for example, to a weather front that passed over node i, then passed over node j and then over node k. Clearly, for this episode $\mathcal{C}_\mathcal{T}T(i,j) = \{b\}$, $\mathcal{C}_\mathcal{T}(j,k) = \{b\}$, and $\mathcal{C}_\mathcal{T}(i,k) = \{b\}$. In this case interval k_1 is inconsistent with the proposed episode since it is impossible that the front passed over node k both before and after passing over nodes i and j.

This simplified example demonstrates two key problems of the "raw" spatiotemporal constraint network. First, since $\mathcal{C}_\mathcal{T}$ is determined only for one edge

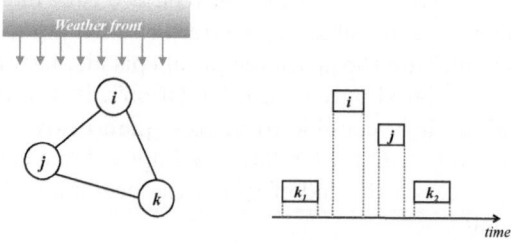

Fig. 4. An example of an inconsistent spatiotemporal constraint network. Left: the network; Right: the time intervals associated with each network node.

at a time, the consistency of temporal constraints along a path in the network can not be assumed. Second, since the Delaunay triangulation is based only on geometric considerations, it does not determine the directionality (or time flow direction) of each edge in the constraint network. Consequently, both directions should be considered for each edge when determining the relationships between the time intervals of the nodes defining the edge. A solution to the first problem will be discussed in Section 4.3. The edge directionality problem will be addressed in Section 4.4.

4.3 Analyzing Path Consistency

As was shown earlier, the spatiotemporal constraint network does not guarantee that a given path will not include any inconsistencies between the participating time intervals. To resolve such inconsistencies, path consistency should be achieved in the constraint network. Following [40] and [41], path consistency is define as:

Definition 4. *A path $p_1, p_2 \ldots p_n$ in a spatiotemporal constrain network $\mathcal{N}$ is path consistent if there exist a sequence of time intervals $i_1, i_2 \ldots i_n$ along this path such that $\mathcal{C}_T(p_1, p_n)$ is satisfied. A spatiotemporal constrain network $\mathcal{N}$ is path consistent iff every path is consistent.*

Path consistency in a constraint network can be achieved by the PC-1 algorithm [42] or the PC-2 algorithm [40] which provide a higher level of efficiency. These algorithms are based on a systematic scan of all possible node triplets in $\mathcal{N}$ and applying a relaxation process. Given a scanning step with three nodes, p_i and p_j and p_k and two sets corresponding constraints $\mathcal{C}_T(p_i, p_j), \mathcal{C}_T(p_j, p_k)$ this process is carried out by intersection over composition [43]:

$$\mathcal{C}_T(p_i, p_k) \leftarrow \mathcal{C}_T(p_i, p_k) \oplus (\mathcal{C}_T(p_i, p_j) \otimes \mathcal{C}_T(p_j, p_k)), \qquad (2)$$

where $\oplus$ is the set–tehoretic intersection. The *composition* ($\otimes$) in Eq. 2 is defined as:

$$\mathcal{C}_T(p_i, p_j) \otimes \mathcal{C}_T(p_j, p_k) = \{c_{ij} \otimes c_{jk} | c_{ij} \in \mathcal{C}_T(p_i, p_j), c_{jk} \in \mathcal{C}_T(p_j, p_k)\}, \qquad (3)$$

where c_{jk} and c_{jk} are two basic temporal relations that is defined by the transitivity table given in [35] (page 836). It should be noted that when the intersection ($\oplus$) in Eq. 2 results in an empty set the constraint network is inconsistent. In the case of the spatiotemporal constraint network proposed here, a more efficient solution to the path consistency problem can be achieved, based on the following Theorem [42]:

Theorem 1. *A triangulated constraint graph $\mathcal{N}$ is path consistent iff every path of length 2 is path consistent.*

Based on this Theorem, it is possible to scan the spatial constraints that are introduced through the Delaunay triangulation and impose path consistency at the *triangle level* rather than the edge level. Our approach is based on two processing steps:

1. **Preprocessing:** In this step the Dealunay triangulation is systematically scanned. For each triangle all possible basic temporal relations are derived for each of the three triangle sides by a cross product between the intervals that were detected at each triangle node. Since the Delaunay triangulation does not define the directionality of the triangle edges, it is assumed that each edge can be directed in both ways, and therefore once all possible interval relations are computed for one direction, the inverse relations are assigned to the opposite direction. At the end of the processing step the initial spatiotemporal constraint network is constructed: each edge in the Delaunay triangulation is labeled with two sets (one for each direction) of basic interval relations.

2. **Path consistency:** Given the initial spatiotemporal constraint network, path consistency is applied by systematically scanning all the Delaunay triangles in the network and applying the intersection over composition operator (Eq. 2) on all three sides of each triangle. The process begins by creating a queue containing all triangles. Eq. 2 is computed for each triangle in the queue. When a triangle side is updated only adjacent triangles that share the updated edge are added back to the queue. The process continues to apply Eq. 2 to all triangles in the queue until it is empty or until inconsistency is detected. This approach is similar to the $\triangle$STP algorithm proposed in [44]. However, $\triangle$STP is further improved here by taking advantage of the known adjacency relations between triangles in the Delaunay triangulation, which limits the number of adjacent triangles that should be checked once inconsistency is detected.

The resulting spatiotemporal constraint network, $\mathcal{N} = (\mathcal{P}, \mathcal{C}_\mathcal{S}, \mathcal{C}_\mathcal{T})$, is a network in which $\mathcal{P}$ is the set of nodes (representing the in-situ sensor locations), $\mathcal{C}_\mathcal{S}$ is a set of edges that were constructed between nodes based on the Dealunay triangulation (each network edge is comprised of two edges, one in each direction), and $\mathcal{C}_\mathcal{T}$ is the set of consistent temporal constraints associated with the edges. more specifically, a given network edge between nodes i and j is comprised of *two* spatial constraints, $\mathcal{C}_\mathcal{S}(p_i, p_j)$ and $\mathcal{C}_\mathcal{S}(p_j, p_i)$, and two corresponding temporal constraint sets (i.e., sets of basic time interval relations), $\mathcal{C}_\mathcal{T}(p_i, p_j)$ and $\mathcal{C}_\mathcal{T}(p_j, p_i)$.

4.4 Discovering Possible Episodes

Given a path–consistent spatiotemporal constraint network, $\mathcal{N} = (\mathcal{P}, \mathcal{C}_\mathcal{S}, \mathcal{C}_\mathcal{T})$, discovering spatiotemporal episodes amounts to searching for paths within the constraint network that are consistent. Since the episodes sought represent spatiotemporal phenomena, each path should be consistent in terms of its spatial and temporal directionality. Spatial directionality should be consistent in terms of the direction of connected edges along the path, and temporal directionality should be consistent in terms of the relations between the basic time intervals associated with two connected edges.

To demonstrate this, consider three nodes, i, j and k, in a section of $\mathcal{N}$ (Fig. 5, left). Based on this information, and assuming a single spatiotemporal process passes through these nodes, two possible episodes may arise: $k \rightarrow j \rightarrow i$ or $i \rightarrow j \rightarrow k$. In the first episode, as we assume the process moves along the path and forward in time, a possible time interval relations would be $\mathcal{C}_\mathcal{T}(k, j) = \{b\}$ and $\mathcal{C}_\mathcal{T}(j, i) = \{b\}$, which corresponds to intervals $s \rightarrow u \rightarrow v$. If a reversed time flow is assumed, a consistent path would be $\mathcal{C}_\mathcal{T}(k, j) = \{bi\}$ and $\mathcal{C}_\mathcal{T}(j, i) = \{mi\}$, which corresponds to intervals $s \rightarrow r \rightarrow q$. If the path $i \rightarrow j \rightarrow k$ is selected then several possible scenarios may arise, not all of which are temporally valid. For example, consider a single process that moves forward in time, a possible time interval relations would be $\mathcal{C}_\mathcal{T}(i, j) = \{m\}$ and $\mathcal{C}_\mathcal{T}(j, k) = \{b\}$. Here, the m relation may exist between two interval combinations: $q \rightarrow r$ and $t \rightarrow u$. However, when applying $\mathcal{C}_\mathcal{T}(j, k) = \{b\}$ next, only interval r has an interval in node k that has this relation (interval s), while interval u does not have an interval that belongs to node k for which the b relation can be found. Consequently, this option should be eliminated. If a reversed time flow is assumed a consistent path would be $\mathcal{C}_\mathcal{T}(i, j) = \{bi\}$ and $\mathcal{C}_\mathcal{T}(j, k) = \{bi\}$. Here, the bi relation may exist between two interval combinations: $v \rightarrow u$ and $t \rightarrow r$. However, when applying $\mathcal{C}_\mathcal{T}(j, k) = \{bi\}$ next, only interval u has an interval in node k that has this relation (interval s), while interval r does not have an interval belonging to node k for which the b relation can be fond. Consequently, this option should be eliminated in this case.

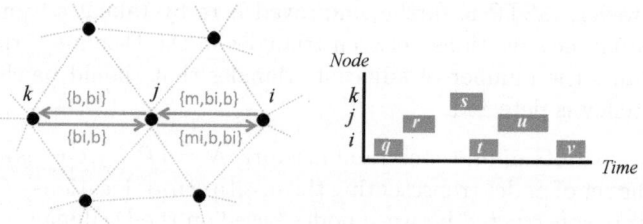

Fig. 5. Maintaining consistency in a path search. Left: a section of a spatiotemporal constraint network. The possible basic time interval relations are indicated next to each edge.; Right: the time intervals associated with each network node.

This example demonstrates the principle of the episode search process. Given a starting point and a time flow direction, an episode can be constructed by searching a path–consistent spatiotemporal constraint network for the longest path that is consistent with the time flow direction constraint. In this case, the time flow constraint serves as a filter for eliminating edges that are not consistent with the time flow constraints, thus serving as a mechanism for determining the correct edge directions along the path. Based on this principle, the spatiotemporal episode mining process is carried out as the following:

1. **Initialization:** Given path–consistent spatiotemporal constraint network, $\mathcal{N}$, the user provides two input parameters: a starting node $\widehat{p_{start}}$ in $\mathcal{N}$, and a set of required interval constraints $\widehat{\mathcal{C}_T}$ (e.g., $\widehat{\mathcal{C}_T} = \{b, o\}$). In addition to these qualitative constraints, a qualitative time difference constraint between time intervals can be defined. For example, a user may constrain the b or the o relation such that given two time intervals, i_a and i_b, the difference between the end of i_a and the beginning of i_b is smaller than a time difference $\triangle t$.

2. **Edge elimination:** A systematic search is carried out on all the edges in $\mathcal{N}$. A list $\mathcal{L}$ is constructed, and each edge that is labeled by $\widehat{\mathcal{C}_T}$ is added to an edge list $\mathcal{L}$. In addition, the time intervals associated with the nodes of the added edge that comply with $\widehat{\mathcal{C}_T}$ are stored in the edge list together with the edge. It should be noted that since edge directionality has not been determined yet, $\mathcal{L}$ may include two edges in opposite directions for a pair of nodes in $\mathcal{N}$.

3. **Path search:** Based on the resulting list of edges, $\mathcal{L}$, a search for the shortest path is carried out from $\widehat{p_{start}}$ using Dijkstra's algorithm [45]. If more than one shortest path is found in $\mathcal{N}$, then the episode is taken as the union of all shortest paths in $\mathcal{N}$.

It should be noted that the process outlined above will produce paths constructed of *directed* edges, and can therefore be seen as a partial ordering of the nodes according to the time interval available at each node and the time constraints imposed. It should also be noted that the results of the process depend on the starting node that is of interest to the user.

5   An Example: Analyzing a Storm in the Gulf of Maine

To illustrate the proposed framework and its benefits this example explores the "Patriot's Day Storm", a recent meteorological event that occurred between April 15 to 17, 2007 [11]. The storm, which dropped close to 150 millimeters of rain in less than 3 days, caused extensive tidal flooding along the northern U.S. coast, resulting in widespread power outages, downed trees, and numerous road closures [12]. During the storm winds of close to 95 kilometers per hour and waves of more than 9 meters were measured by the GoMOOS system [10] (Fig. 1(a)). The analysis of the 2007 Patriot's day storm was carried out in two parts: in Section 5.1 the event schema mining process is demonstrated, then, based on these results the process of spatiotemporal episode construction is demonstrated in Section 5.2.

5.1   Mining Event Schemas

Following the example in Section 1.1, we first focus on the wave height as it was observed during the storm by four different GoMOOS buoys: A01, B01, C02 and J02. These buoys were selected since they provided virtually uninterrupted monitoring of wave heights during the storm, and due to their distribution along the shoreline – while buoys A01, B01 and C02 are located in a region of the gulf that more is open to the ocean, buoy J02 is located in an area that is relatively protected by land and is therefore expected to have a different wave regime. This can be clearly observed in the data set by examining the amplitude of the wave hight during the storm: buoys A01, B01 and C02 recorded heights of more than 9 meters while buoy J02 recorded heights of less than 1.5 meters (see Fig. 6(a)).

Data consisting of the average wave height over 30 minute periods was downloaded from the GoMOOS website [11] for a total of 8 days – covering the period between April 12, 2007 (14:00 UTC) and April 20, 2007 (10:00 UTC), during which the "Patriot's Day Storm" occurred. As the data included several small gaps (typically 1–2 missing values) a monotone piecewise interpolation was applied [46] to close any data gaps. As this interpolation method is monotone, it guarantees that no new extremum points will be generated as a result of the interpolation process, and consequently, no artifacts will be created in scale space. The four data sets can be seen in Fig. 6(a). Following this, the scale–space and granularity tree of each buoy was computed and stored in a GraphGrep database. To query the data set we consider the simple event schema represented by the depth 2 granularity tree pattern Q_1 depicted in Fig. 6(b). This pattern qualitatively describes a subsidence event in the storm: an overall wave height decrease event comprised of a decrease, an increase and a decrease, where the initial decrease is itself comprised of a decrease, a small increase, and a decrease in wave

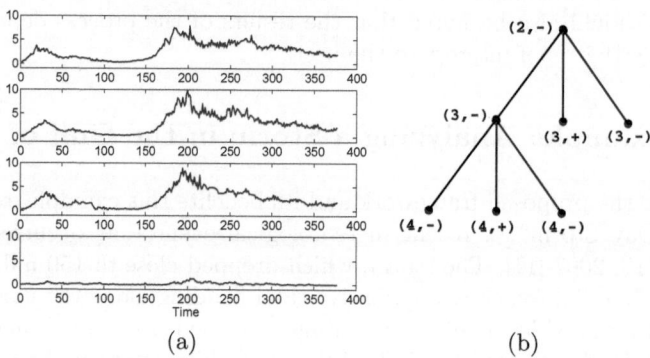

(a)   (b)

Fig. 6. The "Patriot's Day Storm" data: (a) the observation data for buoys A01, B01, C02, and J02 (from top to bottom respectively); (b) The query pattern tree for Q_1

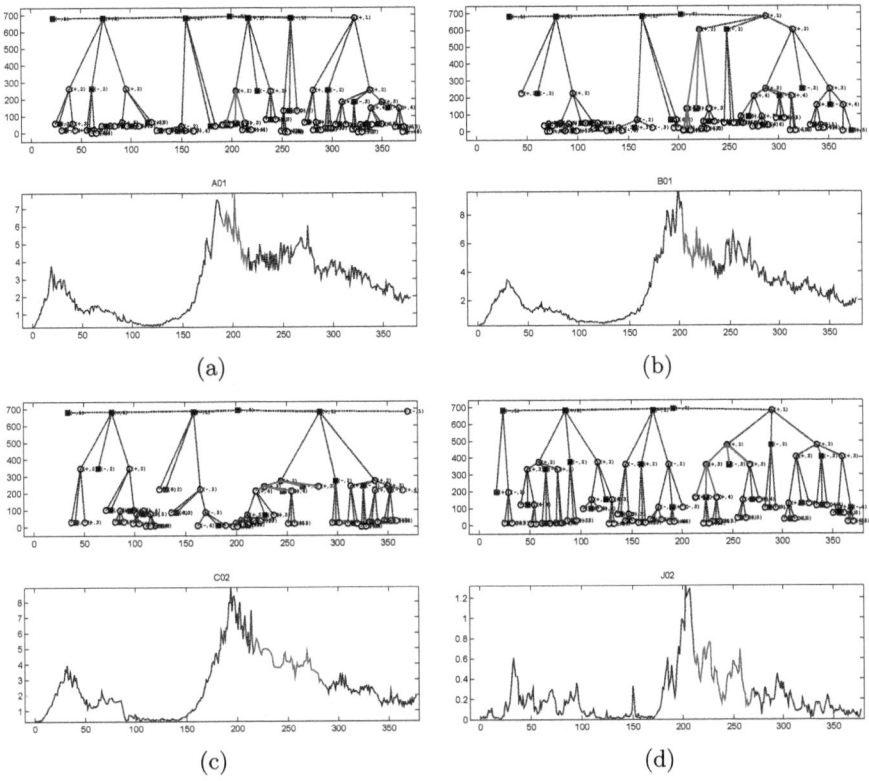

Fig. 7. Event schema mining results: (a) A01; (b) B01; (c) C02; (d) J02. Each of the section highlighted in red indicates the pattern discovered.

height. The results of the matching process for Q_1 using GraphGrep are shown in Fig. 7. Overall, a total of 12 matching patterns were found, of which four sample matches for buoys A01, B01, C02 and J02 are shown in Fig. 7 (due to space constraints). It is important to note that 10 out of the 12 patterns that were detected match *across* buoys and thus represent the same event in the storm.

5.2   Reconstructing Episodes

In order to demonstrate the reconstruction of possible spatiotemporal episodes the data used in Section 5.1 was augmented by data from 4 buoys (44004, 44008, 44005, and 44027) in the Gulf of Maine area (Fig. 8 (a)). The data was obtained from the National Oceanic and Atmospheric Administration (NOAA) national buoy data center [47], and consisted of the average wave height over one hour periods between April 12, 2007 (00:00 UTC) and April 20, 2007 (23:00 UTC). Similarly to the GoMOOS data, the data included several small gaps (typically 1–2 missing values) that were filled using a monotone piecewise interpolation [46].

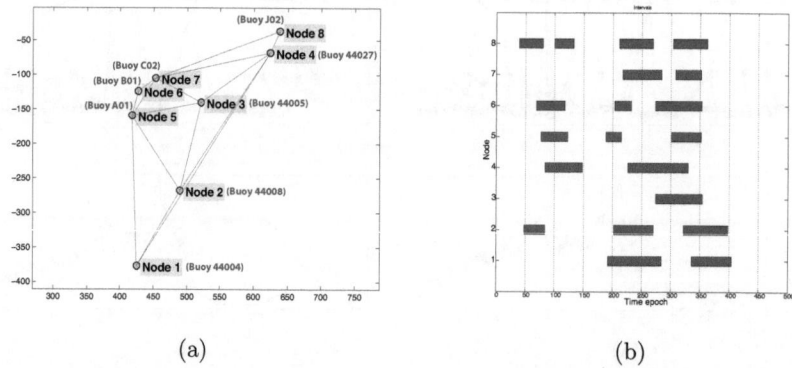

Fig. 8. The reconstruction episode data: (a) the set of 8 buoys from which data was processed and their Delaunay triangulation. Nodes 1–8 correspond to buoys 44004, 44008, 44005, 44027, A01, B01, C02, and J02 respectively.; (b) The time intervals in which the query Q_1 was detected in the 8 buoy data sets.

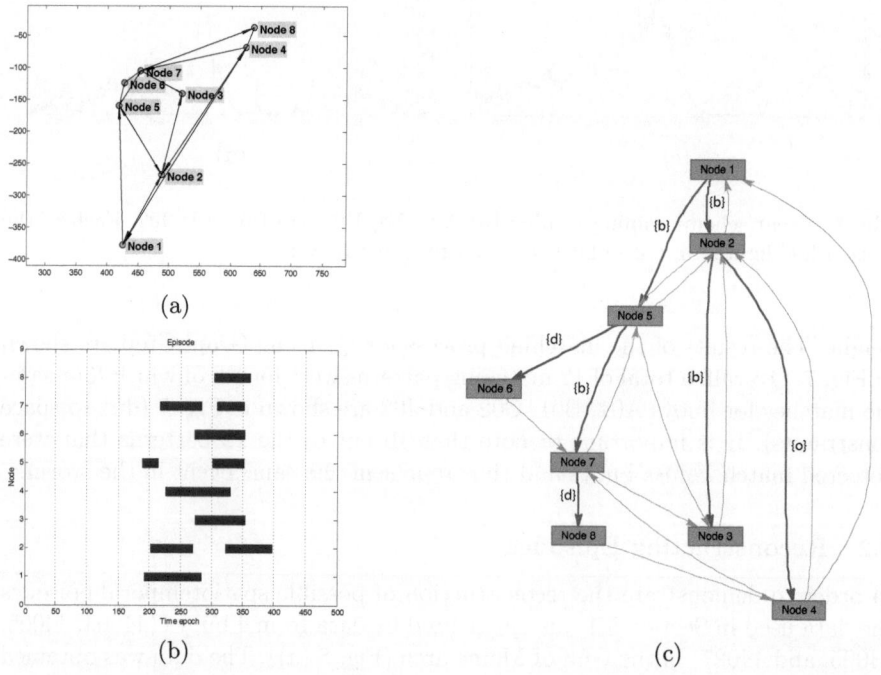

Fig. 9. Reconstructing a possible episode for the 2007 "Patriot's Day Storm": (a) the constraint network; (b) the resulting consistent event time intervals; (c) a graph representation of the episode

The augmented data set was first preprocessed. Using the buoy locations, a triangulation was created using Delaunay triangulation (Fig. 8 (a)). The event schema mining process (Section 3.4) was applied to the data set of each buoy using the query Q_1 (Fig. 6(b)). This resulted in a total of 20 possible event schemas in the entire data set. The time intervals corresponding to these event schemas are shown in Fig. 8 (b).

The episode was constructed with all time intervals using buoy 44004 as $\widehat{p_{start}}$, and by constraining the temporal relations using $\widehat{\mathcal{C}_T} = \{b, o, d\}$. In addition to these qualitative constraint, a quantitative constraint was applied between time intervals such that the time difference between the end and beginning of two consecutive time intervals is smaller than 25 hours (50 time epochs). Then, path consistency end edge elimination was applied (Fig. 9 (a)), followed by a path search process. The search resulted in a total of 5 paths that construct the episode corresponding to 11 time intervals (Fig. 9 (b)). The entire episode (union of all 5 paths paths), along with the relations between the time intervals, is depicted in (Fig. 9 (c)). As can be seen, the hypothesize episode from node 1 shows the progression of the President's Day storm as it moved along the northeast coast of the U.S. from south to north.

6   Conclusion and Future Work

In this work, a new framework for data mining of spatiotemporal events and processes from in-situ sensor data is introduced. The framework is based on two complementary tools. First, inspired by hierarchical event schemas as a fundamental cognitive construct, a tool for segmenting and mining event schemas in a data-driven bottom-up approach is proposed. Second, a tool for ordering event schemas into spatially and temporally consistent episodes is outlined. The combination of these two tools can assist in developing an understanding of spatiotemporal dynamic phenomena as it allows drawing inferences about event sequences, which may lead to a better understanding of causality relations within a dynamic phenomenon.

The work presented here can be further extended in several directions. First, while we have focused primarily on qualitative event schema mining, the implementation of quantitative event schema mining could be further developed. In addition, the inexact event scheme matching problem, in which partial event schema mismatches are permitted, should be further explored. Furthermore, the proposed framework has not been expanded to further process the event schemas that were discovered in order to infer additional knowledge about the geometry, topology, and kinematics of the observed phenomenon. In the context of spatiotemporal episode mining, further work is needed for developing better measures for event–based similarity estimation, as well as the incorporation of more quantitative constraints between events. Finally, the issue of space–time concurrency should be further explored.

References

1. Zacks, J.M., Tversky, B., Iyer, G.: Perceiving, remebering, and communicating structure in events. Journal of Experimental Psychology 130(1), 29–58 (2001)
2. Hard, B.M., Tversky, B., Lang, D.S.: Making sense of abstract events: Building event schemas. Memory and Cognition 34(6), 1221–1235 (2006)
3. Worboys, M., Hornsby, K.: From objects to events: Gem, the geospatial event model. In: Egenhofer, M.J., Freksa, C., Miller, H.J. (eds.) GIScience 2004. LNCS, vol. 3234, pp. 327–343. Springer, Heidelberg (2004)
4. Yuan, M., McIntosh, J.: Assessing similarity of geographic processes and events. Transactions in GIS 9(2), 223–245 (2005)
5. Peuquet, D.J., Duan, N.: An event-based spatiotemporal data model (estdm) for temporal analysis of geographical data. International Journal of Geographical Information Science 9(1), 7–24 (1995)
6. Reistma, F., Bittner, T.: Scale in object and process ontologies. In: Kuhn, W., Worboys, M.F., Timpf, S. (eds.) COSIT 2003. LNCS, vol. 2825, pp. 13–27. Springer, Heidelberg (2003)
7. Grenon, P., Smith, B.: Towards dynamic spatial ontology. Spatial Cognition and Computation 4(1), 69–103 (2004)
8. Gerevini, A., Schubert, L.K.: Efficient algorithms for qualitative reasoning about time. Artificial Intelligence 74(2), 207–248 (1995)
9. Ghallab, M., Nau, D., Traverso, P.: Automated planning. Morgan Kaufmann Publishers, Boston (2004)
10. Wallinga, J.P., Pettirew, N.R., Irish, J.D.: The gomoos moored buoy design. In: Proceedings of OCEANS 2003, September 22-26, vol. 5, pp. 2596–2599 (2003)
11. GoMOOS: Gulf of maine ocean observing system, http://www.gomoos.org (Last visited November 21, 2008)
12. NOAA: National weather service forecast office - gray/portland, http://erh.noaa.gov/gyx/patriot_day_storm_2007.html (Last visited February 21, 2008)
13. Clark, W.C.: Scales in climate impacts. Climage Change 7, 5–27 (1985)
14. Holling, C.S.: Cross-scale morphology, geometry, and dynamics of ecosystems. Ecological Monographs 6(4), 447–502 (1992)
15. Peuquet, D.J.: Making space for time: Issues in space-time data representation. GeoInformatica 5(1), 11–32 (2001)
16. Yuan, M.: Geographic information systems (gis) approaches for geographic dynamics understanding and event prediction. In: Suresh, R. (ed.) Defense Transformation and Net-Centric Systems. 65781(A) of SPIE, vol. 6578 (2007)
17. Peuquet, D.J.: It's about time: A conceptual framework for the representation of temporal dynamics in geographic information systems. Annals of the Association of American Geographers 84(3), 441–461 (1994)
18. Whigham, P.A.: Hierarchies of space and time. In: Frank, A.U., Campari, I. (eds.) COSIT 1993. LNCS, vol. 716, pp. 190–201. Springer, Heidelberg (1993)
19. Yuan, M.: Representing complex geographic phenomena in gis. Cartography and Geographic Information Science 28(2), 83–96 (2001)
20. Höppner, F.: Learning dependencies in multivariate time series. In: Proceedings of the ECAI 2002 Workshop on Knowledge Discovery in (Spatio-) Temporal Data, Lyon, France, pp. 25–31 (2002)

21. Höppner, F.: Discovery of temporal patterns – learning rules about the qualitative behavior of time series. In: Proceedings of the 5th European Conference on Principles and Practice of Knowledge Discovery in Databases, Freiburg, Germany, pp. 192–203 (2001)
22. Gerevini, A.: Processing qualitative temporal constraints, pp. 247–276. Elsevier, Amsterdam (2005)
23. Schwalb, E., Dechter, R.: Processing temporal constraint networks. Artificial Intelligence 93, 29–61 (1995)
24. Schwalb, E., Vila, L.: Temporal constraints: a survey. Constraints: an International Journal 2, 129–149 (1998)
25. Hall, S., Hornsby, K.: Ordering events for dynamic geospatial domains. In: Cohn, A.G., Mark, D.M. (eds.) COSIT 2005. LNCS, vol. 3693, pp. 330–346. Springer, Heidelberg (2005)
26. Croitoru, A.: Deriving and mining spatiotemporal event schemas in in-situ sensor data. In: Gervasi, O., Murgante, B., Laganà, A., Taniar, D., Mun, Y., Gavrilova, M.L. (eds.) ICCSA 2008, Part I. LNCS, vol. 5072, pp. 740–755. Springer, Heidelberg (2008)
27. Lindeberg, T.: Scale-Space Theory in Computer Vision. The Springer International Series in Engineering and Computer Science, vol. 256, 444 pages. Springer, Heidelberg (1994)
28. Witkin, A.P.: Scale-space filtering. In: International Joint Conference on Artificial Intelligence, pp. 1019–1023 (1983)
29. Mokhtarian, F., Mackworth, A.: Scale-based description and recognition of planar curves and two-dimensional shapes. IEEE Transactions on Knowledge and Data Engineering 8(1), 34–43 (1986)
30. Yuille, A.L., Poggio, T.: Scaling theorems for zero-crossings. A.I. Memo 722, Massachusettes Institute of Technology, June 1983, 23 pages (1983)
31. Wu, L., Xie, Z.: Scaling theorems for zero-crossings. IEEE Transactions on Pattern Analysis and Machine Intelligence 12(1), 46–54 (1990)
32. Anh, V., Shi, Y., Tsui, H.T.: Scaling theorems for zero-crossings of bandlimited signals. IEEE Transactions on Pattern Analysis and Machine Intelligence 18(3), 309–320 (1996)
33. Yuille, A.L., Poggio, T.: Fingerprints theorems for zero-crossings. Journal of the Optical Society of America A 2(5), 683–692 (1985)
34. Wada, T., Sato, M.: Scale-space tree and its hierarchy. In: 10th International Conference on Pattern Recognition, pp. 103–108 (1990)
35. Allen, J.F.: Maintaining knowledge about temporal intervals. Communications of the ACM 26(11), 832–843 (1983)
36. Bille, P.: A survey on tree edit distance and related problems. Theoretical Computer Science 337(2005), 217–239 (2005)
37. Shasha, D., Zhang, K.: Approximate tree pattern matching. In: Pattern Matching Algorithms, pp. 341–371. Oxford University Press, Oxford (1997)
38. Giugno, R., Shasha, D.: Graphgrep: A fast and universal method for querying graphs. In: Proceedings of the 16th International Conference on Pattern Recognition (ICPR 2002), Quebec, Canada, August 2002, vol. 2, pp. 112–115 (2002)
39. Shasha, D., Wang, J., Giugno, R.: Algorithmics and applications of tree and graph searching. In: 21st ACM Symposium on Principles of Database Systems (SIGMOD-PODS 2002), Madison, Wisconsin, USA, June 3-6, 2002, pp. 39–52. ACM, New York (2002)
40. Dechter, R., Meiri, I., Pearl, J.: Temporal constraints networks. Artifical Intelligence 49(1991), 61–95 (1991)

41. Bliek, C., Sam–Haroud, D.J.: Path consistency on triangulated constraint graphs. In: Dean, T. (ed.) In Proc. of the Sixteenth International Joint Conference on Artificial Intelligence, pp. 456–461. Morgan Kaufmann, San Francisco (1999)

42. Mackworth, A.K.: Consistency in networks of relations. Artifical Intelligence 8(1977), 99–118 (1977)

43. Dechter, R.: Constraint processing, 481 pages. Morgan Kaufmann, San Francisco (2003)

44. Xu, L., Choueiry, B.: A new efficient algorithm for solving the simple temporal problem. In: 10th International Symposium on Temporal Representation and Reasoning and Fourth International Conference on Temporal Logic (TIME-ICTL 2003), pp. 212–222. IEEE Computer Society Press, Los Alamitos (2003)

45. Dijkstra, E.W.: A note on two problems in connexion with graphs. Numerische Mathematik 1(1959), 269–271 (1959)

46. Fritsch, F.N., Carlson, R.E.: Monotone piecewise cubic interpolation. SIAM Journal of Numerical Analysis 17(2), 238–246 (1980)

47. NOAA: National data buoy center, http://www.ndbc.noaa.gov/.html (Last visited November 23, 2008)

Predictive Indexing for Position Data of Moving Objects in the Real World

Yutaka Yanagisawa

NTT Communication Science Laboratories, NTT Corporation

Abstract. This paper describes a spatial-temporal indexing method for moving objects with a technique to predict future motion positions of moving objects. To build efficient index structure, we had an experiment to analyze practical moving objects, such as people walking in a hall. As the result, we found that any moving objects can be classified to just three types of motion characteristics; 1) staying, 2) straight moving, and 3) random walking. Indexing systems can predict accurate future positions of each object based on our found characteristics, moreover, the index structure can reduce the cost to update MBRs in spatial-temporal data structure. To show an advantage of our prediction method to previous works, we had an experiment to evaluate performance of each prediction method.

1 Introduction

Recently, we can use highly accurate positioning devices to track moving objects, such as pedestrians and cars. The position is one of the most significant data for extracting contexts from the real world. Many context-aware services use the position data for providing services [1], [2]. The Moving Object Database (MoDB) [3] is a database system that can manage position data of real moving objects. Cost reductions in managing such trajectories are one of the most significant challenges for applications using position data. Various types of efficient data structures have been proposed [4] [5] [6] for managing trajectories.

In general, a position is denoted as $p = \{o, t, x, y\}$, which means object o is located at point $\langle x, y \rangle$ at time t, and trajectory λ of a moving object is also denoted as a sequence of positions $\langle p_0, \ldots, p_n \rangle$. Obviously, trajectory can be represented as a model of spatial and temporal data. Thus, most previous MoDBs adapt traditional tree-based indexing mechanisms, such as R-tree [7], for managing trajectories and the positions of each moving object. In the case that a MoDB adapts a traditional spatial index structure, each moving object must send its position to the server for constructing index structures on the server. The database can process quickly a spatial query on the server with this index structures but moving objects must consume much battery to send position data to the server continuously with wireless networks. Because a moving object only has a small batteries to locate it and to send data with wireless networks, the increase in communication cost to update index structure is one of the most serious problems with MoDBs.

M.L. Gavrilova and C.J.K. Tan (Eds.): Trans. on Comput. Sci. VI, LNCS 5730, pp. 77–94, 2009.

To solve this problem, several MoDBs adapt a *predictive* indexing mechanism [8] [9] that predict future positions of a moving object. In this mechanism, the server and each moving object have the same function to predicte the range where the object will move to. The server builds an index structure for a moving object with the predicted range at each time as long as the moving object does not send any position data to the server. When an object moves out of the predicted range at a time, the mobile device on the object sends the position data and the server modifies the index structure with the received position data. After modification of the index structure, both the server and the mobile device calculate again the future range with the same prediction function.

The introduction of the mechanism enables MoDBs to manage positions without frequently communication between the server and each device to update index structures. To greatly reduce the communication cost, the mechanisms must predict the future positions of objects as accurately as possible. In this paper, therefore, we propose a new technique to accurately predict the future ranges where objects will move to. Our proposed technique can also predict accurate future ranges with limited computational resources in a mobile device on a moving object.

To improve prediction accuracy, we investigated the features of the real trajectories obtained in our experiments. From the investigation results, two special motion patterns are found from trajectories: "staying," and "straight-moving." Staying means that an object almost comes to a stop at a point for a period; on the other hand, straight-moving means an object moves in a straight line. Thus, we present a prediction technique based on these two motion patterns and "random-moving," which can represent any motion of objects.

Section 2 describes both the problem and solutions in updating index structures for management of moving objects. We also mention related works in the section. In Section 3, we show several types of trajectories obtained in our experiments and Section 4 explains both the found motion pattern and prediction function to calculate the future positions of moving objects. Moreover, in Section 5, we cite the performance of our proposed prediction technique by comparison with previous existing index structures. Finally, Section 6 concludes our work.

2   Problems and Approach

This section describes the problems on updating index structures for moving object database systems in 2.1. After the explanation of the related works in 2.2, we explain our approach to manage index structure for moving objects in 2.3.

2.1   Problem on Updating Index Structure for MoDB

Here we consider suitable index structures for a standard MoDB. Generally, an MoDB is consists of a server and small mobile devices on each moving object. A mobile device on a moving object has a triplet of a positioning device such as GPS, small storage to store trajectory data of the moving object, and a wireless communication device to send data to the server. The server manages an index structure for moving objects and process queries with the index structure. To construct index structure, the server gathers position data from each mobile device by wireless communication.

We focus on *spatial range query* that is an essential spatial query used in traditional MoDBs. Spatial range query is defined as a query in which a user is interested in certain spatial objects that are related to others and their distances are within a certain threshold distance. In general, a basic range query is represented as a rectangle $R = \langle x, y, width, height \rangle$.

R-tree[7] is one of the most popular spatial index structures using both minimum bounding rectangles and a tree structure. The many improved index structures of R-tree have been proposed, for example, R$^+$-tree[10], R*-tree, M-tree[11], and so on. These traditional spatial index structures are available to manage trajectory data because a trajectory is a type of spatial data. A simple method to process a range query is that the server builds a large index structure with all the gathered trajectory data from every mobile device at each time. The server can process any quests quickly in this method since the server has all the trajectory data. This method, however, increases the communication cost between each mobile device and the server.

To send position data from a mobile device to the server, the mobile device must consume much battery for wireless communication. Every mobile device must send continuously data to the server that gathers all the position data at every time but a small mobile device has little battery to send data. If a mobile device sends data frequently, the battery on the device will be exhausted in a short time. To avoid this problem, it is necessary to reduce the frequency of sending data from a mobile device to the server. One of the simplest methods to reduce communication cost is that each mobile device sends no data to the server while the object stops at a point. This method can reduce the cost but the method is available in limited cases such that most of objects do not move.

Several techniques have been proposed for reduction of the cost with a technique to predict a range where an object will move at a future time. The server and every mobile device have the same prediction model in this method. A mobile device sends position data only at the time when the object moves out of the predicted range. Obviously, this method can reduce more cost than the previous method. In general, the prediction technique extracts motion patterns from the past position of a moving object. TPR-tree and STP-tree are most popular index structure using this prediction technique. Next, we describe prediction techniques used in these index structures.

2.2 Related Works

Here, we mention previous techniques that predict the future positions of moving objects for constructing an effective data structure: TPR-Tree [9], TPR*-Tree [12], and STP-Tree [8]. To compare our technique with these previous works in our experiments, we briefly explain these schemes.

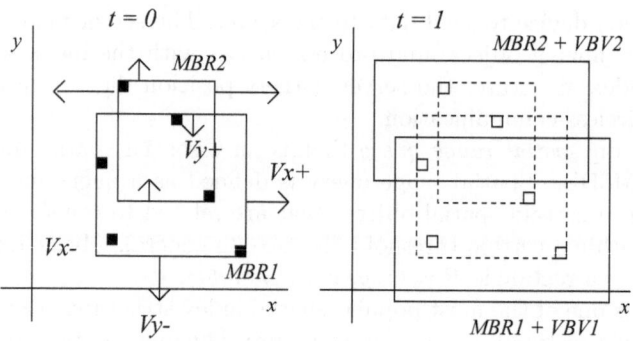

Fig. 1. Prediction on TPR-Tree

Prediction Technique in TPR-Tree. In TPR-Tree and TPR∗-Tree, a database system predicts the future positions of objects using velocities from each axis. To predict a position, the system calculates each maximum velocity of objects in an MBR by positive and negative x- and y-axes from time $t - m$ to t. These velocities are denoted as V_{x+}, V_{y+}, V_{x-}, and V_{y-}, as shown in Figure 1; for example, V_{x+} is the maximum velocity of all objects in an MBR by the positive x-axis during a period. When no object moves in a direction, for instance, no object moves toward the positive y-axis, as illustrated in Figure 1, value V_{y+} has a negative value. In TPR-Tree, the set of V_{x+}, V_{y+}, V_{x-}, and V_{y-} is called the Velocity Bounding Vector (VBV). A database calculates future MBRs from VBV; concretely, each corner point of a future MBR is given by the following equations:

$$P_{x-}(t + j) = V_{x-} \times j \tag{1}$$
$$P_{y-}(t + j) = V_{y-} \times j$$
$$P_{x+}(t + j) = V_{x+} \times j$$
$$P_{y+}(t + j) = V_{y+} \times j$$

In this technique, the MBR validation rate is lower than other techniques because it only uses maximum velocity; however, its reconstruction rate is lower than others because the predicted MBR is always larger than the ideal MBR. When an object completely stops at a point or moves straight at the same velocity, in this technique a database must accurately predict the object's future point. Even if objects are moving randomly, the reconstruction rate is lower than others since the predicted MBR must be larger than the ideal MBR. On the other hand, when objects are moving randomly, the predicted MBR area tends to be much larger than the ideal MBR: in other words, the MBR validation rate becomes low. Similarly, when trajectories have much noise, MBR validation rates also become lower than other techniques.

In the original TPR-Tree, each MBR includes objects that are close to each other at time t. A database does not check overlaps between areas of constructed MBRs. In TPR*-Tree, a database checks for overlaps and reconstructs MBRs so that no MBR overlaps with other MBRs and the VBV of objects in an MBR is similar to each other, after constructing MBRs based on the TRP-tree technique. Since the prediction accuracy of the TPR*-Tree is possibly higher than the original TPR-Tree, in our experiment we compared our methods with it.

For evaluations, we apply our enhanced MBR construction methods, as mentioned in 3.4. After a database temporally stores all points of moving objects at each time, the database constructs optimal MBRs at the time by clustering techniques. For results, we use the same MBRs for evaluation in any tree structure by comparing reconstruction and MBR validation rates.

Prediction Technique in STP-Tree. The prediction technique in STR-Tree uses a nonlinear predictive function represented by the past positions of an object. The essential idea is based on the calculation of approximate predictive functions using several past points through which an object has already passed. To calculate approximate function, STR-Tree uses SVD techniques, which are traditional signal processing techniques.

In STR-Tree, a database system makes a sequence of positions, such as $x(t-m), y(t-m), x(t-m+1), y(t-m+1), \ldots, x(t)$, and $y(t)$ from time $t-m$ to t. We denote a sequence of position vectors from time $t-m$ to t as $\boldsymbol{k}(t)_m$; similarly, a vector sequence from $t-m-1$ to $t-1$ is denoted as $\boldsymbol{k}(t-1)_m$. For predictions, a database system makes $2m \times n$ matrix denoted as $\boldsymbol{K}(t)_{m,n}$ such that the top row of the matrix is given as $\boldsymbol{k}(t)_m$; also n-th row is given as $\boldsymbol{k}(t-n)_m(n < m)$. Another sequence, $\boldsymbol{x}(t)_n = \langle x(t-n), \ldots, x(t) \rangle$, is a sequence of an object's x-axis from time t to $t-n$; similarly, $\boldsymbol{y}(t)_n$ can be defined. Hence, we can calculate the approximate answer sequence of vector $\boldsymbol{w}_x = \langle wx_1, wx_2, \ldots, wx_{2m} \rangle$, $\boldsymbol{w}_y = \langle wy_1, wy_2, \ldots, wy_{2m} \rangle$ in the following equations:

$$\boldsymbol{x}(t)^T = \boldsymbol{K}(t-1)_{m,n} \bullet \boldsymbol{w}_x^T \qquad (2)$$
$$\boldsymbol{y}(t)^T = \boldsymbol{K}(t-1)_{m,n} \bullet \boldsymbol{w}_y^T.$$

An approximate answer can be calculated by Singular Value Decomposition (SVD). Matrix $\boldsymbol{w}_x$, $\boldsymbol{w}_y$ and vector $\boldsymbol{k}_m(t) = \langle x(t-m), \; y(t-m), x(t-m+1), y(t-m+1), \ldots, x(t), y(t) \rangle$, introduces position $\boldsymbol{p}(t+1) = \langle x(t+1), y(t+1) \rangle$ at $t+1$ as the following equation:

$$x(t+1) = \boldsymbol{w}_x \bullet \boldsymbol{k}(t)_m^T \qquad (3)$$
$$y(t+1) = \boldsymbol{w}_y \bullet \boldsymbol{k}(t)_m^T.$$

Positions $\boldsymbol{p}(t+2), \ldots$ after $t+2$, can be calculated by these recursive equations.

In this method, the system predicts the future point of an object based on the affine transformation on the coordinate system; the system accurately predicts future positions if an object moves in an arc, a straight line, or a sign curve. On the other hand, frequent turns by an object decrease prediction accuracy.

Note that in our experiments we also adapt our method to construct MBRs, as in the case of TPR-Trees.

2.3   Approach

Our primal goal is presenting a prediction technique for the index structure to mange trajectories of moving objects without large updating cost. Moreover, we focus on the two following issues:

1. Our prediction technique should use less computational resources on small mobile devices than existing techniques.
2. The prediction model has robustness to noises and measurement errors in practical trajectory data.

The prediction technique in TPR-Tree uses a simple linear prediction function. The technique generates often much larger MBRs than the ideal MBRs in order to avoid the updating data but the performance decreases in processing a range query. The MoDB must check many MBRs to process a query with large MBRs.

On the other hand, the prediction technique in STP-Tree can generate smaller sizes of MBRs than TPR-Tree as long as the motion pattern of an object can be represented as a certain function. The technique requires much more computational resources to predict future positions since the technique uses SVD technique for the prediction. Also the noises have strong influence to the accuracy of the prediction.

Thus, we design the model to predict future motion in the following approaches:

1. Analyzing the practical trajectory data and finding simple motion patterns appears frequently in the practical data,
2. Constructing the prediction model based on found motion patterns,
3. Considering measurement errors on constructing the model.

3   Trajectory Data

3.1   Moving Objects in Real World

For presenting prediction techniques, we experimentally analyzed the trajectories of various types of moving objects, such as cars, people, and parts of human bodies. In this section, we focus on the following characteristics of three types of moving objects.

Trajectories of Moving Vehicles (Vehicle Data). For such data, we obtained the trajectories of working rickshaws in Nara, a famous former capital of Japan. We placed a GPS receiver on each rickshaw that recorded the points where the rickshaw was located every second. The errors of the GPS receivers are within 5 m. The average trajectory length is about 18 km, and the average trajectory period is about nine hours. Example rickshaw trajectories are illustrated in Figure 2. The lines in the figure show the trajectories of rickshaws moving in the northern part of Nara for nine hours a day.

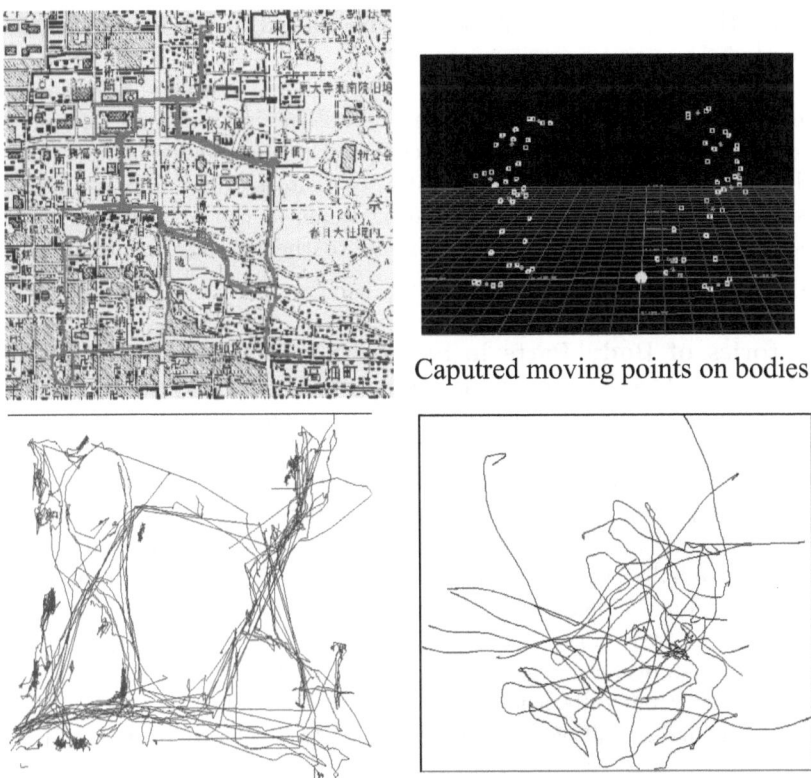

Caputred moving points on bodies

Trajectories of visitors in a forum   Caputred moving points on a hand

Fig. 2. Captured Moving Points

The shape of this trajectory includes many *long straight lines* because a rickshaw moves along streets whose shape is almost a straight line. On the other hand, since a rickshaw waits at intersections for passengers, it tends to stay long at one place. Generally, the trajectories of such moving vehicles as taxis, buses, and trucks have the same characteristics as the trajectories of rickshaws.

Trajectories of Wandering Visitors (Visitor Data). We did an experiment to obtain the trajectories of about 500 visitors and 30 staves to an exhibition that had about 100 booths. The size of the exhibition hall was 40 m × 50 m. We set 10 Lazar sensors and five video cameras for tracking visitors in the hall, and each sensor could obtain the locations of people every 1/2 of a second. The average duration of all trajectories was about 15 minutes, and the average geographical length of the trajectories was about 20 m. The maximum error of a Lazar sensor is less than 1 m.

Because Lazar sensors lose visitors hidden by other people, we completed the incomplete trajectories with video data captured by hand. Figure 2 also illustrates the trajectories of 49 people who walked in the hall during harf a hour period.

Three types of characteristic shapes were found in this type of trajectory: a gentle curved line, a short line, and so on. Because a visitor often walks and stops at booths, trajectory shapes tend to include such characteristic shapes. We also found another characteristic: the velocities of visitors differ, since various types of visitors meander looking for interesting items in the hall. The trajectories of visitors in an exhibition hall are similar to the visitors in museums, large shopping malls, art galleries, and so on.

Trajectories of Body Parts in Sports (Sport Data). We obtained the trajectory data of track points on the bodies of two soccer players using an optical motion capture system. Each player had 36 track points on his/her body and the soccer ball had two track points, and the motion capture system tracked 74 points every 1/120 of a second. The time of all trajectories was two minutes, and the average geographical length of trajectories was about 2 m. The top right image in Figure 2 shows example trajectories of the left legs of the two soccer players when fighting for the ball. This figure shows a projection of trajectories from 3D space to a 2D plane, but the features of the data are the same in each dimension. This figure has eight trajectories because we obtained four sets of trajectories from two players. Each player moves in an area 6×6 m^2. These trajectories have many curves and turns but only a few straight lines. The velocities of these moving points are not fixed, and each point can suddenly accelerate or decelerate.

3.2   Motion Patterns

We found several characteristic motion patterns of practical moving objects in the trajectories we obtained. The motion patterns suggest that "when an object moves in a particular manner, we can predict its future motion." In our experiments, we found two basic patterns: staying and straight-moving.

Staying: When an object almost stops at a place for a period, we describe it as *staying*. We did not find staying objects in the sport data, but 2/5 objects in the vehicle data were staying, and 9/10 objects of the visitor data were staying.

Moving Straight: When an object is moving in a straight line and its velocity is almost fixed, we say the object is *moving straight*. We found this motion pattern in all types of moving objects. Especially since 3/10 objects in the vehicle data are moving straight, the ratio is higher than in other data.

We can classify 7/10 – 9/10 of the objects in any data into these two patterns; however, the rest of the objects cannot be classified into any patterns. To classify all objects, we define one more motion pattern called "Moving Randomly" as follows.

Moving Randomly: When an object is continuously moving in unfixed directions at unfixed velocities, we say the object is *moving randomly*. In practical data, most such objects move in unfixed directions at almost fixed velocities. Such objects are found in visitor and sport data, but rarely in vehicle data.

We can classify any object based on our three defined patterns; moreover, the future motion of any object can be predicted by the definition of motion patterns.

3.3   Noise in Moving

The trajectories of moving objects often have low frequency noise because of positioning errors. Generally, existing databases deal with noiseless data but it is difficult to clean up practical noisy trajectory data. To apply database systems to practical data, in this paper we describe the prediction of the future position of a moving object with such low frequency noise. Trajectory noise has two principal sources: positioning devices and the size of moving objects. Because no positioning device can specify an object's position without errors, trajectories inevitably have errors. The size of the object, moreover, causes noise because devices generally cannot decide where an object's center point is on its surface. For example, errors of laser sensors when tracking walking people are less than 50 cm because the sensor rarely decides the person's center point, and the horizontal area of a person is a circle whose radius is less than 50 cm.

We define maximum noise as the sum of these two errors; for instance, if positioning error is 1 m and error size is 50 cm, maximum error is calculated as 1.5 m. Maximum error is denoted as θ_p, which is an actual measurement.

4   Prediction Model

In this section, we describe functions that predict an object's future point and how to apply prediction techniques to practical moving objects. Here, we consider our indexing technique is applied to both the nearest-neighbor query and the spatial-temporal range query.

4.1   Formalization of Motion Patterns

Before describing functions, we define motion patterns using mathematical equations. We denote a trajectory that includes the points of a moving object from time $t - m$ to t as $\lambda_{t,-m}$. If trajectory $\lambda_{t,-m}$ satisfies condition C, the trajectory's moving object has motion pattern C from time $t - m$ to t. We define three conditions, C_{st}, C_{sw}, and C_{rw}, for each motion pattern mentioned in Section 3.

Staying (C_{st}). We denote a position vector in $\lambda_{t,-m}$ at time i as $p(i)$ for defining 'staying' condition C_{st} as $|p(i) - p(t)| = 0$ where $i = t - m, t - m + 1,$ ..., $t - 1, t$. Condition C_{st} means that the maximum velocity of a moving object equals 0 from time $t - m$ to t.

Actually, practical data have noise θ_p, as mentioned in the previous section, so no practical objects completely stop at a place in the data. We introduce the influence of θ_p to the condition with the next extended equation:

$$C_{st} : |p(i) - p(t)| < \theta_p \tag{4}$$
$$p(i) \in \lambda_{t,-m}.$$

Here $|p|$ means the vector length of p. If an object moves less than distance θ_p from point $p(t)$ during period $t - m$ to t, the object has satisfied the "staying" condition.

Moving Straight(C_{sw}): We denote a velocity vector of an object in $\lambda_{t,-m}$ at time $i(0 \le i \le m)$ as $v(i) = p(i) - p(i - 1)$ for defining the 'moving straight' condition C_{sw} as $v(i) = v(t)$. This condition means that the difference between every velocity vector in $\lambda_{t,-m}$ and velocity vector $v(t) = p(t) - p(t - 1)$ at t always equals 0. Similar to the staying condition, we also considered the influence of noise θ_p. The actual conditions can be defined by:

$$C_{sw} : |v(i) - v(t)| < \theta_p \tag{5}$$
$$v(i) = p(i) - p(i - 1).$$

Moving Randomly(C_{rw}). We classify an object to this pattern when it does not satisfy the previous two conditions: C_{st} and C_{sw}. But objects 'moving randomly' do not move freely on a plane since physical restrictions limit their maximum velocity; for example, no person can walk at 50 m/s. In our method, we must define the maximum velocity of objects as v_{max} for the 'moving randomly' condition.

Obviously, an object moves within a circle such that its center is $p(t)$ and its radius is iv_{max}, where maximum velocity is v_{max} and the end point of the object at t is fixed to $p(t)$. Therefore, ideal 'moving randomly' condition C_{rw} can be defined as $C_{rw} : |v(i)| \le iv_{max}(t - m \le i \le t)$, where maximum velocity is v_{max} in trajectory $\lambda_{t,-m}$. Condition C_{rw}, including the influence of θ_p, is defined by:

$$C_{rw} : |v(i)| \le v_{max} + \theta_p. \tag{6}$$

4.2   Predictive Function

To predict the future position of moving objects, we define functions for each condition.

Notation $R(j)$ means an area where an object of $\lambda_{t,+n}$ will move from time current t to future time $j = t + n$. If an object in trajectory $\lambda_{t,-m}$ satisfies condition C, we can calculate $R(j)$ for the object using the equations in C. $R(j)$ is a closed area, and the shape of $R(j)$ is either a rectangle, a circle, or a combination of such diagrams. Hence, the calculation of $R(j)$ can be represented as function f using $\lambda_{t,-m}$ and time j, for example, $f(\lambda_{t,-m}, j) = R(j)$. We call f a "predictive function."

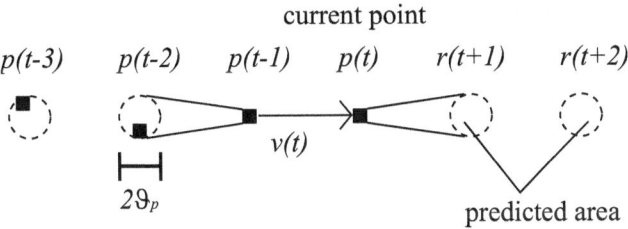

Fig. 3. Prediction for straight moving

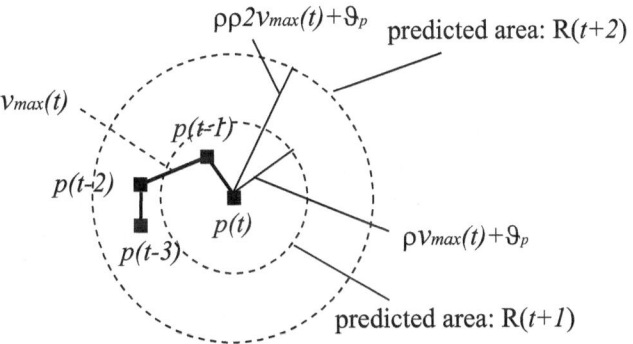

Fig. 4. Prediction for random moving

Note that previous work defined functions that give a future point of a moving object; however, it is actually impossible to determine a future place of an object in just a point. Therefore, we define f as a function that gives an area.

– Staying
   Because $C_{st} : |\boldsymbol{p}(i) - \boldsymbol{p}(t)| < \theta_p$, we define function f_{st} for staying as $R_{st}(j) = f_{st}(\lambda_{t,-m}, j)$, which is a circle such that its center point is $\boldsymbol{p}(t)$ and the radius is θ_p. The predicted area's actual shape is a rectangle because rectangles are available to build indexes based on MBRs.
– Moving Straight
   The function for moving straight is denoted as $R_{sw}(j) = f_{sw}(\lambda_{t,-m}, j)$. Figure 3 shows an example prediction area such that its shape is a circle whose center point is $\boldsymbol{p}(t) + (j - t)\boldsymbol{v}(t)$ and whose radius is θ_p. Similar to staying, we use a rectangle as the actual shape prediction area.
– Moving Randomly
   Based on the definition of condition C_{rw}, prediction area $R_{rw}(j) = f_{rw}(t + n)$ is a circle whose center point is $\boldsymbol{p}(t)$ and whose radius is $nv_{max} + \theta_p$, as illustrated in Figure 4. But this function often gives redundant area, especially because maximum velocity greatly increases the area. To avoid this problem, we reduce the outside of the area where the object seldom reaches.

When an object moves randomly at a velocity less than v_{max} during period n seconds, accuracy ρ in which the object exists in a circle whose center point is p and whose radius is $r(0 \leq r \leq nv_{max})$ is given by the following equation:

$$\rho = \left(\frac{r}{nv_{max}} \right)^n. \tag{7}$$

To predict the area at $t + n$ within accuracy ρ, we decide the radius of the predicted circle as $nv_{max} \sqrt[n]{\rho} + \theta_p$. For example, where $\rho = 0.7$ and $n = 5$, the radius is decided as $0.93 \times 5v_{max}$. In practical trajectory data, v_{max} is larger than the effective velocity, so we use accuracy $\rho = 0.7$ in our evaluation, as mentioned in Section 5.

4.3   Motion Prediction

For calculating the future area of an object at time $t + n$, the prediction system examines trajectory $\lambda_{t,-m}$, whose C_{st}, C_{sw}, or C_{rw} conditions are satisfied by the trajectory. Next, the system applies a function that corresponds to the condition. Because an object can be classified into either condition as mentioned, we can calculate its predicted area. If an object can be classified into both conditions C_{st} and C_{sw}, then the system applies condition C_{st} since the size of area $R_{st}(j)$ is less than $R_{sw}(j)$.

4.4   Construction of MBRs

To manage trajectory data, we adapt a traditional spatial data structure based on R-Tree in multidimensional space, which is similar to TPR- and STP-Trees. In traditional databases, because a set of data is added into a database continuously and randomly, the database must reconstruct the data structure every time a data set is added. Generally, such databases cannot construct the optimal data structure but moving object data are periodically and simultaneously added to a database because positioning devices periodically obtain an object's position. We consider a moving object database that can construct the optimal data structure. A database can calculate the optimal tree-based structure if all data sets of moving objects are simultaneously added to the database.

In the rest of this section, we explain the processes of constructing optimal MBRs.

1. At time t, the database temporally holds all positions $p_0(t)$, ..., $p_n(t)$ of moving objects that will be added to the database.
2. Positions are classified into each class by an average grouping method, a traditional hierarchical clustering method. As a result of the clustering process, each class has objects that are close to each other. In this clustering, the number of classes is indicated by a system administrator before string data. A database calculates MBRs for each class using the position of moving objects included in the class. These MBRs are used as leaf nodes of a tree-based data structure.

3. After the calculation of all leaf nodes, a database constructs each non-leaf MBR in the tree structure from the lower layer to the root.
4. Finally, a database calculates predictive MBRs from t to $t+i$ for every MBR in the tree.

These processes enable databases to calculate effective MBRs that are smaller than MBRs calculated by traditional algorithms for tree construction. Since this technique can be applied to previous proposed tree structures, we will use it to compare our method with previous methods.

5   Evaluation

This section describes the results of experiments that compared our method with previous prediction techniques. In our evaluations, we compared two indicators that show the performance of prediction mechanisms: the reconstruction rate and the MBR validation rate. Reconstruction rate rec_{t+i} at time $t+i$ is given as $\sum_{\tau=t+1}^{t+i} r_\tau/(n_{MBR}*i)$, where n_{MBR} is the number of all original, non-predicted MBRs and r_τ is the number of reconstructed MBRs at time τ according to prediction errors. Whenever a prediction error occurs, a database must possibly reconstruct the tree structure. If no prediction errors occur from time t to $t+i$, the reconstruction rate becomes 0. The other indicator, MBR validation rate val_{t+i}, is given as σ_{t+i}/s_{t+1}, where s_τ is the area of an ideal MBR at τ and σ_τ is the area of a predicted MBR at τ. If no prediction errors occur at time $t+i$, val_{t+i} becomes 1 because the predicted MBR must equal the ideal MBR. In this experiment, we calculate both rates for 'leaf' MBRs, which include the point data directly because non-leaf MBRs includes duplicate prediction errors occurred in leaf MBRs. Note that val_{t+i} is not larger than 1 since val_{t+i} is calculated after the reconstruction of MBRs, so that at least the size of the MBR equals the size of an ideal MBR.

5.1   Experiment Setting

We evaluated our proposed method with practical trajectory data, as mentioned in Section 3. For evaluation, we implemented three prediction techniques on Windows XP and C♯ Language: TPR-base, STP-based, and our proposed method. In each method, we focused on two indicators, reconstruction rate and MBR validation rate, which use 10 points from time $t=-4$ to $t=0$ to predict points from $t=1$ to $t=30$. If a method can completely construct future MBRs at time $t=n$, the reconstruction rate equals 0 at $t=n$. The reconstruction rate is 0.5 if the half of the objects exist outside of constructed MBRs, In other words, a low reconstruction rate means high prediction accuracy. The MBR validation rate is also an indicator of prediction accuracy. The validation rate is given as the ratio of the predicted MBR area at time t to ideal MBR at time t. A high MBR validation rate means high accuracy in contrast to the reconstruction rate. An ideal MBR is constructed such that the MBR completely includes real (not predicted) points of objects at time t, so the predicted MBR equals the ideal MBR if a system can completely predict the future positions of all objects. The rate of the complete predicted MBR becomes 1 since the predicted MBR, which is larger than the ideal MBR, will be reconstructed in a construction algorithm as an ideal MBR, as mentioned above.

We have three experiments using three sets of trajectory data mentioned in Section 3. The first set is a trajectory data set of 39 moving rickshaws, which drive within a central area in Nara city. We use almost 10000 points in this trajectory data set for the first experiment. For this experiment, we set $\theta_p = 10$ (m).

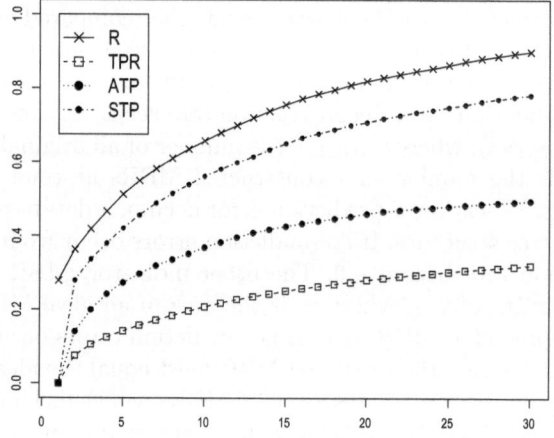

Fig. 5. Comparison results for reconstruction rate (data of moving rickshaws)

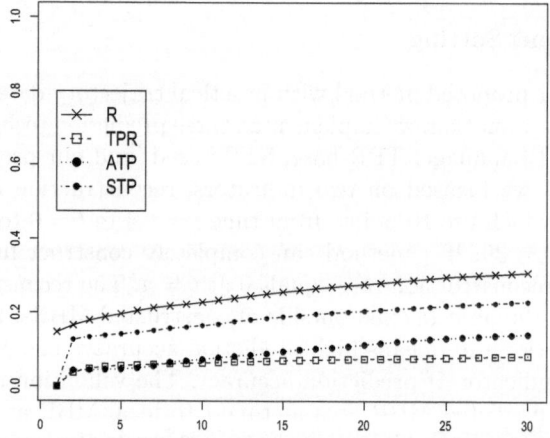

Fig. 6. Comparison results for reconstruction rate (data of walking visitors)

The second set is a trajectory data set of moving feet of two players in playing soccer. We captured trajectory of 18 scene each includes 7200 points during 30 seconds. For this experiment, we set $\theta_p = 2$ (cm).

The third set is a data set of 49 visitors walking in an exhibition hall during half an hour. Each trajectory has 3000–6000 points. For this experiment, we set $\theta_p = 1.5$ (m).

Figures 5, 6, and 7 compare reconstruction rates, and Figures 8, 9, and 10 compare MBR validation rates. The results of our technique are displayed as "adaptive temporal prediction" (ATP) in these figures. The horizontal axis of each figure denotes the past time ($t = 1$ to $t = 10$) from when a system predicted future positions. The time scale depends on the sampling time of each data. The

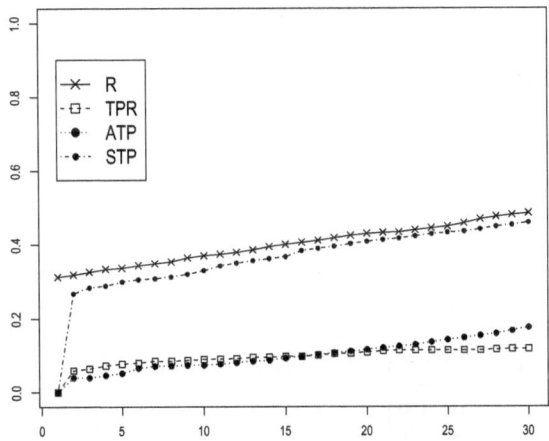

Fig. 7. Comparison results for reconstruction rate (data of sport motion)

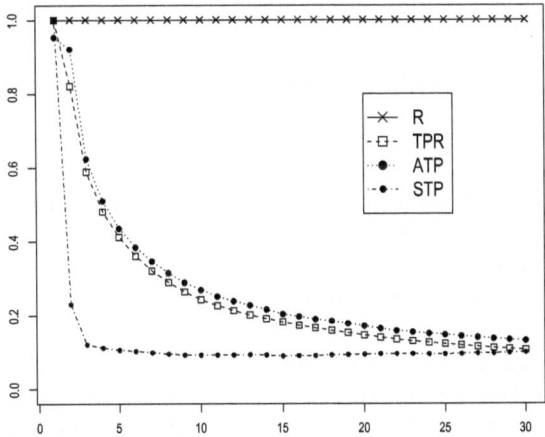

Fig. 8. Comparison results for MBR validation rate (data of moving rickshaws)

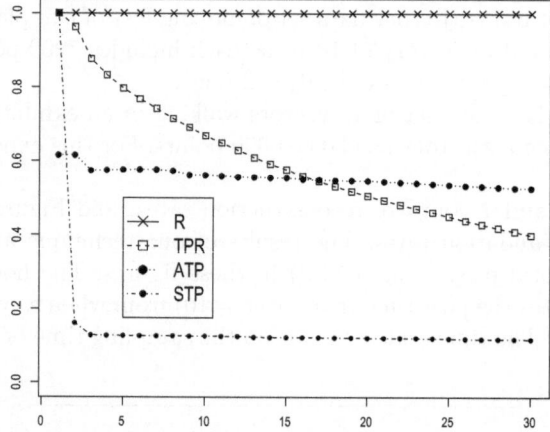

Fig. 9. Comparison results for MBR validation rate (data of walking visitors)

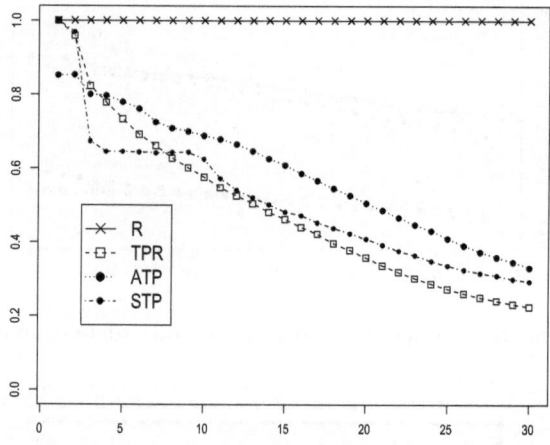

Fig. 10. Comparison results for MBR validation rate (data of sport motion)

vertical axis shows either reconstruction or the MBR validation rate for forty moving objects selected randomly from each data set.

The results of R-Tree [7] shown in the figures can be considered the rate without any prediction, since MBRs in R-Tree will always be reconstructed to an ideal MBR whenever an object moves outside of the original MBR. In other words, if a method's value at a time is lower than the R-Tree value, the performance of the prediction method is worse than no prediction. On the other hand, the MBR validation rate is equals to 1 because MBRs in R-Tree will be reconstructed to the optimal size anytime. Thus, we consider the MBR validation rate of R-Tree the optimal rate at each time.

5.2 Results

The results of experiments show that our technique has the advantages to previous methods in comparison of the validation rate. Especially, our method can generate smaller MBRs than others at 'far' future time. On the other hands, the prediction technique in TPR-Tree shows good performance in comparison of the reconstruction rate. In our experiment, the prediction technique in STR-Tree, which is an improved TPR-Tree method, has disadvantages compared to other methods. The prediction technique of STR-Tree is too sensitive to noise; as a result the predicted points are often different from real points in practical data.

All results of the prediction technique in TPR-Tree will show low reconstruction rates but our technique has almost the same performance in both the sport data and visitor's data. Each reconstruction rate from rickshaw data is higher than rates from the other two types of data because rickshaw trajectories have large noise and the maximum velocity of moving objects are higher than objects in the other data set. For visitor data, the validation rate of our technique is lower than TPR-based technique in the near future but the rate in the far future is higher. Because many visitors stop at a point during long time, our technique generates redundant MBRs in the near future.

MBR validation rates of STP-based prediction technique decreases more rapidly than other methods for predicted time. For the distant future, the STP-based prediction technique constructs redundant predicted MBRs, which are much larger than ideal MBRs because few anomaly trajectory data that have large noise make the prediction technique generate redundant large MBRs. Our proposed method can construct accurate predicted MBRs where include such an anomaly trajectory.

We also examined the performance of an enhanced prediction technique in STP-Tree, which constructs MBRs larger than the original MBRs, the same as θ_p. In other words, we introduced θ_p into the prediction technique to evaluate the effectiveness of θ_p. The introduction of θ_p certainly improved STP-based prediction performance. But we basically only found slight improvement because the influence of the sensitive function is stronger than the improvement of θ_p. If we can reduce noise from the practical trajectory data, STR-based prediction performance will possibly be improved more. Actually, it is difficult to reduce noise from several points of moving objects, so we conclude that our proposed method is better than other methods even if θ_p is introduced to those other methods.

6 Conclusion

In this paper, we proposed a motion prediction method based on three motion patterns: staying, moving straight, and moving randomly to make predictive indexes for moving objects. Moreover, evaluation results showed the advantages of our methods in experiments that compared previous prediction techniques using practical trajectory data.

In our method, we suppose the trajectory data can be obtained accurately and completely; however, we should introduce a complementary method for missing trajectories. In the future, we will apply our method in application systems using trajectories and evaluate its performance in these systems with a complementary method. For applying practical application systems, we will also enhance our prediction technique based on geographic conditions; for example, when an object moves up a slope, its velocity probably decreases.

References

1. Lester, J., Choudhury, T., Borriello, G.: A practical approach to recognizing physical activities. In: Fishkin, K.P., Schiele, B., Nixon, P., Quigley, A. (eds.) PERVASIVE 2006. LNCS, vol. 3968, pp. 1–16. Springer, Heidelberg (2006)
2. Hightower, J., Consolvo, S., LaMarca, A., Smith, I.E., Hughes, J.: Learning and recognizing the places we go. In: Beigl, M., Intille, S.S., Rekimoto, J., Tokuda, H. (eds.) UbiComp 2005. LNCS, vol. 3660, pp. 159–176. Springer, Heidelberg (2005)
3. Wolfson, O., Sistla, P., Xu, B., Zhou, J., Chamberlain, S.: DOMINO: Databases fOr MovINg Objects tracking. In: SIGMOD 1999 Conference Proceedings, pp. 547–549 (1999)
4. Mokhtar, H., Su, J., Ibarra, O.H.: On moving object queries. In: PODS 2002 Symposium Proceedings, pp. 188–198 (2002)
5. Kollios, G., Gunopulos, D., Tsotras, V.J.: On indexing mobile objects. In: PODS 1999 Symposium Proceedings, pp. 261–272 (1999)
6. Kollios, G., Tsotras, V.J., Gunopulos, D., Delis, A., Hadjieleftheriou, M.: Indexing animated objects using spatiotemporal access methods. IEEE Transactions on Knowledge and Data Engineering 13(5), 758–777 (2001)
7. Guttman, O.: R-trees: a dynamic index structure for spatial searching. In: SIGMOD 1984 Conference Proceedings, pp. 47–57 (1984)
8. Tao, Y., Faloutsos, C., Papadias, D., Liu, B.: Prediction and indexing of moving objects with unknown motion patterns. In: SIGMOD 2004 Conference Proceedings, pp. 611–622. ACM Press, New York (2004)
9. Šaltenis, S., Jensen, C.S., Leutenegger, S.T., Lopez, M.A.: Indexing the positions of continuously moving objects. In: SIGMOD 2000 Conference Proceedings, pp. 331–342. ACM Press, New York (2000)
10. Sellis, T., Roussopoulos, N., Faloutsos, C.: The R⁺-tree: A dynamic index for multidimensional objects. In: VLDB 1987 Conference Proceedings, pp. 3–11 (1987)
11. Ciaccia, P., Patella, M., Zezula, P.: M-tree: An efficient access method for similarity search in metric spaces. In: VLDB 1997 Conference Proceedings, pp. 426–435 (1997)
12. Tao, Y., Sun, J., Papadias, D.: Analysis of predictive spatio-temporal queries. ACM Trans. Database Syst. 28(4), 295–336 (2003)

SHWMP: A Secure Hybrid Wireless Mesh Protocol for IEEE 802.11s Wireless Mesh Networks*

Md. Shariful Islam[1], Md. Abdul Hamid[2], and Choong Seon Hong[1,**]

[1] Department of Computer Engineering, Kyung Hee University, Republic of Korea
sharif@networking.khu.ac.kr, cshong@khu.ac.kr
[2] Department of Information and Communications Engineering,
Hankuk University of Foreign Studies, Republic of Korea
hamid@hufs.ac.kr

Abstract. In recent years, mesh networking has emerged as a key technology for the last mile Internet access and found to be an important area of research and deployment. The current draft standard of IEEE 802.11s has defined routing for Wireless Mesh Networks (WMNs) in layer-2 and is termed as Hybrid Wireless Mesh Protocol (HWMP). However, security in routing or forwarding functionality is not specified in the standard. As a consequence, HWMP in its current from is vulnerable to various types of routing attacks such as flooding, route disruption and diversion, spoofing etc. In this paper, we propose SHWMP, a secure HWMP protocol for WMN. The proposed protocol uses cryptographic extensions to provide authenticity and integrity of HWMP routing messages and prevents unauthorized manipulation of mutable fields in the routing information elements. We show via analysis that the proposed SHWMP successfully thwarts all the identified attacks. Through extensive ns-2 simulations, we show that SHWMP provides higher packet delivery ratio with little increase in end-to-end delay, path acquisition delay and control byte overhead.

Keywords: Wireless Mesh Network, Secure Hybrid Wireless Mesh Protocol, Authentication, Merkle Tree.

1 Introduction

Wireless mesh networking (WMN) has emerged as one of the most promising concept for self-organizing and auto-configurable wireless networking to provide adaptive and flexible wireless Internet access solutions for mobile users. Potential application scenarios for wireless mesh networks include backhaul support for cellular networks, home networks, enterprise networks, community networks, and intelligent transport system networks [1]. The increased interest in WMN has reflected in producing a standard named IEEE 802.11s, which is in progress and expected to be finalized by mid 2009. Our work is based on the current draft version D2.02 [2] of

---

* "This research was supported by the MKE, Korea, under the ITRC support program supervised by the NIPA)"(NIPA-2009-(C1090-0902-0016)).
** Corresponding author.

M.L. Gavrilova and C.J.K. Tan (Eds.): Trans. on Comput. Sci. VI, LNCS 5730, pp. 95–114, 2009.

IEEE 802.11s that introduces the concept of embedding routing in layer-2 named Hybrid Wireless Mesh Protocol (HWMP). HWMP has been developed to ensure interoperability between devices of different vendors and is the key reason for integrating routing in MAC layer.

Wireless mesh networks (WMNs) consist of mesh clients and mesh routers, where the mesh routers form a wireless infrastructure/backbone and interwork with the wired networks to provide multihop wireless Internet connectivity to the mesh clients. The network architecture of a 802.11s WMN is depicted in Fig. 1. A mesh point (MP) is an IEEE 802.11s entity that mainly acts as a relay node. A mesh access point (MAP) is an MP but can also work as an access point. Legacy wireless mobile stations (STA) are connected to an MAP through generic WLAN protocols. Thus, configuration of an MAP allows a single entity to logically provide both mesh functionalities and AP functionalities simultaneously. A mesh portal (MPP) is a logical point and has a bridging functionality and connects the mesh network to other networks such as a traditional 802.11 WLAN or a non-802.11 network and act as the gateway to the WMN infrastructure. In a WMN, traffic flows between MP to MP for intra-mesh traffic and between MP-MPP or vice-versa for traffic to / from outside mesh.

The security in routing or forwarding functionality is not specified in IEEE 802.11s [3]. Our study identifies that existing HWMP is vulnerable to various types of routing attacks such as flooding, route disruption and diversion, spoofing etc. The main reason is that the intermediate nodes need to modify mutable fields (i.e., hop count, TTL, metric etc) in the routing element before forwarding and re-broadcasting them. Since other nodes will act upon those added information, these must also be protected somehow from being forged or modified. However, only source authentication does not solve this problem, because the information are added or modified in intermediate nodes. This motivates us to devise a hop-by-hop authentication mechanism in our proposal. An earlier version of this work can be found in [4].

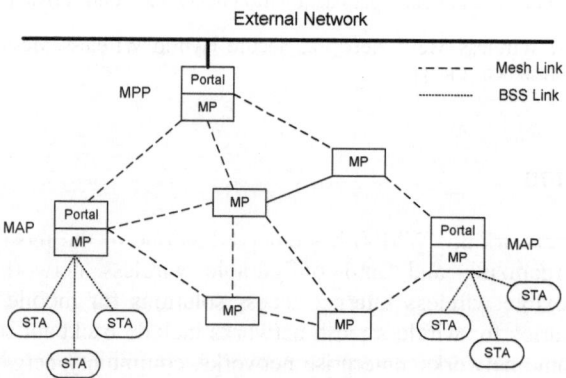

Fig. 1. Network architecture of IEEE 802.11s WMN

The contributions of this paper are as follows. We identify the security vulnerabilities of HWMP. Particularly we show that flooding, route disruption, route diversion, routing loop formation through spoofing are the major threats posed by an adversary in HWMP. We propose a Secure Hybrid Wireless Mesh Protocol (SHWMP) for IEEE 802.11s based WMN. The fields of the routing information elements are classified as mutable and non-mutable fields. The proposed SHWMP protects the non-mutable part using symmetric key encryption and authenticates mutable information exploiting the concept of Merkle tree [5]. Results from analysis and simulation demonstrate that SHWMP is robust against the identified attacks, provides higher packet delivery ratio, requires no extra communication cost and incurs little path acquisition delay, end-to-end delay, control byte, computation and storage overhead.

The rest of the paper is organized as follows. Following section discusses some of the related works. We give a brief overview of the cryptographic primitive used in this work in Section 3. Section 4 briefly introduces HWMP in 802.11s. Possible attacks are shown in section 5. Section 6 shows the key distribution in 802.11s. We have proposed our idea in section 7 followed by security analysis in section 8. We have evaluated network performance through simulation in section 9. Finally, section 10 concludes our work.

2 Related Works

Research on layer-2 routing is still in its early age. As of now, there is no state-of-the-art solution exists in the literature for securing layer-2 routing. In [6], the authors have just summarized the proposed routing from IEEE 802.11s draft D0.01 [7]. However, the optional routing protocol RA-OLSR, that was described in [6] is no longer considered as a candidate protocol for IEEE 802.11s routing and is omitted from current draft D.2.02 [2]. In [8], the author has described just an update of layer-2 routing in the current draft. IEEE 802.11s's framework and research challenges are summarized in [3]. A hybrid centralized routing protocol is presented in [17] that incorporates tree-based routing with a root-driven routing protocol.

Apart from these, there has been some research on securing layer 3 routing. Ariadne in [9] ensures a secure on-demand source routing. Authentication is done using TESLA [10], digital signatures and standard Message Authentication Code (MAC). However, as the route request is not authenticated until it reaches the destination, an adversary can initiate route request flooding attack. A variant of Ariadne named endairA is proposed in [11] with a difference that instead of signing a route request, intermediate nodes sign the route reply. It requires less cryptographic computation, but still vulnerable to malicious route request flooding attack. SAODV [12] is a secure variant of AODV. Operations are similar to AODV, but uses cryptographic extensions to provide authenticity and integrity of routing messages. It uses hash chains in order to prevent manipulation of hop count field. However, an adversary can always increase the hop count. Another secure on-demand distant vector protocol, ARAN (Authenticate Routing for Ad hoc Networks), is presented in [13]. Just like SAODV, ARAN uses public key cryptography to ensure integrity of routing message. However, a major drawback of ARAN is that it requires extensive signature generation and verification during the route request flooding.

In our proposed scheme, we will use the existing keying hierarchy specified in current 802.11s specification. So, there is no extra burden for enforcing external keying mechanism (like PKI, KDC etc.). That is, we are not assuming that a pairwise key exists between any two nodes in the networks as path security can not be assured in 802.11s. We have used Merkle Tree in our scheme for authenticating mutable fields in the routing information elements. Our secure routing employs symmetric crypto-graphic primitives only and uses Merkle tree-based hop-by-hop authentication mechanism by exploiting existing key-hierarchy of 802.11s standard.

3   Cryptographic Primitives

In this section, we briefly describe two of the important cryptographic primitives Merkle tree and Authentication Path used in our proposed scheme.

3.1 Merkle Tree

Merkle Tree is a useful technique to build secure authentication and signature schemes from hash functions. A Merkle tree [4][14][15] is a complete binary-tree that has equipped with a function *hash* and an assignment function F such that for any interior node n_{parent} and two child node n_{left} and n_{right}, the function F satisfies: $F(n_{parent}) = hash(F(n_{left}) \| F(n_{right}))$.

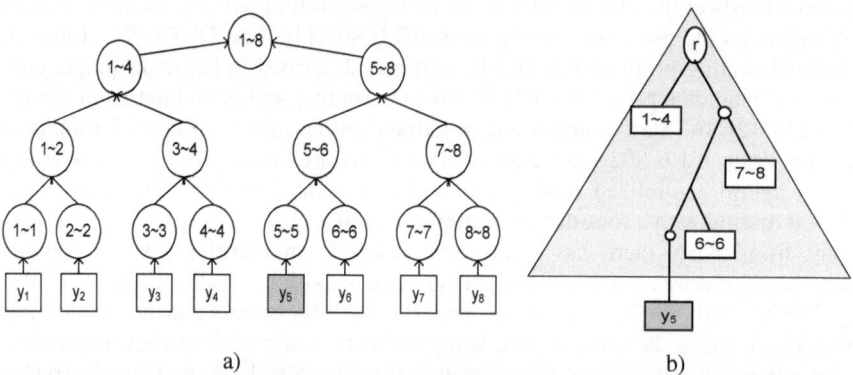

a)   b)

Fig. 2. a) A Merkle tree with 8 leaves and 8 leaf pre-images $(y_1, y_2,...,y_8)$. Each leaf node is a hash of its corresponding pre-image and each internal node is the hash of the concatenation of two child values. y_5 is the value of the pre-image needs to be verified and the root of the tree is known to be public. b) The set of white rectangular nodes makes up the authentication path for the leaf pre-image y_5. Node r represents the root and value of y_5 is verified if both the publicly known root (1~8) (in Fig. 2a) and computed root r (in Fig. 2b) are same.

The hash function used is a candidate one-way function such as SHA-1[16]. Fig. 2a depicts a Merkle tree with eight leaf nodes, each being a hash of a leaf pre-image denoted by a box. Then, function F is used to assign the values of each internal node. The value of the root is considered public while all the values associated with a leaf pre-image are known by the owner of the tree.

3.2 Authentication Path

The authentication path of a leaf pre-image consists of values of all the nodes that are siblings of the nodes on the path between the leaf pre-image and the root. Fig. 2b shows the authentication path of the marked leaf pre-image y_5. To verify the value of a leaf pre-image a receiver needs to compute the potential values of its ancestors by iteratively using of the F function shown in previous section. Note that, computing the F function α times, where α denotes the number of leaf nodes in the authentication path, will result in getting the value of the root. A leaf pre-image is authenticated and accepted as correct if and only if the computed root value is equal to the already known root value.

4 Overview of HWMP for IEEE 802.11s WMN

The Hybrid Wireless Mesh Protocol (HWMP) has combined the flavor of reactive and proactive routing strategy by employing both on-demand path selection mode and proactive tree building mode. On-demand mode allows two MPs to communicate using peer-to-peer paths. This mode is mainly used by nodes that experience a changing environment and when there is no root MP configured. On the other hand, proactive tree building mode can be an efficient choice for nodes in a fixed network topology. The mandatory routing metric used in HWMP is the airtime cost metric [2] that measures the link quality (e.g. amount of channel resource consumed by transmitting a frame over a particular link). In HWMP, both on demand and proactive mode can be used simultaneously.

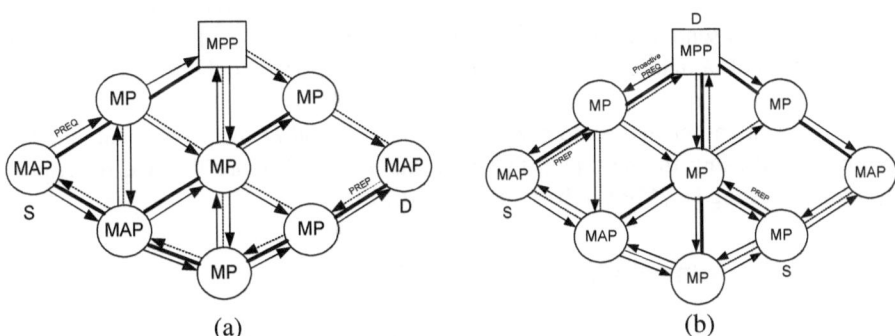

Fig. 3. a) On-demand mode. b) Proactive mode.

4.1 On-Demand Mode

In an On-demand mode a source MP broadcast *path request* (PREQ) message requesting a route to the destination. The PREQ is processed and forwarded by all intermediate MPs and sets up the reverse path from the destination to the source of route discovery. The destination MP or any intermediate MP with a path to the destination may unicast a *path reply* (PREP) to the source MP that creates the forward path to the destination. As shown in Fig. 3a, source MAP *S* broadcasts PREQ message and intermediate nodes re-broadcasts PREQ after updating its path to the source as shown in solid lines. The destination MAP *D*, unicast a PREP message to the source MAP *S* using the reverse path shown in solid lines.

4.2 Proactive Mode

In *Proactive Tree Building* mode, the MP that configured as a root MP (i.e usually the MPP) can initiate route discovery process in two ways. *Firstly,* it announces its presence by periodically broadcasting a *root announcement* RANN message that propagates metric information across the network. Upon reception of a RANN message, an MP that has to create or refresh a path to the root MP sends a unicast PREQ to the root MP. The root MP then unicast a PREP in response to each PREQ. The unicast PREQ creates the reverse path from the root MP to the originating MP, while the PREP creates the forward path from the MP to the root MP. *Secondly*, the root MP proactively disseminates a proactive PREQ message to all the MPs in the networks with intent to establish a path as shown in Fig. 3b. An MP after receiving a proactive PREQ, creates or updates its path to the root MP by unicasting a proactive PREP, if and only if the PREQ contains a greater sequence number, or the sequence number is the same as the current path and the PREQ offers a better metric than the current path to the root MP. Thus, a routing tree is created with the MPP being the root of the tree.

4.3 Hybrid Mode

HWMP also allows both on-demand and proactive mode to work simultaneously. This hybrid mode is used in situations where a root MP is configured and a mesh point *S* wants to send data to another mesh point *D* but has no path to *D* in its routing table. Instead of initiating on-demand mode, S may send data to the root portal, which in turns delivers the data to *D* informing that both *S* and *D* are in the same mesh. This will trigger an on-demand route discovery between *S* and *D* and subsequent data will be forwarded using the new path that performs better. A more detailed description regarding existing HWMP can be found in [2][6][8].

5   Security Vulnerabilities of HWMP

The existing HWMP routing mechanism relies on the fact that all participating mesh entities cooperate with each other without disrupting the operation of the protocol. Without proper protection, the routing mechanism is susceptible to various kind of attacks. In the following, we identify and describe possible attacks that can be launched in the existing HWMP routing mechanism.

5.1 Flooding

The simplest of attacks that a malicious node can launch is by flooding the network with a PREQ messages destined to an address which is not present in the network. As the destination node is not present in the network, every intermediate node will keep forwarding the PREQ message. As a result, a large number of PREQ message in a short period will consume the network bandwidth and can degrade the overall throughput. As shown in Fig. 4a, the malicious node M initiates route discovery with a PREQ for a destination that is not in the network. So that intermediate nodes re-broadcasts PREQ and within a short time the network is flooded with fake requests.

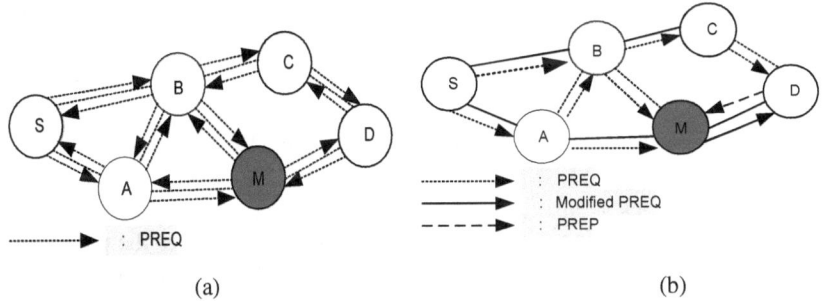

(a)   (b)

Fig. 4. a) Flooding. b) Route Disruption.

5.2 Route Disruption

By launching a route disruption attack, an adversary can prevent discovering a route between two legitimate nodes. In other word, if there exist a route between the two victim nodes, but due the malicious behavior of the attacker, the routing protocol can not discover it. In HWMP, route-disruption attacks can easily be launched by a malicious node as shown in Fig. 4b. The malicious node M can prevent the discovery of routes between nodes S and D. M can modify the metric field value to zero on the PREQ message it receives from A or B and re-broadcast. So, after receiving the modified PREQ, D will choose M as the next hop in the reverse path and unicast PREP to M. Now, M can prevent the route discovery by dropping the valid PREP message destined for S.

5.3 Route Diversion

A malicious node can launch a route diversion attack by modifying mutable fields in the routing information elements such as hop count, sequence number and metric field. A malicious node M can divert traffic to itself by advertising a route to a destination with a Destination Sequence Number (DSN) greater than the one it received from the destination. For example, the malicious node M in the Fig. 5a receives a PREQ from A which was originated from S for a route to node D. As HWMP allows intermediate PREP, M can unicast a PREP to A with a higher destination sequence number than the value last advertised by D. After getting the PREQ message D will also unicast PREP to the source S. At some point, A will receive both the PREPs and consider the PREP with higher destination sequence number as the valid one and discards the original PREQ as if it was stale. So, A will divert all subsequent traffic destined for D to the malicious node M.

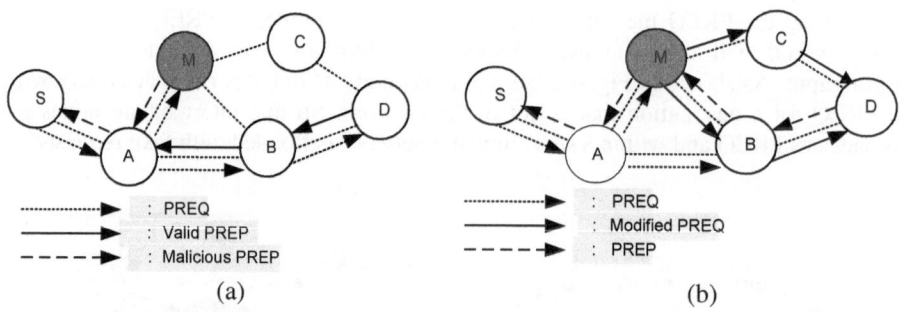

Fig. 5. Route diversion attacks. a) Increasing DSN. b) Decreasing Metric.

Route diversion attack is also possible by modifying the metric field used in the HWMP PREQ messages. Value of the metric field is used to decide the path from a source to a destination, which is a cumulative field contributed by each node in the path. A malicious node can modify the mutable metric field to zero to announce a better path to a destination. As depicted in Fig. 5b, *M* can modify the metric field in the PREQ to zero and re-broadcasts it to the network. So, the reverse path created should go through the malicious node *M*. As a result, all traffics to the destination *D* will be diverted through the attacker.

5.4 Routing Loops

A malicious node can create routing loops in a mesh network by spoofing MAC addresses and modifying the value of the metric field. Consider the following network (Fig. 6) where a path exists between the source *S* and destination *X* that goes through node *B* and *C*. Also, there is a path exists from *A* to *C* through *D*. Assume that a malicious node *M* as shown in Fig. 6a is present in the vicinity where it can listen to the PREQ/PREP messages pass through *A*, *B*, *C* and *D* during route discovery process. It can create a routing loop among the nodes *A*, *B*, *C* and *D* by impersonation combined with the modification of metric field in PREP message.

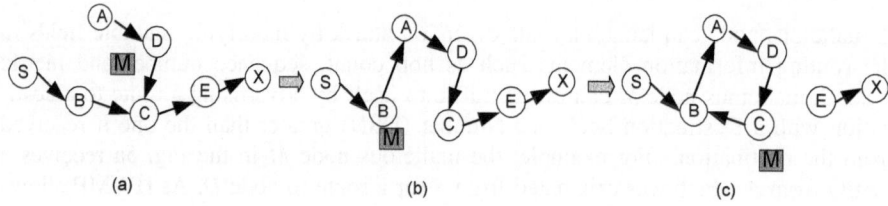

Fig. 6. Formation of Routing loops

First it impersonates node *A*'s MAC address and moved out of the reach of node *A* and closer to node *B*. And then it sends a PREP message to node *B* indicating a better metric value than that of the value received from *C*. So, node *B* now re-establishes its route to *X* that should go through *A* as shown in Fig. 6b. At this point, the malicious node impersonates node *B* and moves closer to node *C* and sends a PREP to node *C* indicating a better metric than the one received from *E*. So, node *C* will now choose *B* as the next hop for its route to the destination *X* as shown in Fig. 6c. Thus a loop has been formed and the destination *X* is unreachable from all the four nodes.

6 Key Distribution in IEEE 802.11s

IEEE 802.11s ensures link security by using Mesh Security Association (MSA) services. 802.11s extends the security concept of 802.11 by a key hierarchy, inherits functions of 802.11i and uses 802.1X for initial authentication [2]. The operation of MSA relies on mesh key holders, which are functions that are implemented at MPs within the WMN.

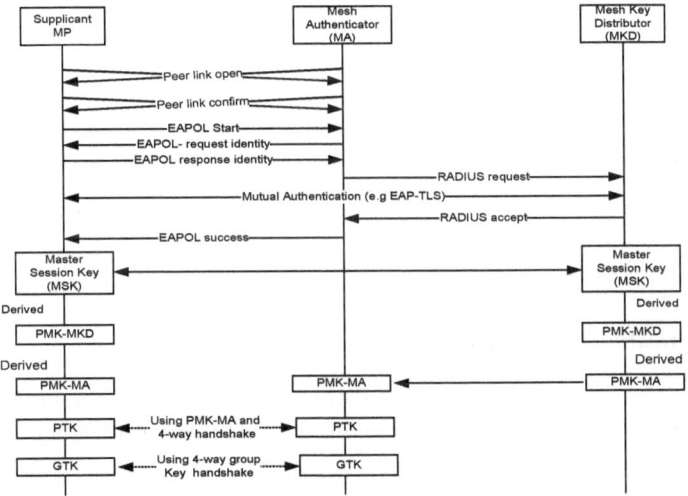

Fig. 7. Key establishment procedure in IEEE 802.11s

Two types of mesh key holders are defined: mesh authenticators (MAs) and mesh key distributors (MKDs). A single MP may implement both MKD and MA key holders, an MA alone and no key holders. Fig. 7 depicts the key establishment procedure between two MPs in IEEE 802.11s.

The first level of link security branch, PMK-MKD is mutually derived by the supplicant MP and MKD, from the Master Session Key (MSK) that is created after the initial authentication phase between supplicant MP and MKD or from a pre-shared key (PSK) between MKD and supplicant MP, if exists. The second level of link security branch PMK-MA is also derived by the supplicant MP and MKD. MKD then delivers the PMK-MA to the MA and thus permits to initiate MSA 4-way handshake which results in deriving a PTK of 512 bits between supplicant MP and MA.

During the MSA 4-way handshake, an MA receives the GTK of the supplicant MP. After the completion of MSA 4-way handshake, a group handshake is used to send the GTK of the MA to the supplicant MP. The GTK is a shared key among all supplicant MPs that are connected to the same mesh authenticator (MA). In our proposed secure routing algorithm, PTK is used for encryption of unicast messages and GTK is used for encrypting broadcast messages.

7 Secure Hybrid Wireless Mesh Protocol (SHWMP)

The routing protocol proposed in this section is a secure version of Hybrid Wireless Mesh Protocol (SHWMP). HWMP routing information elements have a mutable and a non-mutable part. We exploit these existing mutable and non-mutable fields to design a secure layer-2 routing. More specifically, we (i) use the existing key distribution, (ii) identify the mutable and non-mutable fields, (iii) show that mutable fields can be authenticated in hop-by-hop fashion using the concept of Merkle hash tree, and (iv) use symmetric key encryption to protect non-mutable fields. We describe the proposed protocol in details in the following subsections.

7.1 Use of Keys

All the entities in the mesh infrastructure (MP, MAP and MPP) can act as supplicant MP and Mesh Authenticator (MA). Before initiating a route discovery process, all the MPs authenticate its neighboring MPs, send its GTK and establish PTK through key distribution process described in Section 6. We use this GTK for securing broadcast messages such as PREQ, RANN and PTK is used to secure unicast messages such as PREP, proactive PREQ.

7.2 Identification of Mutable / Non-mutable Fields

The information elements in the HWMP contain fields that can be modified in the intermediate routers which we termed as mutable and those that can not be modified termed as non-mutable fields.

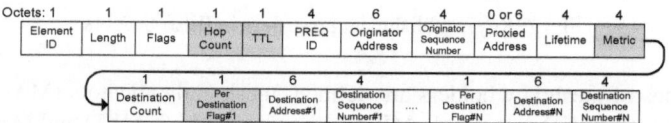

Fig. 8. Format of a PREQ element

Fig. 8 shows the format of a PREQ element where the mutable fields are:

a. Hop count field: Provides information on the number of links in the path, incremented by each intermediate node, but it is not used for routing decision.
b. TTL field: The time-to-leave field defines the scope of the PREQ in number of hops. TTL value is decremented by 1 at each intermediate node.
c. Metric field: HWMP uses an airtime link metric instead of hop count metric as in AODV, to take a decision on path selection. Whenever an intermediate node receives a PREQ that is to be forwarded, it calculates the airtime cost to the current path and adds the value to the existing metric field.

d. Per destination flag: The Destination Only (DO) and Reply and Forward Flag (RF) determine whether the route-reply message (RREP) will be sent by intermediate node or only by destination. If DO flag is not set and RF flag is set, the first intermediate node that has a path to the destination sends PREP and forwards the PREQ by setting the DO flag to avoid all intermediate MPs sending a PREP. In this case, per destination flag field is also a mutable field.

Fig. 9 and Fig. 10 show the format of a PREP and RANN information element. In both the cases, the mutable fields are hop-count, TTL and metric indicated by shadowed boxes.

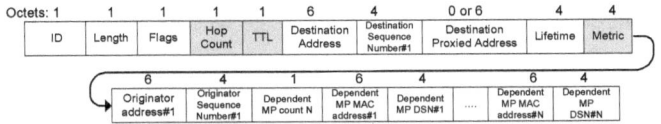

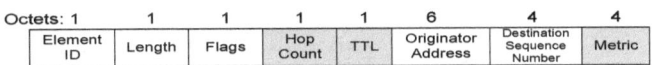

Fig. 9. Format of a PREP element

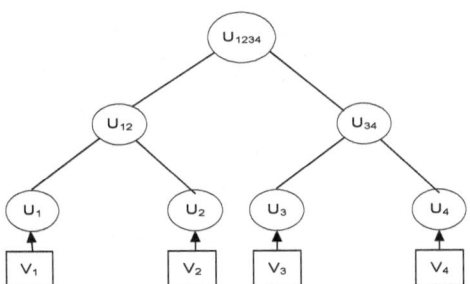

Fig. 10. Format of a RANN element

7.3 Construction of Merkle Tree

Let the mutable fields of routing information elements that need to be authenticated are v_1, v_2, v_3 and v_4. We hash each value v_i into u_i with a one-way hash function such that $u_i = h(v_i)$. Then we assign the hash values to the leaves of the binary tree as shown in Fig. 11.

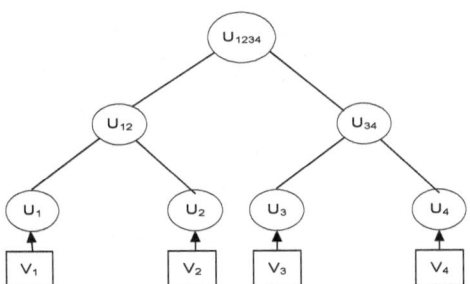

Fig. 11. Construction of Merkle tree

Moreover, to each internal vertex u of this tree, we assign a value that is computed as the hash of the values assigned to the two children of u such as $u_{12} = h(u_1 \| u_2)$. Finally, we found the value of the root and make a message authentication code on the root by using GTK of the sender or by PTK between sender and receiver for authenticating broadcast and unicast messages, respectively. The sender can reveal a value v_i that needs to be authenticated along with the values assigned to the siblings of the vertices along the path from v_i to root that we denote as authentication path, *authpath(v_i)*. The receiver can hash the values of the authentication path in appropriate order (as described in Section 3.2) to compute the root and create a MAC on the root using the derived key. It then compares the values of the two MACs, If these two values match, then the receiver can be assured that the value v_i is authentic.

7.4 Securing on Demand Mode

Consider a source MP S that wants to communicate with a destination MP X as shown in Fig. 12.

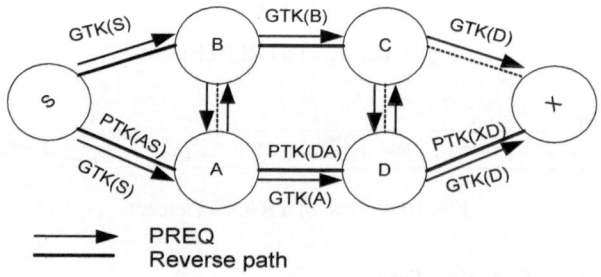

      PREQ
      Reverse path

Fig. 12. Secure on-demand path selection

In order to establish a secure route, source node S, destination node X and set of intermediate nodes (F_1 that includes A, B and F_2 that includes C, D) executes the route discovery process in the following way:

$$S \rightarrow *: \text{MAC}_{\text{GTK}}\text{root}(S), \{v_i, \text{authpath}(v_i)\}, \{\text{PREQ}-\text{MF}\}_{\text{GTK}} \tag{1}$$

$$F_1 \rightarrow *: \text{MAC}_{\text{GTK}}\text{root}(F_1), \{v_i, \text{authpath}(v_i)\}, \{\text{PREQ}-\text{MF}\}_{\text{GTK}} \tag{2}$$

$$F_2 \rightarrow *: \text{MAC}_{\text{GTK}}\text{root}(F_2), \{v_i, \text{authpath}(v_i)\}, \{\text{PREQ}-\text{MF}\}_{\text{GTK}} \tag{3}$$

$$X \rightarrow F_2: \text{MAC}_{\text{PTK}}^{X,F_2}\text{root}(X), \{v_i, \text{authpath}(v_i)\}, \{\text{PREP}-\text{MF}\}_{\text{PTK}}^{X,F_2} \tag{4}$$

$$F_2 \rightarrow F_1: \text{MAC}_{\text{PTK}}^{F_2,F_1}\text{root}(F_2), \{v_i, \text{authpath}(v_i)\}, \{\text{PREP}-\text{MF}\}_{\text{PTK}}^{F_2,F_1} \tag{5}$$

$$F_1 \rightarrow S: \text{MAC}_{\text{PTK}}^{F_1,S}\text{root}(F_1), \{v_i, \text{authpath}(v_i)\}, \{\text{PREP}-\text{MF}\}_{\text{PTK}}^{F_1,S} \tag{6}$$

From the key management of 802.11s, the node S is equipped with one GTK that it shares with its neighbors and set of PTKs for communicating with each neighbor individually. Before broadcasting the PREQ, it first creates a Merkle tree with the leaves being the hash of mutable fields of PREQ message. S then creates a MAC on the root of the Merkle tree it just created. Then, S broadcasts message (1) which includes the MAC of the root created using the GTK, mutable fields v_i s that need to be authenticated along with the values of its authentication path *authpath(v_i)*and encrypted PREQ message excluding the mutable fields.

Any of the neighboring nodes of *S*, after receiving the PREQ, tries to authenticate the mutable fields by hashing the values received in an ordered way, create a MAC on it using the shared GTK and comparing that with the received MAC value of the root. If the two values match, the intermediate MP is ascertain that the values are authentic and came from the same source that created the tree.

Let us consider for example that *B* receives a PREQ from its neighboring node *S* and wants to authenticate the value of the metric field *M*. According to our protocol, *B* and *C* should receive the value *M* along with the values (i.e. U_H and U_{TF}) of the authentication path as shown in Fig. 13. *B* and *C* can now verify the authenticity of *M* by computing $h(h(h(M)\| U_H)\| U_{TF})$ and a MAC on this value using the key GTK. It then compares the received MAC value with the new one, if it found a match, then it can assure that the value *M* is authentic and came from the same entity that has created the tree and computed the MAC on the U_{root}.

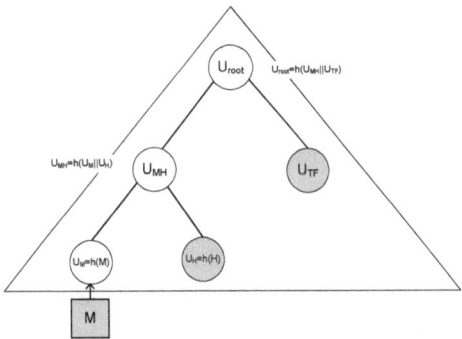

Fig. 13. Circles in grey construct the Authentication path for the metric field M

The intermediate nodes then update the values of the mutable fields like hop count, metric and TTL and create Merkle trees from the modified fields. They also decrypt the non-mutable part of the PREQ message and re-encrypt it with their own broadcast key and re-broadcast (as shown in (2) and (3)) the PREQ message following the same principle. After receiving the PREQ, the destination MP updates the mutable fields; creates its own Merkle Tree and unicasts a PREP message as (4) using the same principle but this time using PTK instead of using GTK. The PREP is propagated as (5) and (6) to the source MP in the reverse path created using PREQ and thus a secure forward path from the source to the destination is established.

7.5 Securing Proactive Mode

In the *Proactive RANN* mode, the RANN message is broadcasted using the group transient key as shown in Equation (7) to (9) to protect the non-mutable fields and authenticate the mutable fields (hop count, TTL and metric) using the Merkle tree approach. As there are only three mutable fields in the RANN message a node requires generating a random number to construct the Merkle tree. After receiving the

RANN message an MP that needs to setup a path to the root MP unicast a PREQ to the root MP as per Equation (10) to (12). On receiving each PREQ the root MP replies with a unicast PREP to that node as described in Equation (13) to (15). *Proactive PREQ* mode can also be secured by transmitting proactive PREQ and PREP in the same way discussed above.

$$R \to *: MAC_{GTK} root(R), \{v_i, authpath (v_i)\}, \{RANN\text{-}MF\}_{GTK} \tag{7}$$

$$F_1 \to *: MAC_{GTK} root(F_1), \{v_i, authpath (v_i)\}, \{RANN\text{-}MF\}_{GTK} \tag{8}$$

$$F_2 \to *: MAC_{GTK} root(F_2), \{v_i, authpath (v_i)\}, \{RANN\text{-}MF\}_{GTK} \tag{9}$$

$$D \to F_2: MAC_{PTK}^{D,F_2} root(D), \{v_i, authpath (v_i)\}, \{PREQ\text{-}MF\}_{PTK}^{D,F_2} \tag{10}$$

$$F_2 \to F_1: MAC_{PTK}^{F_2,F_1} root(F_2), \{v_i, authpath (v_i)\}, \{PREQ\text{-}MF\}_{PTK}^{F_2,F_1} \tag{11}$$

$$F_1 \to R: MAC_{PTK}^{F_1,R} root(F_1), \{v_i, authpath (v_i)\}, \{PREQ\text{-}MF\}_{PTK}^{F_1,R} \tag{12}$$

$$R \to F_1: MAC_{PTK}^{R,F_1} root(R), \{v_i, authpath (v_i)\}, \{PREP\text{-}MF\}_{PTK}^{R,F_1} \tag{13}$$

$$F_1 \to F_2: MAC_{PTK}^{F_1,F_2} root(F_1), \{v_i, authpath (v_i)\}, \{PREP\text{-}MF\}_{PTK}^{F_1,F_2} \tag{14}$$

$$F_2 \to D: MAC_{PTK}^{F_2,D} root(F_2), \{v_i, authpath (v_i)\}, \{PREP\text{-}MF\}_{PTK}^{F_2,D} \tag{15}$$

Notations used in Equation (7) to (15) are as follows: R is considered as the root MP and D is the MP that needs to setup a path to R. F_1 and F_2 are the intermediate nodes in the path. $MAC_k root(X)$, represents the MAC of the Merkle tree's root created by node X using a shared key k. {RANN/PREQ/PREP-MF} represents the routing information elements without the mutable fields. v_i and *authpath*(v_i) denote the fields that need to be authenticated and the values assigned to the authentication path from v_i to root of the tree, respectively.

7.6 Securing Hybrid Mode

The hybrid mode is the combination of both on-demand and proactive mode. As described in Section 4.3, the hybrid mode is first initiated with the proactive mode where an MP can establish route to the destination via the proactively built tree with the MPP as the root, using our secure proactive routing explained in previous section. After getting the notification from the root MPP that the destination is within the mesh, the source MP initiates the secure on-demand node as described in Section 7.4. The path that performs better will be chosen for subsequent communication. Therefore, individually securing on-demand and proactive mode can ensure security for the hybrid mode too.

8 Analyses

In this section, we will analyze the proposed SHWMP in terms of robustness against the attacks presented in Section 4 and also the overhead required for ensuring secure routing.

8.1 Security Analysis

1) Preventing Flooding: In the proposed SHWMP, a node can participate in the route discovery process only if it has successfully establishes a GTK and PTK through key distribution mechanism of 802.11s. Thus it will not be possible for a malicious node to initiate a route discovery process with a destination address that is not in the network. Again, as the PREQ message is encrypted during transmission, a malicious node can not insert new destination address.

2) Preventing Route Disruption: This type of attack is caused by the malicious behavior of a node through modification of a mutable field and dropping routing information elements. Note that, in our proposed scheme only authenticated nodes can participate in the route discovery phase. Moreover, routing information elements are authenticated and verified per hop. So, it is not possible to launch a route disruption attack in SHWMP.

3) Preventing Route Diversion: The root cause of route diversion attack is the modification of mutable fields in routing messages. These mutable fields are authenticated in each hop. If any malicious node modifies the value of a field in transit, it will be readily detected by the next hop while comparing the new MAC with the received one. It will find a miss-match in comparing the message authentication code (MAC) and the modified packet will be discarded.

4) Avoiding Routing Loops: Formation of routing loops requires gaining information regarding network topology, spoofing and alteration of routing message. As all the routing information is encrypted between nodes, an adversary will be unable to learn network topology by overhearing routing messages. Spoofing will not benefit the adversary as it will require authentication and key establishment to transmit a message with spoofed MAC. Moreover, fabrication of routing messages is detected by integrity check. So, proposed mechanism ensures that routing loops can not be formed.

8.2 Overhead Analysis

1) Computation cost: The computation cost of a sender and receiver are defined by following equations:

$$k \times h + m + e \text{ (sender)} \tag{16}$$
$$\alpha \times h + m + d \text{ (receiver)} \tag{17}$$

Where, k is the number of hash operations required to construct a Merkle tree. Cost of computing a hash function is defined by h. m is the cost involved in computing the MAC of the root, whereas e and d are encryption and decryption cost. To authenticate a particular value, a receiver needs to compute the root by calculating α hash operations, where α defines the number of leaf nodes in the authentication path as described in Section 3.2. Note that, the computation cost to verify a single mutable field requires $O(\log_2 \alpha)$ hash evaluations.

2) Communication Overhead: It is defined by the number of routing messages required to establish a secure path and defined by (18), (19) and (20).

$$(n\text{-}1)\times broadcast + h\times unicast \quad (on\text{-}demand) \qquad (18)$$

$$n\times broadcast + h\times unicast \quad (proactive\ PREQ) \qquad (19)$$

$$n\times broadcast + 2h\times unicast \quad (proactive\ RANN) \qquad (20)$$

Where, n is the number of nodes in the network, h is the number of hops in the shortest path. The number of messages required for establishing a path in HWMP is same as our proposed one. So, our protocol does not incur any extra communication overhead.

3) Storage Requirements: A node needs to store the number of fields that need to be authenticated, hashed values of the Merkle tree and the MAC of the root value. So, storage requirement of a node is given by (21)

$$\sum_{i=1}^{n} d_i + (k \times l) + S_M, \qquad (21)$$

where, d_i is the size of a mutable field, k is the number of hashes in the Merkle tree, l is the size of a hashed value and S_M is the size of the MAC.

9 Performance Evaluation

We use ns-2 [18] to simulate our proposed secure routing (SHWMP) protocol and compare that with the existing HWMP. We have simulated 50 static mesh nodes in a $1500 \times 1500\ m^2$ area. We use 5 to 10 distinct source-destination pairs that are selected randomly. Traffic source are CBR (constant bit-rate). Each source sends data packets of 512 bytes at the rate of four packets per second during the simulation period of 900 seconds.

In order to compare HWMP with SHWMP, both protocols were run under identical traffic scenario. Both on-demand and proactive mode were simulated. We consider the following performance metrics:

1. **Packet delivery ratio:** Ratio of the number of data packets received at the destinations to the number of data packets generated by the CBR sources. It in turn determines the efficiency of the protocol to discover routes successfully.

2. **Control overhead (in bytes):** Ratio of the control overhead bytes to the delivered data bytes.

3. **Path acquisition delay:** Time required to establish a route from source to destination which actually measures the delay between sending a PREQ/proactive PREQ to a destination and the receipt of corresponding PREP.

4. **End-to-end delay:** Average delay experienced by a data packet from a source to destination. Note that, end-to-end delay includes all the delays including medium access delay, processing delays at intermediate nodes etc.

5. **Average path length:** Average length of a path (in terms of hop count) discovered by the routing protocol. We calculate this by averaging the number of hops taken by each packet to reach the destination.

As shown in Fig. 14, the packet delivery ratio is better in SHWMP for both on-demand and proactive mode than that of HWMP. We assume that 10% misbehaving nodes are present in the network. Since the misbehaving nodes participates in the route discovery process, in HWMP sometimes packets are intentionally dropped by the misbehaving nodes. But, in the proposed protocol, misbehaving nodes can not participate in the route discovery process and thus always achieve a higher packet delivery ratio.

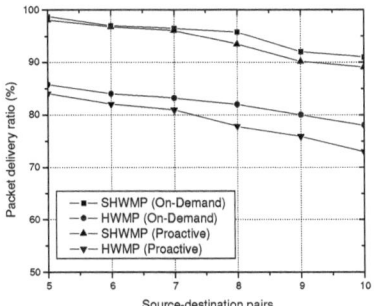

Fig. 14. Packet delivery ratio

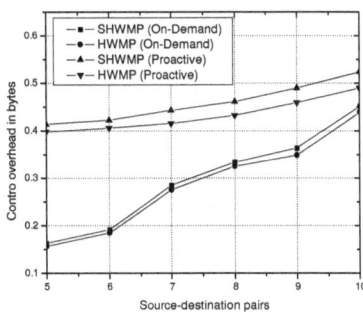

Fig. 15. Control overhead

Fig. 15 shows that the control overhead in bytes for the two protocols are almost identical both in the case of on-demand mode and proactive tree building mode. However, SHWMP has a little more overhead as it needs to include MAC values along with the routing messages that increases the size of a control message. However, the number of control packets transmitted by the two protocols is roughly equivalent.

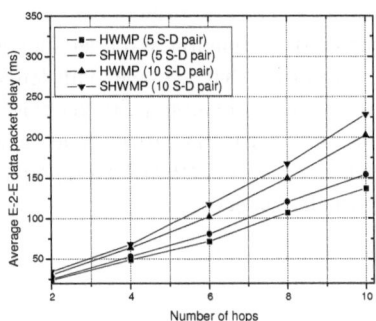

Fig. 16. End-to-end delay for data

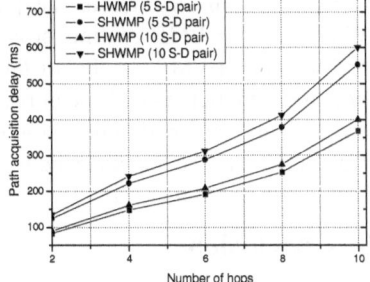

Fig. 17. Path acquisition delay

Fig. 16 depicts that the average end-to-end delay of data packets for both protocols are almost equal. We run the simulation using 5 and 10 source-destination pairs, and as the traffic load increases, end-to-end delay also increases. It is also evident that the effect of route acquisition delay on average end-to-end delay is not significant. Average route acquisition delay for the proposed SHWMP scheme is much higher than that of the HWMP mechanism as shown in Fig. 17. Because, in addition to normal routing operation of HWMP, the proposed SHWMP scheme requires computing hash and MAC values to verify the authenticity of a received packet, which require extra processing delay.

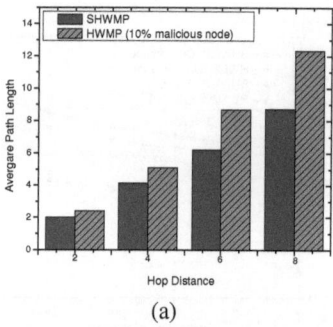
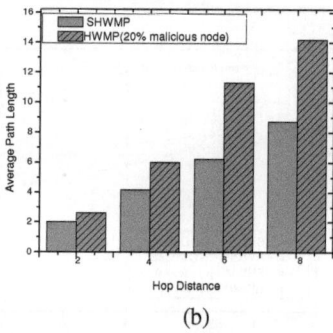

(a)   (b)

Fig. 18. Average path length, a) 10% malicious node. b) 20% malicious node.

In Fig. 18a and Fig. 18b, it is shown that the average path length is less with the proposed SHWMP protocol compared to HWMP protocol in the presence of 10% and 20% malicious nodes, respectively. Since, malicious nodes, when, participate in the route discovery process, can increase the value of the metric field while observing a small hop-count values in the PREQ message. This in turn restricts a node to select a shorter path in the HWMP protocol. With secure HWMP, a malicious node cannot participate in the route discovery process until it is authenticated by the neighbour node(s). We chose source-destination pairs those are 2, 4, 6 and 8 hops away. We derive the hop distance between each source-destination pair using Euclidean geometry. However, in simulation, the path from source to destination is created using HWMP and SHWMP routing protocol. The result shows that average path-length increases for HWMP as the distance between source-destination pairs increases. Since, more malicious nodes can participate in the route discovery process and cause increase in average path length. On the other hand, in case of SHWMP average path length is close to the hop distance measured using Euclidean geometry.

10   Conclusion

The goal of this paper is to develop a secure routing mechanism for wireless mesh networks. We have identified that several security attacks can be launched by the adversary with the existing HWMP routing protocol. Then, we have designed SHWMP, a secure extension of layer-2 routing protocol for IEEE 802.11s. Through analysis, it is shown that all the identified attacks can be defended with the proposed

SHWMP protocol. Furthermore, through extensive simulations we have compared different performance metrics for the existing HWMP and the proposed SHWMP. Our simulation results show that the proposed protocol outperforms existing one in terms of packet delivery ratio and average path length. Due to cryptographic extension (which is a must to provide secure routing), our protocol incurs little overhead in terms of control overhead in bytes and path acquisition delay. In future work, we intend to address the attack mounted by colluding adversarial nodes against the secure routing protocol.

References

1. Akyildiz, I.F., Wang, X., Wang, W.: Wireless Mesh Networks: a Survey. Computer Networks 47(4) (2005)
2. IEEE 802.11s Task Group, Draft Amendment to Standard for Information Technology Telecommunications and Information Exchange Between Systems – LAN/MAN Specific Requirements – Part 11: Wireless Medium Access Control (MAC) and Physical Layer (PHY) specifications: Amendment: ESS Mesh Networking, IEEE P802.11s/D2.02 (September 2008)
3. Wang, X., Lim, A.O.: IEEE 802.11s Wireless Mesh Networks: Framework and Challenges. In: AdHoc Networks, pp. 1–15 (2007), doi:10.1016/j.adhoc.2007.09.003
4. Islam, M.S., Yoon, Y.J., Hamid, M.A., Hong, C.S.: A Secure Hybrid Wireless Mesh Protocol for 802.11s Mesh Network. In: Gervasi, O., Murgante, B., Laganà, A., Taniar, D., Mun, Y., Gavrilova, M.L. (eds.) ICCSA 2008, Part I. LNCS, vol. 5072, pp. 972–985. Springer, Heidelberg (2008)
5. Merkle, R.C.: A Certified Digital Signature (subtitle: That Antique Paper from 1979). In: Brassard, G. (ed.) CRYPTO 1989. LNCS, vol. 435, pp. 218–238. Springer, Heidelberg (1990)
6. Bahr, M.: Proposed Routing for IEEE 802.11s WLAN Mesh Networks. In: 2nd Annual International Wireless Internet Conference (WICON), Boston, MA, USA (2006)
7. IEEE P802.11s™/D0.01, Draft amendment to standard IEEE 802.11™: ESS Mesh Networking. IEEE (March 2006)
8. Bahr, M.: Update on the Hybrid Wireless Mesh protocol of 802.11s. In: Proc. of IEEE International Conference on Mobile Adhoc and Sensor Systems, 2007. MASS, pp. 1–6 (2007)
9. Hu, Y.-C., Perrig, A., Johnson, D.B.: Ariadne: A Secure On-Demand Routing Protocol for Ad Hoc Networks. In: Proc. MobiCom 2002, Atlanta, GA (2002)
10. Perrig, A., Canetti, R., Tygar, J.D., Song, D.: Efficient Authentication and Signing of Multicast Streams over Lossy Channels. In: Proc. of IEEE Symposium on Security and Privacy, 2000, pp. 56–73 (2002)
11. Gergely, A., Buttyan, L., Vajda, I.: Provably Secure On-demand Routing in Mobile Ad Hoc Networks. IEEE transactions on Mobile Computing 5(11), 1533–1546 (2006)
12. Zapata, M.G., Asokan, N.: Securing Adhoc Routing Protocols. In: Proc. of ACM Workshop of Wireless Security(Wise), pp. 1–10 (2002)
13. Sangiri, K., Dahil, B.: A Secure Routing Protocol for Ad Hoc Networks. In: Proc. of 10th IEEE International Conference on Network Protocols, ICNP 2002 (2002)
14. Szydlo, M.: Merkle Tree Traversal in Log Space and Time. In: Cachin, C., Camenisch, J.L. (eds.) EUROCRYPT 2004. LNCS, vol. 3027, pp. 541–554. Springer, Heidelberg (2004)

15. Jakobsson, M., Leighton, T., Micali, S., Szydlo, M.: Fractal Merkle Tree Representation and Traversal. In: Joye, M. (ed.) CT-RSA 2003. LNCS, vol. 2612, pp. 314–326. Springer, Heidelberg (2003)
16. FIPS PUB 180-1, Secure Hash Standard, SHA-1,
    http://www.itl.nist.gov/fipspubs/fip180-1.htm
17. Lim, A.O., Wang, X., Kado, Y., Zhang, B.: A Hybrid Centralized Routing Protocol for 802.11s WMNs. Journal of Mobile Networks and Applications (2008)
18. The Network Simulator – ns-2, http://www.isi.edu/nsnam/ns/index.html

EDAS: Energy and Distance Aware Protocol Based on SPIN for Wireless Sensor Networks

Jaewan Seo[1], Moonseong Kim[2,*], Hyunseung Choo[1], and Matt W. Mutka[2]

[1] School of Information and Communication Engineering
Sungkyunkwan University, Korea
{todoll2,choo}@skku.edu
[2] Department of Computer Science and Engineering
Michigan State University, USA
mkim@msu.edu, mutka@cse.msu.edu

Abstract. Energy-efficient is a challenge in designing effective dissemination protocols for Wireless Sensor Networks (WSNs). Several recent studies have been conducted in this area, and SPMS, which outperforms the well-known SPIN protocol, is a particularly representative protocol. One of the many characteristics of SPMS is its use of the shortest path to minimize energy consumption. However, since it repeatedly uses the same shortest path and, hence, reduces energy consumption, it is impossible to maximize the network lifetime. In this paper, a novel data dissemination protocol is proposed, called **E**nergy and **D**istance **A**ware protocol based on **SPIN** (EDAS), which guarantees energy-efficient data transmission, as well as maximizing network lifetime. EDAS solves the network lifetime problem by taking account of both the residual energy and the most efficient distance between the nodes, to determine a path for data dissemination. Simulation results show that the EDAS guarantees energy-efficient transmission and moreover increases the network lifetime by approximately 69% than that of SPMS.

Keywords: Wireless Sensor Networks, Data Dissemination Protocol, Energy Efficiency, Lifetime, SPIN, SPMS.

1   Introduction

Wireless Sensor Networks (WSNs) have recently emerged as a core technology to be applied in smart homes and interactive human environments [1] [2]. Their implementations have several challenges such as the limitations of sensor nodes, and more importantly battery power. If certain nodes deplete their battery earlier than others when data is transmitted, the network lifetime decreases. Here, the network lifetime is defined as the time from network initialization to the first node failure. WSNs should be fully connected to achieve the main goal of the sensor node that is to sense and transmit data back to the sink where the end users can access the data and perform further processing. In the worst case scenario, failures may result in network disconnection. Therefore,

---

* Corresponding author.

M.L. Gavrilova and C.J.K. Tan (Eds.): Trans. on Comput. Sci. VI, LNCS 5730, pp. 115–130, 2009.

low power consumption and network lifetime are important issues in WSNs and the problems have been studied widely [3].

Several solutions to maximize network lifetime have been recently proposed. One is control of the transmission power of wireless nodes such as [4] [5]. Some clustering algorithms have been designed to focus on efficient data aggregation to reduce the volume of information [6] [7]. Another method causes nodes to sleep in order to spare energy [8] [9]. The last one uses an energy-efficient data dissemination protocol [10] [11] [12] [13]. In this paper, we focus on this energy-efficient data dissemination protocol, particularly a flooding scheme, for following reasons.

1. Data dissemination is a fundamental feature in WSNs, since thousand of sensor nodes collect and transmit data;
2. Energy-efficient data dissemination with multi-hop distance considerations would increase network lifetime;
3. Selecting nodes with a high residual energy would contribute to a prolonged network lifetime;
4. Optimizing flooding would reduce the number of redundant transmissions.

The two well-known schemes of flooding and Sensor Protocols for Information via Negotiation (SPIN) are employed for earlier data dissemination [14]. First, in flooding, each node retransmits the received data from all its neighbors. This simple, stateless protocol is used at each node when data is transmitted. It rapidly disseminates data; however, it also quickly consumes energy when data implosion occurs [15] [16]. Second, SPIN solves this weakness by exchanging information and negotiation at each node. However, although SPIN exchanges high-level metadata, data descriptors, to prevent data duplication, there is the issue of efficiency as this method transmits all data at the same power level and thus does not consider the distance to a neighbor.

Khanna, *et al.* recently proposed a protocol, Shortest Path Minded SPIN (SPMS), to minimize energy consumption using the shortest path and multi-hop to reach the destination [17]. Each node executes the Bellman-Ford algorithm within its zone to use the shortest path. The zone is defined as the area a node can reach by transmitting at maximum power [18]. Thus, SPMS has to maintain a routing table. Since it uses the shortest path, energy use is minimized. However, since specific nodes in the shortest path would be used repeatedly, network lifetime may rapidly decrease. Moreover, once a node failure occurs, parts of the network along the shortest path may be rendered useless.

As mentioned above, energy-efficient transmission and network lifetime are essential in WSNs. The network lifetime is of particular importance to reduce the number of network disconnections. In this paper, a novel data dissemination protocol is proposed, called Energy and Distance Aware protocol based on SPIN (EDAS), which significantly increases network lifetime and guarantees energy-efficient transmission when the sensing data is disseminated throughout the entire network. EDAS avoids the network lifetime problem of SPMS by selecting a path to transmit data according to its residual energy such as [19]. Furthermore, it flexibly selects paths based on two attributes; the first is residual energy, where EDAS selects a high energy node to prevent a certain node from being selected repeatedly, the second is a specified distance, since the path is selected based on the reasonable transmission distance so that EDAS can decrease the energy consumption rather than SPMS.

The remainder of this paper is organized as follows. In Section 2, the disadvantages of SPIN and SPMS are discussed. The proposed protocol design and its description are given in Section 3 and simulation results are presented in Section 4. Section 5 summarizes the main conclusions of the paper.

2   Related Work

2.1   SPIN

Since a node transmits data to its neighbors irrespective of whether the neighbor already received the data via flooding, a more efficient approach such as SPIN is proposed to solve the data implosion problem of flooding. Nodes negotiate with their neighbors before transmitting data in SPIN to overcome this problem. This guarantees only required data will be transmitted. SPIN uses metadata that describes the data to negotiate successfully.

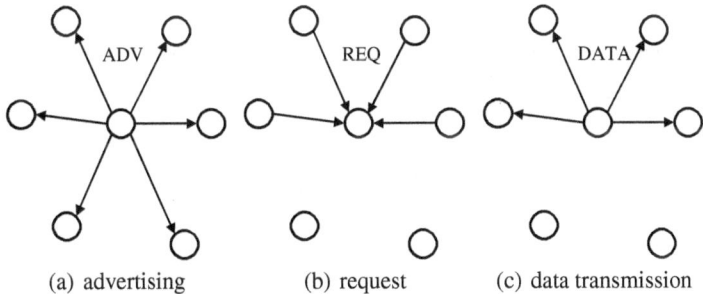

(a) advertising   (b) request   (c) data transmission

Fig. 1. SPIN algorithm

SPIN involves the following steps:

1. Step 1 (Fig. 1 (a)) is the advertisement phase wherein a sensor node, having data, broadcasts an advertisement message including metadata to its neighbor nodes.
2. Step 2 (Fig. 1 (b)) is the request phase where the neighboring nodes that have received the advertisement message check the metadata to determine whether they need the data. If the data is required, the neighbors send a request message to the sensor node that sent the advertisement.
3. Step 3 (Fig. 1 (c)) is the data transmission phase where the sensor node that receives the request message transmits the data to the neighbors that have sent the requests.

SPIN disseminates data throughout the entire network by the repeated use of these three steps.

The strength of the SPIN approach is in its simplicity and the nodes make uncomplicated decisions whenever data is received; hence little energy is consumed in the computation. SPIN is able to distribute 50% more data per unit energy than flooding, because an advertisement message is low compared to the sensing data [14]. However, SPIN transmits the data at the same power level, since it does not consider the distance to the neighbors. Of course energy consumption generally increases exponentially with distance, and so SPIN is incapable of performing energy-efficient transmissions. This work considers, including the provision of reasonable transmission distance by adopting an energy model in WSNs.

2.2 SPMS

SPMS employs the metadata concept used in SPIN and uses a multi-hop model for data transmission to avoid the exponential increase in energy consumption with distance [20]. The next hop should be known before transmitting data for multi-hop routing. It is infeasible to maintain a routing table for each node, since sensor networks comprise thousands of nodes. Thus, SPMS maintains a routing table for a zone, which is defined as the maximum power level of a node. This reduces the cost of building the routing table. Each node can build a routing table with the shortest path using the Bellman-Ford algorithm. After building the routing table as shown in Fig. 2. SPMS begins data transmission. Each link has a cost associated with it which represents the transmission power needed to reach the neighbor.

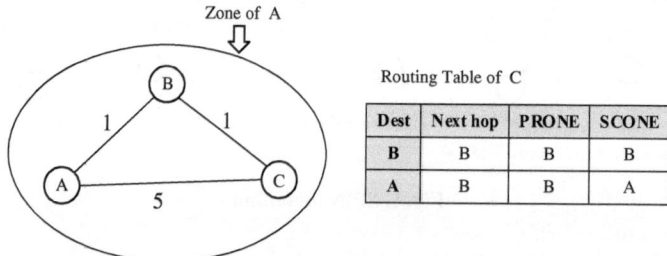

Fig. 2. Routing table in SPMS

SPMS involves the following steps:

1. Step 1 involves a metadata exchange, similar to that in SPIN, whereby a node having data to send, broadcasts an advertisement message to its neighbor nodes in the same zone.
2. Step 2 is the request phase where the neighbors receiving the advertisement message check the metadata and decide whether or not the data are necessary. If necessary, the neighbor sends a request message, but sends the data through the shortest path.

To send the data through the shortest path, if the node that sent the advertisement message is not its next hop neighbor in the computed shortest path, then it should wait for a predetermined fixed period before sending the request message. Therefore, after each node sends a request message, it receives the data through the shortest path. If there is no advertisement message during the fixed period, the node sends the request message to the source node.

SPMS sends data through the shortest path except in the case of a node failure. When the predecessor node on the computed shortest path fails, the current node sends a request message to the node, from which it initially received an advertisement message instead of using the shortest path. However, it may not guarantee energy-efficient transmission when the distance between the node that sent the advertisement message and the current node is too long. Further, if the advertisement node fails, there is no way to receive the data. To solve this problem, each node in SPMS maintains a Primary Originator Node (PRONE) and a Secondary Originator Node (SCONE). PRONE is an energy-efficient source from which data can be obtained and SCONE is an alternative node that can be used if PRONE fails. As mentioned, SPMS has the mechanism to overcome its weakness due to only using the shortest path, however the network lifetime problem persists.

2.3   SPMS-Rec

SPMS-Recursive (SPMS-Rec) is an enhanced SPMS protocol designed to reduce energy consumption and delay [21]. In SPMS, if a relay node, e.g., node R, which is between the source and destination, transmits a request message and finds its next hop node to be failed, it simply drops the packet, and the destination node tries to request alternative routes. This wastes energy unnecessarily and increases the delay, since the request message may already have arrived at the relay node R. The reason for this problem is that the relay node does not store any state; thus, there is no way to detect or recover from failures. To solve this problem, in SPMS-Rec, when each relay node requests a message, it stores a request table and sets its timer. If the relay node does not receive data during a fixed period, it assumes that failure has occurred. Therefore, relay nodes can try to find another route locally instead of destination node.

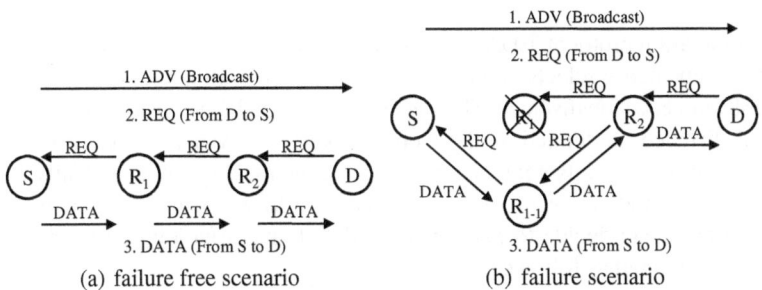

(a) failure free scenario   (b) failure scenario

Fig. 3. Operation of SPMS-Rec

The operation of SPMS-Rec is generally identical to SPMS (Fig. 3 (a)), yet the failure operation differs as described next. In the case of failure, SPMS-Rec invokes the following steps. The destination node D sends a request message to its PRONE, node S, via relay node R_2, and sets its timer $T_{out}^{dat}(D)$. The relay node R_2 first tries to send the request message through the shortest path and sets timer $T_{out}^{dat}(R_2)_1$. Once $T_{out}^{dat}(R_2)_1$ expires due to failure of node R_1, the relay node R_2 tries to send the request message through an alternative path and sets its timer $T_{out}^{dat}(R_2)_2$ as shown in Fig. 3 (b). If relay node R_2 receives the data from node R_{1-1} then R_2 will forward it to node D. Otherwise, when $T_{out}^{dat}(R_2)_2$ expires, the relay node R_2 decides whether to send the request message directly or to drop it. Thus SPMS-Rec is able to reduce energy consumption and time delays since it can detect or recover from failures locally using timer.

3   The Proposed Dissemination Protocol: EDAS

3.1   Motivation

SPMS aims to maximize energy efficiency by transmitting data through the shortest path. However, this may result in a specific node that is used repeatedly, since the shortest path has already been selected, which is inefficient in terms of the network lifetime. Further, SPMS does not guarantee maximal energy-efficiency in the case of a node failure, since parts of the network can no longer use the shortest path. SPMS-Rec improves energy efficiency in comparison to SPMS, although it still has network lifetime problems. Therefore, a protocol that has to guarantee energy-efficient transmission as well as a long network lifetime. In this paper, a data dissemination protocol, called Energy and Distance Aware protocol based on SPIN, namely EDAS, is proposed. The objective of EDAS is to propose a dissemination protocol that simultaneously guarantees these requirements. EDAS differs from SPMS-Rec in that it changes the path every time an event occurs. To select an appropriate path, EDAS strikes a balance between energy-efficient transmission and network lifetime aspects based on two attributes, namely, the residual energy and the most efficient transmission distance.

3.2   Node Marker N^m

As a dissemination protocol, EDAS should spread the sensing data throughout the entire network. It repeatedly selects neighbor nodes that have both a high energy level and a reasonable transmission distance. The state of the node, which considers residual energy and ideal distance, is defined as the Node marker N^m. When a node broadcasts an advertisement message, its neighbors that receive the message independently calculate N^m, since each node is aware of its location via a location finding system [22] [23] [24]. Each node can decide to participate in data transmission based on the value of N^m which is defined as follows:

$$N^m(w_e, w_d) = w_e N_e + w_d N_d \tag{1}$$
$$w_e, w_d \geq 0; w_e + w_d = 1 \tag{2}$$

where, N_e, N_d denote the residual energy for each node and the level of harmonic energy-consumption distance between the two nodes, respectively. Each attribute can be adjusted by weight factor w_e and w_d.

The first factor for the node's energy level is:

$$N_e = 1 - min\{1, -\log_{10}(E_{res}/E_{ini})\} \tag{3}$$

where, E_{res} and E_{ini} are defined as the residual energy and initial energy, respectively. N_e represents the energy level sensitively using logarithmic properties. In Fig. 4, the abscissa axis is the normalized residual energy and the ordinate axis is Eq. 3. The logarithm function rapidly decreases the selected probability when the normalized residual energy is low. EDAS is able to select a node with a high energy level using the property of logarithm function. In our simulation, we assume that the logarithm is base 10. Hence, random selection is achieved when the residual normalized energy is 10%: this value can be regulated.

The second factor N_d is for the ideal transmission distance. Since the energy consumption increases exponentially with the distance when data is transmitted, it may consume less energy when relaying data over shorter distances, as opposed to directly transmitting data to the destination node. However, if the data is relayed too many times, it would consume even more energy. Therefore, considering the ideal transmission distance is important to achieve energy efficient transmission. In our simulation environment, we use the following energy model [25]:

$$E_{tx} = \alpha_{11} + \alpha_2 d^p, \ p > 1 \tag{4}$$

$$E_{rx} = \alpha_{12} \tag{5}$$

where, E_{tx} and E_{rx} denote the energy consumed for transmitting and receiving a bit over a distance d, respectively; α_{11}, the energy/bit consumed by the transmitter electronics; α_2, the energy dissipated in the transmit op-amp; and α_{12}, the energy/bit consumed by the receiver electronics.

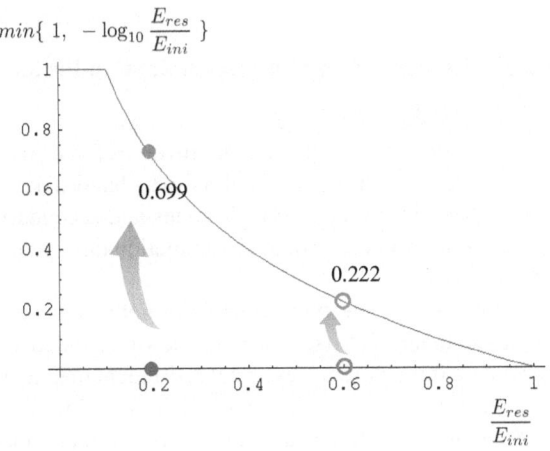

Fig. 4. The characteristic of the logarithm function

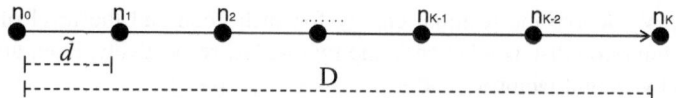

Fig. 5. Ideal data dissemination through relay nodes

Fig. 5 shows the data dissemination through the ideal distance between nodes n_0 and n_K. The path length between nodes a and b is defined as $P(a, b)$. Therefore, when data is transmitted, the expectation of energy consumption through the path $P(n_0, n_K)$ is represented as follows.

$$E(P(n_0, n_K)) = \sum_{r=1}^{K} E(P(n_{r-1}, n_r)) \qquad (6)$$

If the average distance between each relay node is defined as $\tilde{d}$, then the expected number of relay nodes is $\lfloor D/\tilde{d} \rfloor$. Therefore, the energy consumption between nodes n_0 and n_K is as follows:

$$E(P(n_0, n_K)) \approx \frac{D}{\tilde{d}}(\alpha_{11} + \alpha_{12} + \alpha_2 \tilde{d}^p) \qquad (7)$$

The energy consumption $E(P(n_0, n_K))$ is minimized when the value has a local minimum value by parameter $\tilde{d}$. Therefore, $\frac{\partial}{\partial \tilde{d}}E(P(n_0, n_K)) = 0$ and the value of $\tilde{d}$ is $p\sqrt{\frac{\alpha_{11}+\alpha_{12}}{\alpha_2(p-1)}}$ [25]. Therefore, N_d is defined as:

$$N_d = 1 - min\{1, (|d - \tilde{d}|)/\tilde{d}\} \qquad (8)$$

where, d is the distance between the two nodes; N_d represents the degree of adjacency to $\tilde{d}$.

3.3 An Energy and Distance Aware Data Dissemination Protocol Based on SPIN: EDAS(w_e, w_d)

EDAS(w_e, w_d) uses the concept of metadata and involves three steps, namely, advertisement, request, and transmission, through a three-way handshake. Nodes are classified as: primary nodes that transfer data to their zones and secondary nodes that only receive data. Fig. 6 shows an example of data dissemination through the primary (black) and secondary nodes (white).

The order of arrival is used to determine whether or not a node belongs to a primary node. Each node can calculate N^m itself and set the timer based on value N^m in a distributed manner when it initially receives an advertisement message. If the calculated value N^m increases, each node sets a timer with shorter period. In other words, it is highly probable that a node with a large value of N^m is a primary node.

The state transitional graph of this procedure is shown in Fig. 7 for detailed explanation. Once one node receives an advertisement message from the sender node, it

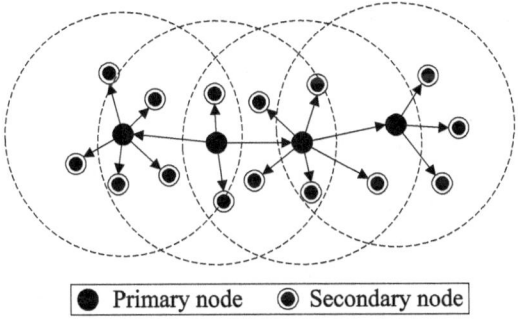

Fig. 6. The basic concept of data dissemination in EDAS

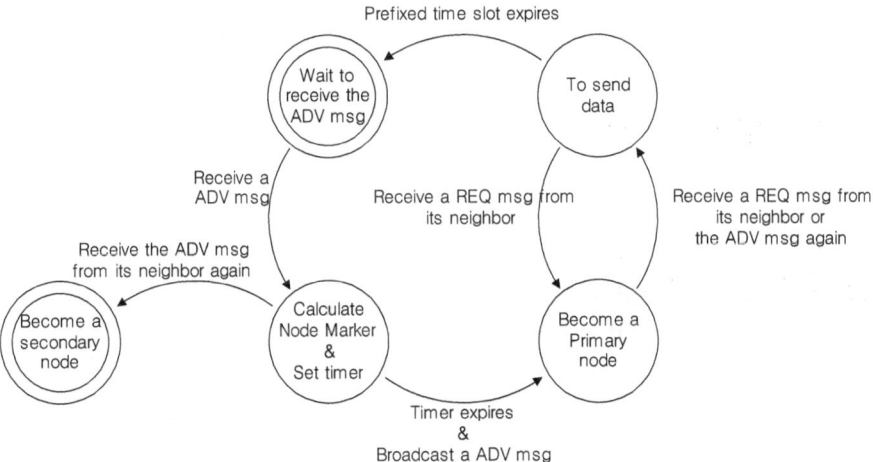

Fig. 7. EDAS node state transitions

calculates N^m and sets the timer. When the timer expires, the node first listens to check if the channel is clear. If it is, it broadcasts the advertisement message to its neighbors and promotes itself to a primary node. If it overhears the advertisement message from its neighbor again, it becomes a secondary node. Since primary nodes should have a high energy level and ideal distance, it is highly probable that nodes with a high value of N^m could be primary nodes.

Fig. 8 describes the algorithm. For convenience, this paper explains the EDAS(w_e, w_d) protocol using diagrams.

1. A primary node broadcasts an advertisement message to its neighbor nodes. (Fig. 8 (a))
2. The neighbors that receive the advertisement message set their timers after calculating N^m. The neighbor nodes that have higher values of N^m set their timers

with shorter periods. That is, the timer value is in inverse proportion to N^m. (see Fig. 8 (a))

3. The neighbor whose timer expires first, it becomes a primary node and broadcasts an advertisement message. (Fig. 8 (b))
4. If the primary node that transmits the original advertisement message receives an advertisement message again, it transmits the data to the node from which it received the message. (Fig. 8 (b) and (c))
5. If a node whose timer has not yet expired receives the advertisement message again, it becomes a secondary node and sends a request message to the node from which it received the message. (Fig. 8 (b) and (c))
6. If a node receives an advertisement message for the first time, go to step 2.
7. All nodes in the entire network repeat the above procedure until they receive the required data.

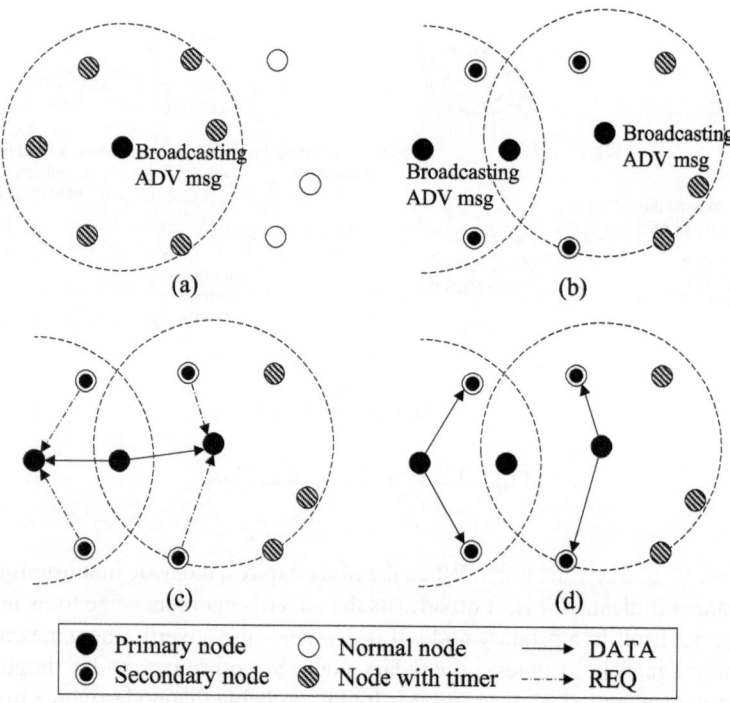

Fig. 8. EDAS(w_e, w_d) Algorithm

4 Performance Evaluation

The protocol EDAS(w_e, w_d) was implemented using JAVA to evaluate its performance compared to SPIN and SPMS-Rec and the main parameters are listed in Table 1. Sensor nodes with a 60m transmission range are randomly distributed in a network over an area of $500m \times 500\ m$. The energy model has values of α_{11}, $\alpha_{12} = 80nJ/bit$ and $\alpha_2 = 100pJ/bit/m^2$. $\tilde{d}$, calculated from the energy model, is 40m. The control messages (advertisement, request) are 15 $bytes$ each, while the data is 500 $bytes$ and p is set to 2. We assume that the wireless links are reliable.

Table 1. Simulation variables

network size	$250m \times 250m$
initial energy	$0.5\ J$
α_{11}, α_{12}	$80\ nJ/bit$
α_2	$100\ pJ/bit/m^2$
packet size (ADV, REQ)	15 $bytes$
packet size (DATA)	500 $bytes$
transmission range	$60\ m$
$\tilde{d}$	$40\ m$
p	2

Fig. 9(a) shows a graph of each scheme's average energy consumption according to the density at the time when the event has occurred 50 times. SPIN has the highest average energy consumption because it transmits all data at the same power level. SPMS-Rec exhibits a better performance than SPIN, although its energy consumption increases with density compared to that of EDAS(w_e, w_d) since all nodes send an advertisement message in SPIN and SPMS-Rec. Conversely, since EDAS(w_e, w_d) differs to the above procedure, its energy consumption does not increase greatly. Fig. 9(b) shows the energy consumption of for each w. The average energy consumption is the lowest value when w_d is 1, because the data is spread by only using the ideal efficient transmission range. When EDAS(w_e, w_d) seeks immediate gains for a residual energy as w_e is 1, the energy consumption is the worst for each w.

Fig. 10 explains the network lifetime based on the density with initial energy of $0.5 J$. We generate source data packets until the first node failure occurs to compare the network lifetime in different schemes. EDAS(w_e, w_d) considers residual energy to distribute data over the entire network; this significantly prolongs network lifetime. When w_e is 1, then EDAS(w_e, w_d) has the longest network lifetime since it only considers the residual energy. When both w_e and w_d are 0.5, the network lifetime is approximately 69% longer than that of SPMS-Rec.

SPMS solves the energy problem in the SPIN protocol, but the node failure issue still exists. Although SPMS-Rec supplements this weakness, there is still the problem of the network lifetime, as depicted in Fig. 11. The proposed EDAS(w_e, w_d) exhibits an approximate 100% success ratio until it reaches 300 events except when w_d is set 1, because it does not consider a residual energy level. EDAS(w_e, w_d) takes into account both the node's residual energy level and its energy-efficient transmission range. That is, EDAS(w_e, w_d) is an especially interesting system for energy-efficient data transmission in networks.

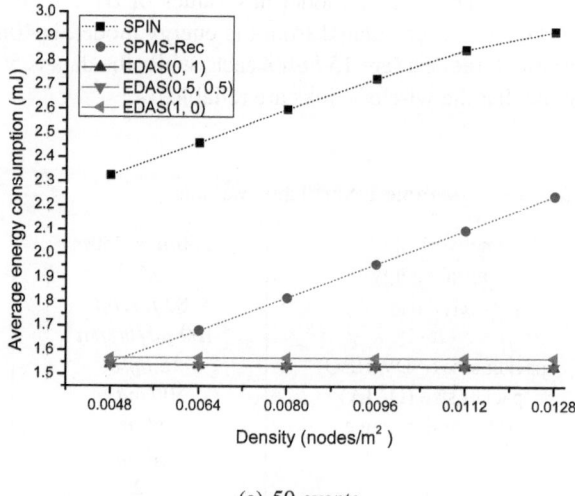

(a) 50 events

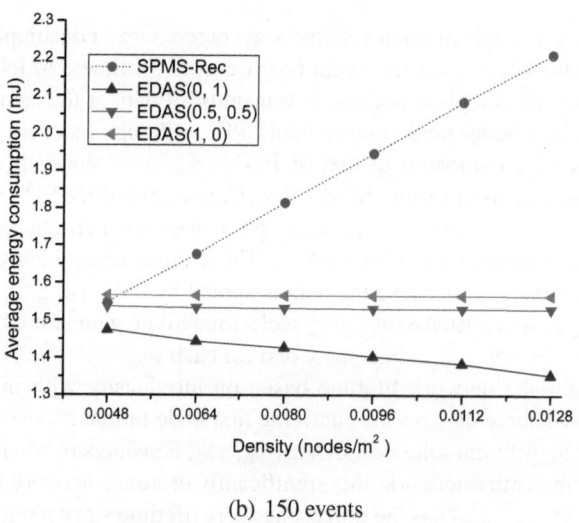

(b) 150 events

Fig. 9. Average energy consumption

To compare the impact of considering residual energy, we report the remaining energy level of the sensor nodes after arbitrary source nodes generate data packets 150 times. Fig. 12 indicates the distribution of the energy consumption in the network's nodes. Even if several node failures occur as shown in Fig. 11 and Fig. 12(a), the nodes having high residual energy are inefficiently distributed when using the SPIN protocol. SPMS-Rec (Fig. 12(b)) also exhibits decreased energy consumption since it repeatedly

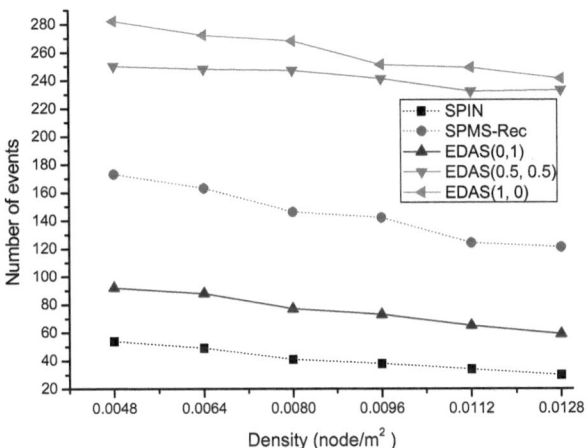

Fig. 10. Network lifetime

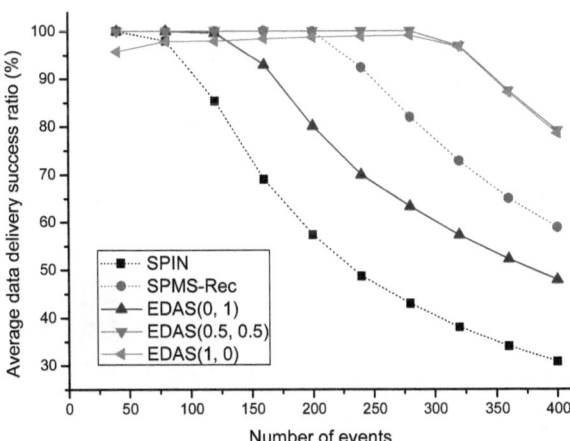

Fig. 11. Average data delivery success ratio

uses the nodes in the shortest path. However, this study confirms that the energy consumption distribution is likely to be uniform and the energy efficiency is excellent for EDAS(w_e, w_d) as shown in Fig. 12(c) and Fig. 12(d).

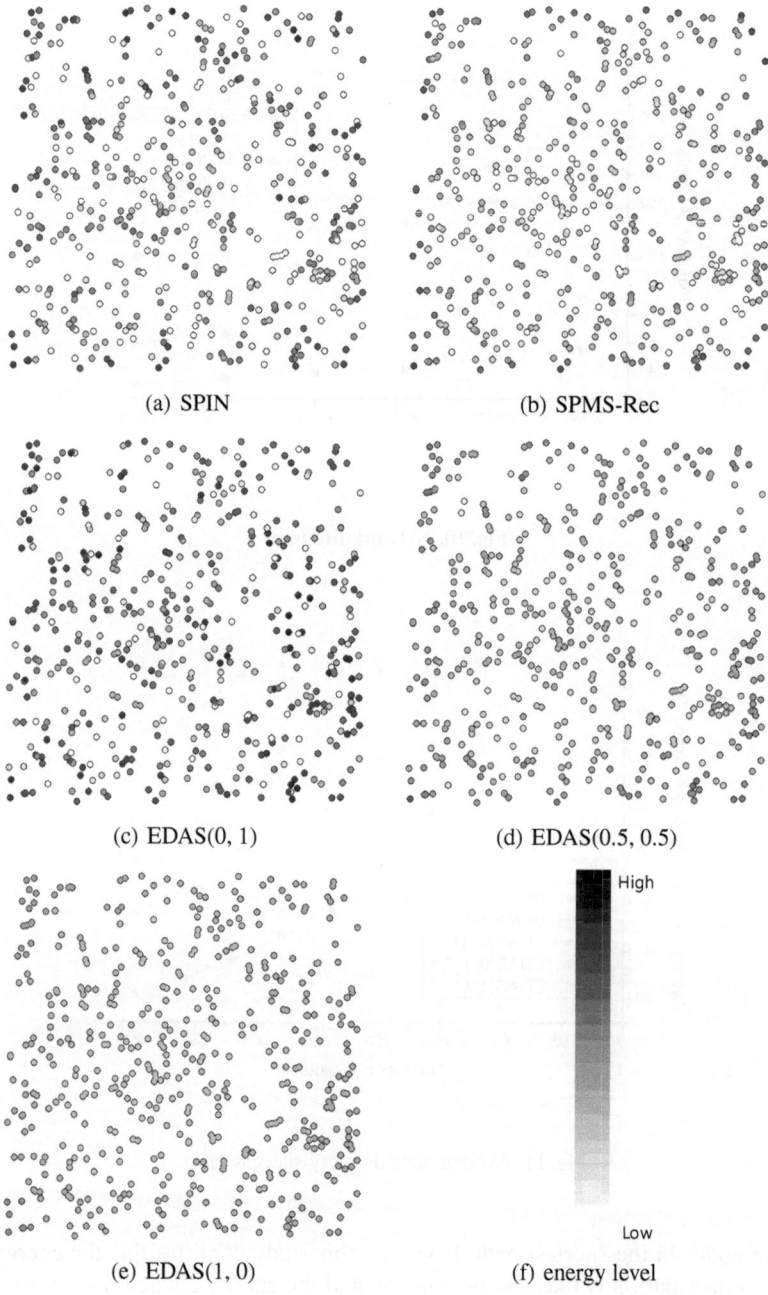

(a) SPIN

(b) SPMS-Rec

(c) EDAS(0, 1)

(d) EDAS(0.5, 0.5)

(e) EDAS(1, 0)

(f) energy level

Fig. 12. Distribution of energy consumption after generating 150 events

5   Conclusion

In this paper, several energy-efficient data dissemination protocols for Wireless Sensor Networks have been studied. A new protocol, called the Energy and Distance Aware protocol based on SPIN (EDAS) has been proposed. Comparisons have been made between SPIN and SPMS-Rec through simulation experiments and these have shown that the performance of the proposed protocol is superior to earlier protocols. In particular, with regard to energy efficiency in terms of density, EDAS is superior to existing schemes as only some of the nodes broadcast the advertisement messages. Hence EDAS guarantees energy-efficient transmission and a longer network lifetime through an established path based on two attributes, appropriate distance and residual energy. Moreover, EDAS provides a more flexible routing protocol since it uses two weighted attributes.

Since wireless links tend to be unreliable due to factors such as interference, attenuation, and fading, we will study how to extend EDAS to unreliable environments. EDAS may need to be developed because when links are unreliable, the link quality may change dynamically, and primary nodes need to be resilient to such changes. In future studies, it is also hoped to consider extending this proposed protocol to cover multiple events.

Acknowledgment

This research was supported by MKE, Korea under ITRC IITA-2009-(C1090-0902-0046) and by the MEST, Korea under the WCU Program supervised by the KOSEF (No. R31-2008-000-10062-0); and, by the National Science Foundation, USA under grants no. CNS-0721441 and OCI-075362.

References

1. Akyildiz, I.F., Su, W., Sankarasubramaniam, Y., Cayirci, E.: Wireless sensor networks: a survey. Computer Networks 38, 393–422 (2002)
2. Al-Karaki, J.N., Kamal, A.E.: Routing Techniques in Wireless Sensor Networks: A Survey. IEEE Wireless Communications 11(6), 6–28 (2004)
3. Kamimura, J., Wakamiya, N., Murata, M.: Energy-Efficient Clustering Method for Data Gathering in Sensor Networks. In: Proceedings of BROADNETS (2004)
4. Ingelrest, F., Simplot-Ryl, D., Stojmenovic, I.: Optimal Transmission Radius for Energy Efficient Broadcasting Protocols in Ad Hoc Networks. IEEE Trans. on Parallel and Distributed Systems 17(6), 536–547 (2006)
5. Cardei, M., Wu, J., Yang, S.: Topology Control in Ad hoc Wireless Networks with Hitchhiking. In: SECON 2004, pp. 480–488. IEEE, Los Alamitos (2004)
6. Heinzelman, W., Chandrakasan, A., Balakrishnan, H.: An Application-Specific Protocol Architecture for Wireless Microsensor Networks. IEEE Transactions on Wireless Communications 1(4), 660–670 (2002)
7. England, D., Veeravalli, B., Weissman, J.B.: A Robust Spanning Tree Topology for Data Collection and Dissemination in Distributed Environments. IEEE Transactions on Parallel and Distributed Systems 18(5), 608–620 (2007)

8. Cardei, M., Thai, M., Li, Y., Wu, W.: Energy-efficient target coverage in wireless sensor networks. In: INFOCOM 2005, vol. 3, pp. 1976–1984. IEEE, Los Alamitos (2005)
9. Cardei, M., Du, D.: Improving wireless sensor network lifetime through power aware organization. Wireless Networks 11(3), 333–340 (2005)
10. Senouci, S.-M., Pujolle, G.: Energy efficient routing in wireless ad hoc networks. In: ICC 2004, vol. 7, pp. 4057–4061. IEEE, Los Alamitos (2004)
11. Kwon, S., Shroff, N.B.: Energy-Efficient Interference-Based Routing for Multi-hop Wireless Networks. In: INFOCOM 2006, pp. 1–12. IEEE, Los Alamitos (2006)
12. Srinivas, A., Modiano, E.: Minimum Energy Disjoint Path Routing in Wireless Ad-Hoc Networks. In: MOBICOM 2003, pp. 122–133. ACM, New York (2003)
13. Ganesan, D., Govindan, R., Shenker, S., Estrin, D.: Highly resilient, energy-efficient multi-path routing in wireless sensor networks. SIGMOBILE Mobile Computing and Communications Review (MC2R) 1(2), 333–340 (2001)
14. Heinzelman, W.R., Kulik, J., Balakrishnan, H.: Adaptive Protocols for Information Dissemination in Wireless Sensor Networks. In: MOBICOM 1999. ACM/IEEE (1999)
15. Ni, S.Y., Tseng, Y.C., Chen, Y.S., Sheu, J.P.: The Broadcast Storm Problem in a Mobile Ad Hoc Network. In: MOBICOM 1999, pp. 151–162. ACM/IEEE (1999)
16. Le, T.D., Choo, H.: Efficient Flooding Scheme Based on 2-Hop Backward Information in Ad Hoc Networks. In: ICC, pp. 2443–2447. IEEE, Los Alamitos (2008)
17. Khanna, G., Bagchi, S., Wu, Y.-S.: Fault Tolerant Energy Aware Data Dissemination Protocol in Sensor Networks. In: Dependable Systems and Networks (DSN), pp. 739–748. IEEE, Los Alamitos (2004)
18. Hass, Z., Pearlman, M.R.: The Performance of Query Control Schemes for the Zone Routing Protocol. IEEE/ACM Transactions on Networking 9(4), 427–438 (2001)
19. Hassanein, H., Luo, J.: Reliable Energy Aware Routing In Wireless Sensor networks. In: DSSNS 2006, pp. 54–64. IEEE, Los Alamitos (2006)
20. Pottie, G.J., Kaiser, W.J.: Embedding the internet: wireless integrated network sensors. Communications of the ACM 43(5), 51–58 (2000)
21. Khosla, R., Zhong, X., Khanna, G., Bagchi, S., Coyle, E.J.: Performance Comparison of SPIN based Push-Pull Protocols. In: WCNC 2007, pp. 3990–3995. IEEE, Los Alamitos (2007)
22. Savvides, A., Han, C., Strivastava, M.: Dynamic fine-grained localization in ad-hoc networks of sensors. In: Proceedings of Mobicom, pp. 166–179. ACM, New York (2001)
23. Huang, T., He, C., Blum, B.M., Stankovic, J.A., Abdelzaher, T.: Range-free localization schemes for large scale sensor networks. In: Mobicom, pp. 81–95. ACM, New York (2002)
24. Nasipuri, A., Li, K.: A directionality based location discovery scheme for wireless sensor networks. In: Proceedings of the first ACM international workshop on Wireless Sensor Networks and Applications, pp. 150–111. ACM, New York (2002)
25. Bhardwaj, M., Garnett, T., Chandrakasan, A.P.: Upper bounds on the lifetime of sensor networks. In: ICC, vol. 3, pp. 785–790. IEEE, Los Alamitos (2001)

Scheme to Prevent Packet Loss
during PMIPv6 Handover*

Seonggeun Ryu and Youngsong Mun

School of Computing, Soongsil University,
Sangdo 5 Dong Dongjak Gu, Seoul, Korea
sgryu@sunny.ssu.ac.kr, mun@ssu.ac.kr

Abstract. Mobile IPv6 (MIPv6) is a presentative protocol which supports global IP mobility. MIPv6 causes a long handover latency that a mobile node (MN) cannot send or receive packets. This latency can be reduced by using Proxy Mobile IPv6 (PMIPv6). PMIPv6 is a protocol which supports IP mobility without participation of the MN, and is studied in Network-based Localized Mobility Management (NETLMM) working group of IETF. There is much packet loss during handover in PMIPv6, although PMIPv6 reduces handover latency. In this paper, to reduce packet loss in PMIPv6 we propose Packet Lossless PMIPv6 (PL-PMIPv6) with authentication. In PL-PMIPv6 a previous mobile access gateway (pMAG) registers to a Local Mobility Anchor (LMA) on behalf of a new MAG (nMAG) during layer 2 handoff. Then, the nMAG buffers packets during handover after registration. Therefore, PL-PMIPv6 can reduce packet loss in MIPv6 and PMIPv6. Also, we use Authentication, Authorization and Accounting (AAA) infrastructure to authenticate the MN and to receive MN's profiles securely. For the comparison with MIPv6 and PMIPv6, detailed performance evaluation is performed. From the evaluation results, we show that PL-PMIPv6 can achieve low handover latency and low total cost.

1 Introduction

In wireless/mobile networks, mobile nodes (MN) can change their attachment points while they communicate with correspondent nodes (CN). Hence, mobility management is essential for tracking the MNs current locations so that their data can be delivered correctly. IP-based mobility management is critical, since the next-generation wireless/mobile networks are anticipated to be unified networks based on IP technology, i.e., all-IP networks. Mobile IPv6 (MIPv6) [1] from the Internet Engineering Task Force (IETF) is standardized for mobility management in IPv6 wireless/mobile networks. Mobile IPv6 requires client functionality in the IPv6 stack of a mobile node. Exchange of signaling messages between the MN and a home agent enables the creation and maintenance of binding between the MN's home address and its care-of address. Mobility as specified in MIPv6 requires the IP host to send IP mobility management signaling messages to the

---

* This work was supported by the Soongsil University Research Fund.

M.L. Gavrilova and C.J.K. Tan (Eds.): Trans. on Comput. Sci. VI, LNCS 5730, pp. 131–142, 2009.

HA, which is located in the network. MIPv6 is an approach of host-based mobility to solve the IP mobility challenge. However, it takes a long time to process handover and there is much packet loss during handover, since there are many signaling messages via wireless link which occurs long delay during handover.

Network-based mobility is another approach to solving the IP mobility challenge. It is possible to support mobility for IPv6 nodes without host involvement by extending MIPv6 signaling messages and reusing the HA. This approach to supporting mobility does not require the MN to be involved in the exchange of signaling messages between itself and the HA. A Mobile Access Gateway (MAG) in the network performs the signaling with the HA and does the mobility management on behalf of the MN attached to the network. This protocol is called by Proxy Mobile IPv6 (PMIPv6) [2] in Network-based Localized Mobility Management (NETLMM) working group of IETF. PMIPv6 can reduce handover latency, since the proxy mobility agent on behalf of the MN performs handover process. That is, there is a little signaling message via wireless link.

There is much packet loss during handover in PMIPv6, although PMIPv6 reduces handover latency. In this paper, to reduce packet loss in PMIPv6 we propose Packet-Lossless PMIPv6 (PL-PMIPv6) with authentication. The similar scheme was studied to reduce packet loss and handover latency in MIPv6, such as fast handovers for MIPv6 (FMIPv6) [3]. In PL-PMIPv6, a previous MAG (pMAG) registers to a Local Mobility Anchor (LMA) on behalf of a new MAG (nMAG) during layer 2 handoff. Then, the nMAG buffers packets during handover after registration. Therefore, PL-PMIPv6 can reduce packet loss in MIPv6 and PMIPv6. Also, we use Authentication, Authorization and Accounting (AAA) infrastructure to authenticate the MN and to receive MN's profiles securely. We show performance of PL-PMIPv6 through comparison of packet loss during handover of MIPv6, PMIPv6 and PL-PMIPv6.

The rest of the paper is organized as follows. Section 2 specifies PMIPv6 protocol as related works. Packet-lossless PMIPv6 (PL-PMIPv6) is proposed in Section 3. Section 4 presents analytical model and the numerical results. In Section 5, we conclude this paper.

2   Related Works

Proxy Mobile IPv6 protocol is intended for providing network-based IP mobility management support to a mobile node, without requiring the participation of the MN in any IP mobility related signaling. The mobility entities in the network will track the MN's movements and will initiate the mobility signaling and then setup the required routing state.

The core functional entities in the NETLMM infrastructure are the Local Mobility Anchor (LMA) and the Mobile Access Gateway (MAG). The LMA is responsible for maintaining the MN's reachability state and is the topological anchor point for the MN's home network prefix. The MAG is the entity that performs the mobility management on behalf of an MN, and it resides on the access link where the MN is anchored. The MAG is responsible for detecting

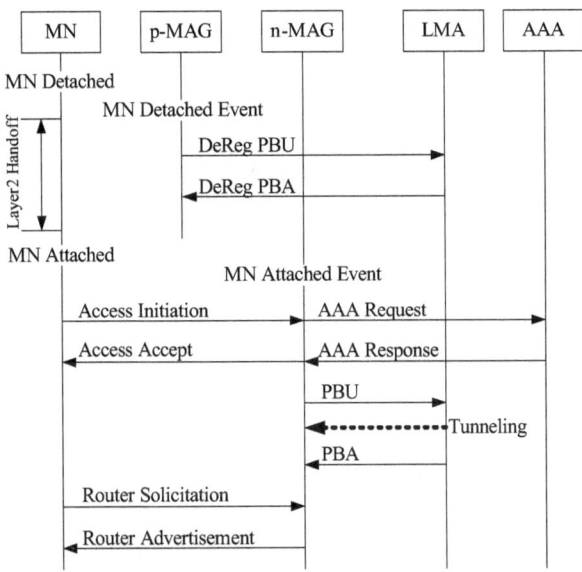

Fig. 1. Signaling flow of PMIPv6 with Authentication

the MN's movements to and from the access link and for initiating binding registrations to the MN's LMA.

Once an MN enters PMIPv6 domain and attaches to an access link, the MAG on that access link, after identifying the MN and acquiring its identity, will determine if the MN is authorized for the network-based mobility management service.

For updating the LMA about the current location of the MN, the MAG sends a Proxy Binding Update (PBU) message to the MN's LMA. Upon accepting this PBU message, the LMA sends a Proxy Binding Acknowledgement (PBA) message including the MN's home network prefix. It also creates the Binding Cache entry and establishes a bi-directional tunnel to the MAG. The MAG on receiving the PBA message sets up a bi-directional tunnel to the LMA and sets up the data path for the MN's traffic. At this point the MAG will have all the required information for emulating the MN's home link. It sends Router Advertisement (RA) messages to the MN on the access link advertising the MN's home network prefix as the hosted on-link-prefix. Figure 1 shows the signaling call flow for the MN's handover from previously attached MAG (pMAG) to the newly attached MAG (nMAG). After obtaining the initial address configuration in the PMIPv6 domain, if the MN changes its point of attachment, the MAG on the previous link will detect the MN's detachment from the link and then signal the LMA and then remove the binding and routing state for that MN. However, the LMA upon accepting the request will wait for some short amount of time before it deletes the binding, for allowing a smooth handover. The MAG on the new access link upon detecting the MN on its access link will signal the LMA for

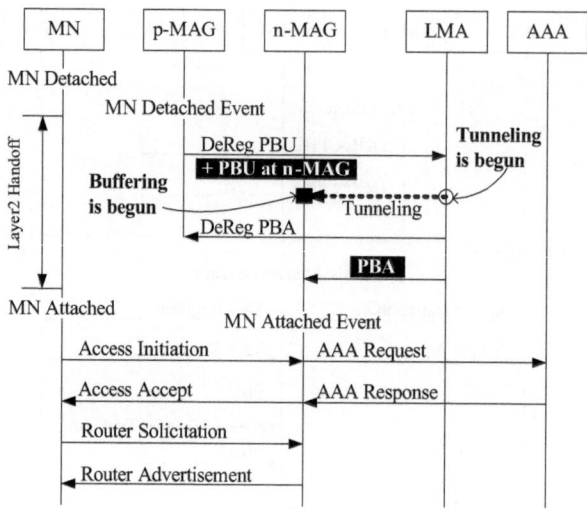

Fig. 2. Signaling flow of the proposed scheme (PL-PMIPv6)

updating the binding state. Once that signaling is complete, the MN will continue to receive the RAs containing its home network prefix. The MN believes that it is still on the same link and will use the same address configuration on the new access link.

3 Proposed Scheme (PL-PMIPv6)

In MIPv6, there are schemes to reduce both handover latency and packet loss, such as FMIPv6 and Hierarchical MIPv6 (HMIPv6 [4]). In FMIPv6, address test is performed during handover to reduce handover latency, and also tunnel between a previous access router and a new access router is established to prevent packet loss. HMIPv6 employs a mobility anchor point (MAP) to handle binding update (BU) for MNs within the MAP domain. In this way, network-wide signaling is only required when the MN roams outside of its current MAP domain and, thus, signaling traffic and handoff latency can be reduced.

Also, in PMIPv6 there are schemes to reduce handover latency and packet loss [5] [6]. Fast handover for PMIPv6 is a scheme that only LMA exchanges signaling with MAGs to set up the fast handover [5]. But, this scheme does not follow the order of signaling flow in PMIPv6. The scheme in [6] is the Extended PMA (EPMA) and LPMA functionalities to reduce signaling cost for intra domain handover and to optimize packet delivery [6]. This scheme is applied HMIPv6 mechanism in PMIPv6. But, this scheme does not prevent packet loss during handover.

In this paper, we propose Packet-Lossless PMIPv6 (PL-PMIPv6) with authentication, to reduce packet loss in PMIPv6. PL-PMIPv6 follows the order of signaling flow in PMIPv6 and reduces packet loss. Figure 2 shows signaling flow

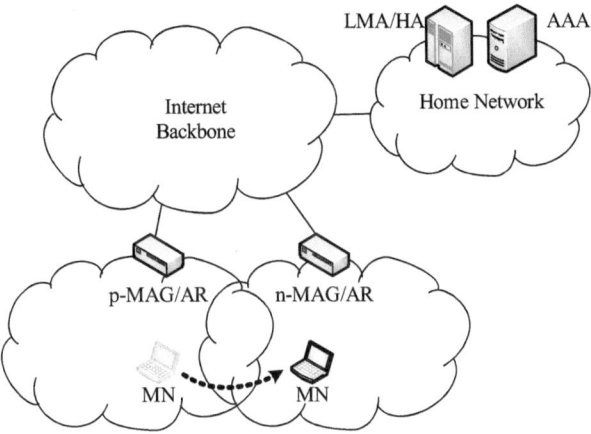

Fig. 3. System Model

of PL-PMIPv6 during handover. After the pMAG is aware of the MN's detachment, it sends the DeReg PBU message to the LMA in PMIPv6. In PL-PMIPv6 when pMAG sends the DeReg PBU message, the PBU message of nMAG is included in DeReg PBU message. That is, the pMAG registers on behalf of the nMAG in advance to reduce handover latency. As a result, the tunnel between the LMA and the nMAG is established in advance. Also, when the nMAG receives the PBA message, it begins to buffer packets to the MN. After layer 2 handoff, the MN sends the Router Solicitation (RS) message and receives the RA message including the MN's home network prefix.

In addition, the PBU message of the nMAG included in the DeReg PBU message is a tentative BU message in TEBU scheme [7], since the MN's movement may not be anticipated, exactly. Then, the general registration in the nMAG will be performed after finish of handover.

In PMIPv6, we use AAA infrastructure to authenticate the MN like in [6]. The nMAG can receive the MN's profile securely using AAA infrastructure.

4   Performance Evaluations

We make a comparison of MIPv6, PMIPv6 and PL-MIPv6 handover in terms of handover latency, costs and total cost ratio. For those comparisons, we use a system model in Fig. 3. In the system model, we evaluate performance of three schemes when an MN moves between MAGs. We assume that a correspondent node generates data packets destined to the MN at a mean rate λ, and the MN moves between MAGs at a mean rate μ. We define packet to mobility ratio (PMR, ρ) as the mean number of packets received by the MN from the correspondent per movement. When the movement and packet generation processes are independent and stationary, the PMR is given by $\rho = \lambda/\mu$. We assume that a cost for transmitting a packet is dependent on the distance between the sender

Table 1. Symbols of delay

Symbol	Explanation
$t_{MN,MAG}$	transfer delay between a MN and a MAG.
$t_{MAG,HN}$	transfer delay between a MAG and a LMA or a AAA server in the home network.
$t_{MAG,MAG}$	transfer delay between adjacent two MAGs.

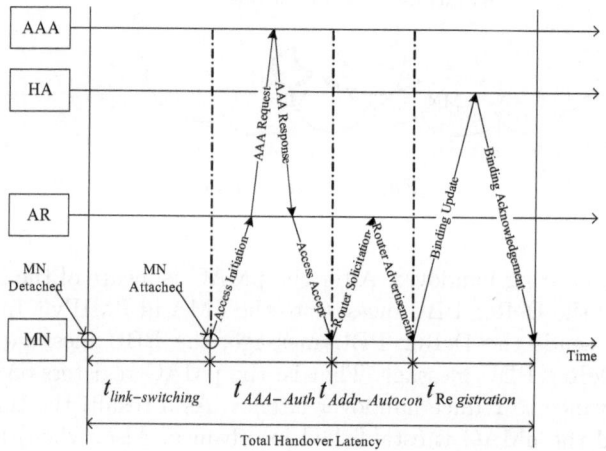

Fig. 4. Handover latency of MIPv6

and receiver. We define that l_d is the average length of a data packet and l_c is the average length of a control packet [8].

4.1 Handover Latency

In this section, we analyze handover latency of MIPv6, PMIPv6 and PL-PMIPv6. To analyze handover latency of three schemes, we define symbols of delay in Table 1 which refers to the system model in Fig. 3.

Handover latency consists of three latencies such as a link switching latency, an IP connectivity latency and a location update latency. The link switching latency is due to a layer 2 handoff. The IP connectivity latency is due to movement detection and new IP address configuration after the layer 2 handoff. The location update latency is due to registrations to the HA and the CNs. An MN can send or receive packets in nMAG after the IP connectivity.

Handover Latency of Mobile IPv6. Figure 4 shows handover latency of MIPv6. We define handover latency of MIPv6 referring to Fig. 4 as,

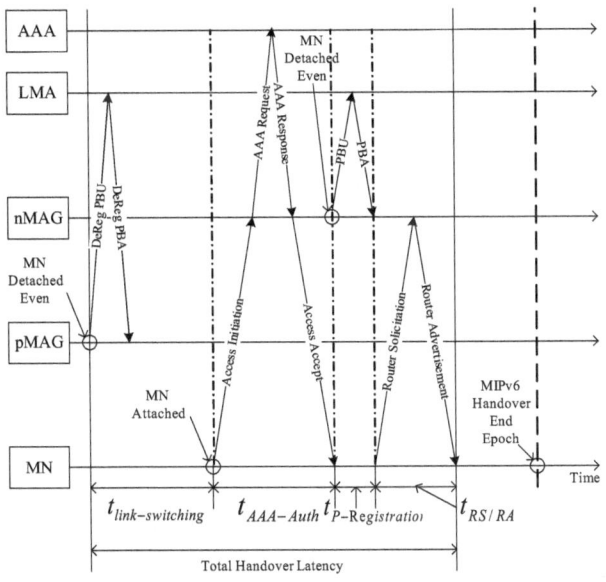

Fig. 5. Handover latency of PMIPv6

$$T_{Latency}^{MIPv6} = t_{link-switching} + t_{AAA-Auth} + t_{Addr-Autoconf} + t_{Registration}. \quad (1)$$

$t_{link-switching}$ is a delay during layer 2 handover. $t_{AAA-Auth}$ is a delay during authentication of an MN through AAA infrastructure $(2 \cdot (t_{MN,MAG} + t_{MAG,HN}))$. $t_{Addr-Autoconf}$ is a delay during process of stateless autoconfiguration $(2 \cdot t_{MN,MAG})$. $t_{Registration}$ is a delay during binding update to a HA $(2 \cdot (t_{MN,MAG} + t_{MAG,HN}))$.

When the MN is detached from the previous AR, it scans new ARs, and then it associates with the new AR. The MN processes AAA procedure to obtain access authentication and authorization from AAA server in the home network. Then the MN performs stateless address autoconfiguration, to setup its care-of address. Finally, the MN send the BU message to the home agent.

Handover Latency of Proxy Mobile IPv6. In Fig. 5, we define handover latency of PMIPv6. Handover latency of PMIPv6 is as follows,

$$T_{Latency}^{PMIPv6} = t_{link-switching} + t_{AAA-Auth} + t_{P-Registration} + t_{RS-RA}. \quad (2)$$

$t_{link-switching}$ and $t_{AAA-Auth}$ are same as those in MIPv6. $t_{P-Registration}$ is a delay during proxy binding update to a LMA $(2 \cdot t_{MAG,HM})$. t_{RS-RA} is a delay during exchanging of a router solicitation (RS) and a router advertisement (RA) messages between the MN and the MAG $(2 \cdot t_{MN,MAG})$. The pMAG deregister its binding to the LMA during $t_{link-switching}$.

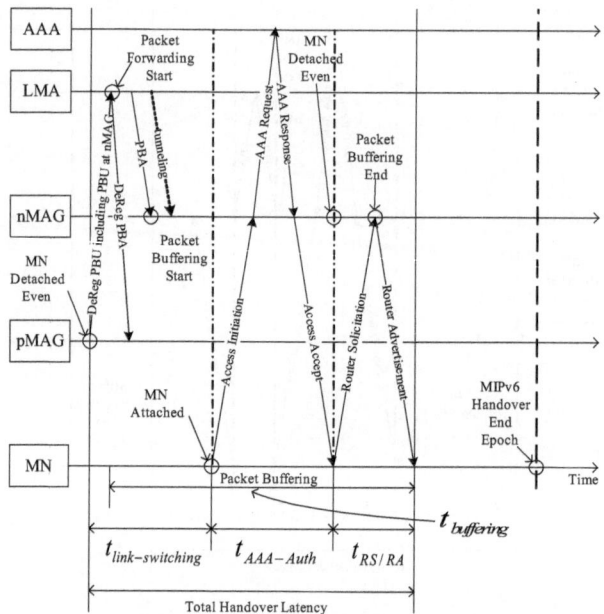

Fig. 6. Handover latency of PL-PMIPv6

When the pMAG detects that the MN is detached, it sends the deregistration PBU message to deregister its current binding. After AAA process is finished, the nMAG sends the PBU for registration message to the LMA. Finally, the MN sends the RS message, and receives the Router Advertisement (RA) message.

Handover Latency of the Proposed Scheme. Handover latency of PL-PMIPv6 can be defined by formula 3 through Fig. 6. In this case, delay of proxy registration is reduced, since proxy registration is performed during layer 2 handoff.

$$T_{Latency}^{PL-PMIPv6} = t_{link-switching} + t_{AAA-Auth} + t_{RS-RA}. \tag{3}$$

When the pMAG finds out detachment of the MN, it sends the deregistration PBU message including the PBU message. Hence, delay that the nMAG sends the PBU message and receives PBA message is eliminated in PL-PMIPv6.

Results about Handover Latencies. We present some results based on the above analysis, and draw handover latencies according to layer 2 handoff, wireless link delay, and delay between the MAG and the home network. Figure 7 shows handover latency of three schemes. In Fig. 7, we use that $t_{MN,MAG} = 10ms$, $t_{MAG,MAG} = 2ms$, $t_{MAG,HN} = 10ms$, referring to [9] [10].

Layer 2 handoff delay generally causes a long handover latency. Then, it may be reduced, however in this paper, we do not consider layer 2 handoff delay. Handover latency is seriously affected by accesses of wireless link during handover.

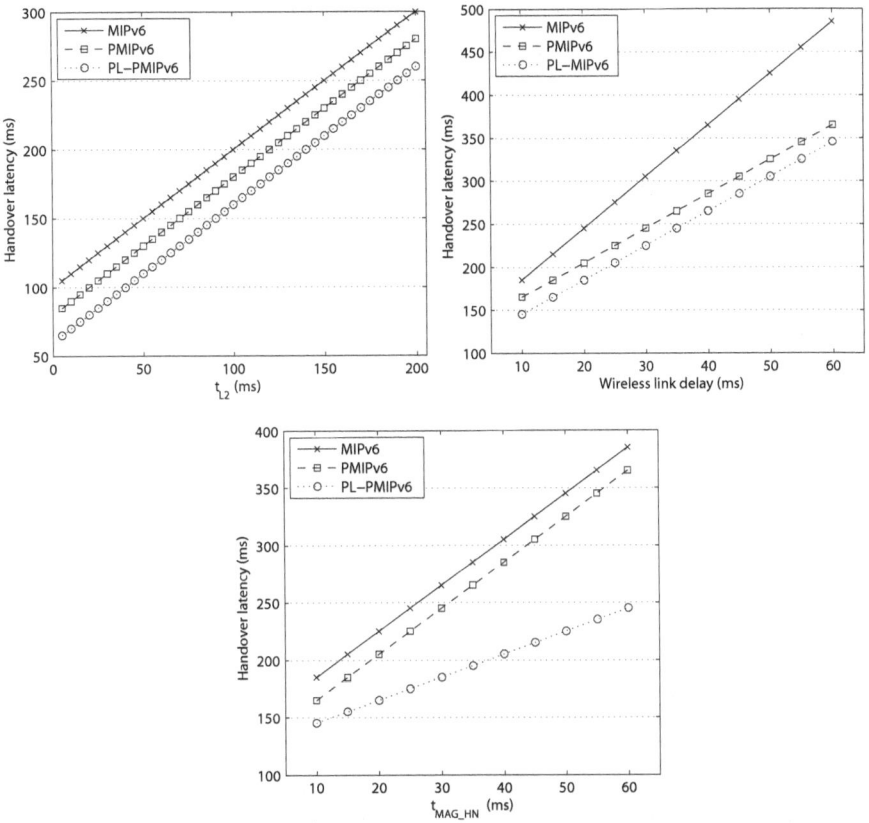

Fig. 7. Handover latency vs wireless link delay

In Fig. 7, handover latencies of PMIPv6 and PL-PMIPv6 are lower latency than that of MIPv6, since wireless link is used less. Also, PL-PMIPv6 reduces more handover latency than PMIPv6, since the registration of nMAG is performed during layer 2 handoff. In the third figure of Fig. 7, Handover latency of PL-PMIPv6 is quite shorter than those of the other schemes.

4.2   Total Cost during Handover

In this section we analyze three scheme in terms of total cost. Total cost consists of signaling cost and packet delivery cost during handover. Signaling cost is cost of messages for signaling, and packet delivery cost is cost of packet transferred from correspondent node.

We assume that both signaling and delivery cost are influenced by transmission delay. That is, signaling cost and delivery cost of a packet consist of an average length of signal and data packet and transmission delay, respectively.

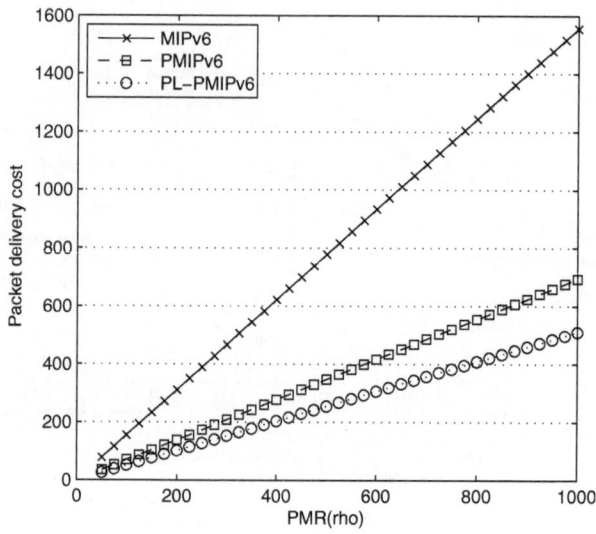

Fig. 8. PMR vs packet delivery cost

The following is signaling costs of three schemes referring to Fig. 4,5,6.

$$C_{Signaling}^{MIPv6} = S_{AAA-Auth} + S_{Addr-Autoconf} + S_{Registration}, \tag{4}$$

$$C_{Signaling}^{PMIPv6} = S_{P-Deregistration} + S_{AAA-Auth} + \\ + S_{P-Registration} + S_{RS-RA}, \tag{5}$$

$$C_{Signaling}^{PL-PMIPv6} = S_{P-Deregistration} + S_{AAA-Auth} + \\ + S_{P-Registration}/2 + S_{RS-RA}. \tag{6}$$

$S_{AAA-Auth}$ is cost for process of AAA authentication $(l_c \cdot t_{AAA-Auth})$. $S_{Addr-Autoconf}$ is cost for stateless address autoconfiguration $(l_c \cdot t_{Addr-Autoconf})$. $S_{Registration}$ is cost for binding update to a HA $(l_c \cdot t_{Registration})$. $S_{P-Registration}$ is cost for proxy binding update to a LMA $(l_c \cdot t_{P-Registration})$. S_{RS-RA} is cost for exchanging a RS and a RA messages $(l_c \cdot t_{RS-RA})$. We have known that signaling costs of three schemes differ little.

We assume that delivery cost consists of packet transmission cost and packet lost during handover. Packet delivery costs of three schemes are as follows :

$$C_{Delivery}^{MIPv6} = \lambda \cdot d_{HN,MN} \cdot T_{Latency}^{MIPv6} \cdot \eta, \tag{7}$$

$$C_{Delivery}^{PMIPv6} = \lambda \cdot d_{HN,pMAG} \cdot T_{Latency}^{PMIPv6} \cdot \eta, \tag{8}$$

$$C_{Delivery}^{PL_PMIPv6} = \lambda \cdot d_{HN,MN} \cdot t_{MAG,HN} \cdot \eta + \\ + \lambda \cdot d_{HN,nMAG} \cdot t_{buffering}. \tag{9}$$

$d_{HN,MN}$ is delivery cost from home network to a MN. $d_{HN,pMAG}$ is delivery cost from home network to a pMAG. $d_{HN,nMAG}$ is delivery cost from home network

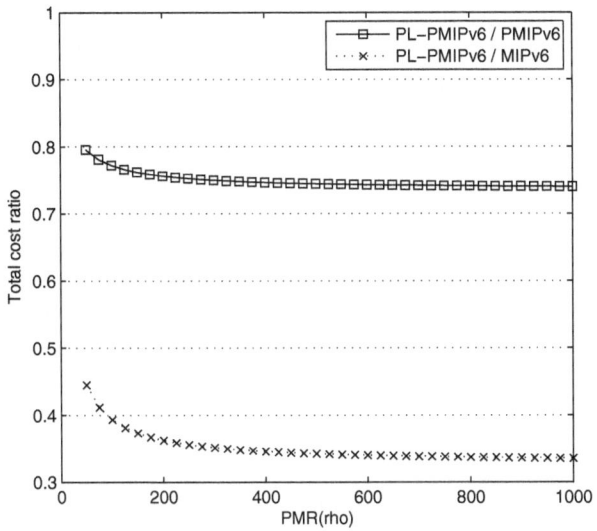

Fig. 9. PMR vs total cost ratio

to a new MAG through the pMAG. In MIPv6 and PMIPv6, all packets from the CN are lost during handover. However, in PL-PMIPv6 most packets are tunneled to nMAG and are buffered. Figure 8 shows delivery costs of three schemes. In Fig. 8, costs of PMIPv6 and PL-PMIPv6 are much lower than that of MIPv6, since handover latencies are shorter in PMIPv6 and PL-PMIPv6. In addition, cost of PMIPv6 is the lowest in three schemes, because packets from the CN are tunneled and buffered in PMIPv6.

We calculated total costs of three schemes using above formulas. Then we make cost ratio formulas to compare performance of three schemes.

$$CostRatio_{MIPv6}^{PL-PMIPv6} = \frac{TotalCost_{PL-PMIPv6}}{TotalCost_{MIPv6}} \qquad (10)$$

$$CostRatio_{PMIPv6}^{PL-PMIPv6} = \frac{TotalCost_{PL-PMIPv6}}{TotalCost_{PMIPv6}} \qquad (11)$$

Figure 9 shows total cost ratios of MIPv6 and PMIPv6 about PL-PMIPv6, respectively. Through this figure, we show that PL-PMIPv6 reduces total cost during handover. That is, PL-PMIPv6 enhances handover performance of 66% compared with MIPv6 and 26% compared with PMIPv6.

5   Conclusions

PMIPv6 is a protocol which supports IP mobility without participation of the MN, and is studied in Network-based Localized Mobility Management (NETLMM) working group of IETF. In PMIPv6 all packets from CNs are lost during handover.

To solve this problem, we propose a scheme to reduce packet loss, such as Packet-Lossless PMIPv6 (PL-PMIPv6). PL-PMIPv6 can prevent packet loss during handover, since process of proxy registration is performed in advance and then packets from CNs are tunneled and buffered to nMAG. Also, we use AAA infrastructure to authenticate a MN and to receive MN's profiles securely.

Finally, we analyze the handover latency and the total cost for PL-PMIPv6. The analysis shows that with PL-PMIPv6, the handover latency and the total cost can be reduced compared to PMIPv6. Recently, PMIPv6 has been standardized in IETF. Then, if PL-PMIPv6 is used together with PMIPv6, PMIPv6 will be a more robust protocol in NETLMM working group of IETF.

References

1. Johnson, D., Perkins, C., Arkko, J.: Mobility Support in IPv6, RFC 3775, IETF (June 2004)
2. Gundavelli, S., Leung, K., Devarapalli, V., Chowdhury, K., Patil, B.: Proxy Mobile IPv6, RFC 5213, IETF (August 2008)
3. Koodli, R. (ed.): Fast Handovers for Mobile IPv6, RFC 4068, IETF (July 2005)
4. Soliman, H., Castelluccia, C., El Malki, K., Bellier, L.: Hierarchical Mobile IPv6 Mobility Management (HMIPv6), RFC 4140, IETf (August 2005)
5. Kim, P., Kim, S., Jin, J.: Fast Handovers for Proxy Mobile IPv6 without Inter-MAG Signaling, draft-pskim-netlmm-fastpmip6-00.txt, IETF (November 2007)
6. Park, S., Kang, N., Kim, Y.: Localized Proxy-MIPv6 with Route Optimization in IP-Based Networks. IEICE Trans. Commun. E90.B(12) (December 2007)
7. Ryu, S., Mun, Y.: The Tentative and Early Binding Update for Mobile IPv6 Fast Handover. In: Jia, X., Wu, J., He, Y. (eds.) MSN 2005. LNCS, vol. 3794, pp. 825–835. Springer, Heidelberg (2005)
8. Jain, R., Raleigh, T., Graff, C., Bereschinsky, M.: Mobile Internet Access and QoS Guarantees using Mobile IP and RSVP with Location Registers. In: Proc. IEEE Intl. Conf. on Communications, June 1998, pp. 1690–1695. IEEE, Atlanta (1998)
9. Fathi, H., Prasad, R., Chakraborty, S.: Mobility Management For Voip In 3G Systems: Evaluation Of Low-Latency Handoff Schemes. IEEE Wireless Communications 56(1), 260–270 (2007)
10. Kwon, T., Gerla, M., Das, S.: Mobility Management for VoIP Service: Mobile IP vs. SIP. IEEE Wireless Communications 9(5), 66–75 (2002)

Wavelet Based Approach to Fractals and Fractal Signal Denoising*

Carlo Cattani

DiFarma, University of Salerno, Via Ponte Don Melillo I-84084 Fisciano (SA)
ccattani@unisa.it

Abstract. In this paper localized fractals are studied by using harmonic wavelets. It will be shown that, harmonic wavelets are orthogonal to the Fourier basis. Starting from this, a method is defined for the decomposition of a suitable signal into the periodic and localized parts. For a given signal, the denoising will be done by simply performing a projection into the wavelet space of approximation. It is also shown that due to their self similarity property, a good approximation of fractals can be obtained by a very few instances of the wavelet series. Moreover, the reconstruction is independent on scale as it should be according to the scale invariance of fractals.

Keywords: Harmonic Wavelets, Weierstrass Function, Scale Invariance, Fractals, Denoising.

1 Introduction

Wavelets are some special functions $\psi_k^n(x)$ [6,11,12,17,22] which depend on two discrete valued parameters: n is the scale (refinement, compression, or dilation) parameter and k is the localization (translation) parameter. These functions fulfill the fundamental axioms of multiresolution analysis so that by a suitable choice of the scale and translation parameter one is able to easily and quickly approximate any localized functions (even tabular) with decay to zero.

Due to this multiscale approach the approximation method by wavelets has some advantages due to the minimum set of coefficients for representing the phenomenon and the direct projection into a given scale, thus giving a direct physical interpretation of the phenomenon. In each scale the wavelet coefficients and, in particular, the detail coefficients β_k^n describe "local" oscillations. Therefore wavelets seems to be the more expedient tool for studying those problems which are localized (in time or in frequency) and/or have some discontinuities.

Among the many different families of wavelets a special attention was paid to the complex Littlewood Paley bases, also called harmonic wavelets [4,6,7,8,9,10, 18,19]. They seems to be one of the most expedient tool for studying processes which are localized in Fourier domain or have high frequency oscillation in time.

---

* Preliminary results presented at the International Conference on Computational Science and Applications (ICCSA 2008), June 30-July 3, 2008 Perugia (It) [7].

M.L. Gavrilova and C.J.K. Tan (Eds.): Trans. on Comput. Sci. VI, LNCS 5730, pp. 143–162, 2009.
© Springer-Verlag Berlin Heidelberg 2009

It has been recently observed [7, 9, 10] that also some kind of highly oscillating functions and random-like or fractal-like functions can be easily reconstructed by using the harmonic wavelets. For this reason they will be used, in the following, to study the regularity of a slightly modified Weierstrass function [2, 3, 21], which has a fractal behavior. Indeed the Weierstrass function is based on periodic functions ranging from $-\infty$ to ∞, therefore a slight modification is needed in order to let its reconstruction, by harmonic wavelet series, converges in $L_2(\mathbb{R})$. Therefore we will focus on the so-called deterministically self-similar signals (see e.g. [22]) which are deterministic self-similar functions with compact support.

The self-similarity property of fractals will be analyzed in terms of wavelet coefficients (see also [22]) and it will be shown that to reconstruct a fractal only a small set of coefficients, at the lower scale, are needed.

In the last part of this paper a signal made by a 2π–periodic function and a localized self-similar function will by analyzed by using harmonic wavelets. A fundamental theorem, which characterizes harmonic wavelets with respect to the periodic functions, will be given. It will be shown that harmonic wavelets are orthogonal to the Fourier basis, for this reason the wavelet reconstruction of a 2π–periodic function will trivially vanish. Based on this, the denoising of the signal will be obtained by a projection into a wavelet space with no error (contrary to what usually happens with ordinary techniques of hard/soft denoising). In this case the only error is due to the scale of wavelet approximation and it can be shrunk to zero by simply increasing the scale.

2   Preliminary Remarks on Self-similar Functions

Self-similarity, or scale invariance, is a characteristic feature of many natural phenomena [1, 3, 14, 13, 15, 16, 20, 21, 22], biological signals, geological deformations, natural phenomena as clouds, tree branches evolution, leaves formation, stocks and financial data etc. All these phenomena can be represented by some kind of singular (nowhere differentiable) functions which are fractals. They are mathematically known through tabular and recursive-analytical formulas both deterministic and stochastic thus being represented by a very large family of functions. Most of them are characterized by the scale invariance (or self-similarity). This, however, is a feature of many functions, including fractals, and behaves as if the function is identical to any of its rescaled instances up to a suitable renormalization of the amplitude. In particular, the following definition holds:

Definition 1: The function $f(x), x \in A \subset \mathbb{R}$ is *scale invariant* with scaling exponent $H, (0 < H < 1)$ (Hausdorff dimension) if, for any $\mu \in \mathbb{R}$, it is

$$f(\mu x) = \mu^H f(x) \ . \tag{1}$$

It is assumed $\mu > 0$ so that

$$\mu = 2^C \Rightarrow C = \log_2 \mu \ , \tag{2}$$

and the scale invariance reads as

$$f\left(2^C x\right) = 2^{CH} f\left(x\right) . \tag{3}$$

Indeed the condition (1) is quite strong and can't be easily fulfilled, in the sense that there are not so many functions for which (1) holds true. For example the Cantor set function is invariant only for $\mu = 3^k, k \in \mathbb{N}$. For this reason the invariance it is assumed to be satisfied by only few values of μ and the corresponding weakened version of scale invariance is called discrete scale invariance [1, 2].

Some examples of such a kind of functions are:

− The Weierstrass function [2, 21]

$$f\left(x\right) = \sum_{k=1}^{\infty} a^k \cos b^k x \left(a > 1, b > 1\right) . \tag{4}$$

− The Mandelbrot-Weierstrass function [2]

$$f\left(x\right) = \sum_{k=-\infty}^{\infty} a^{-kH} \left(1 - \cos a^k x\right) \left(a > 1, 0 < H < 1\right) . \tag{5}$$

− The Cantor function [13, 15, 16]

$$\begin{cases} f_0\left(x\right) &= x \\ \\ f_{n+1}\left(x\right) = \begin{cases} \dfrac{1}{2} f_n\left(3x\right) & , 0 \le x \le \frac{1}{3} \\ \dfrac{1}{2} & , \dfrac{1}{3} < x \le \dfrac{2}{3} \\ \dfrac{1}{2} + \dfrac{1}{2} f_n\left(3x - 2\right) & , \dfrac{2}{3} < x \le 1 . \end{cases} \end{cases}$$

There exist also some functions which are scale invariant but not necessarily fractals like, for example, the Mellin functions [2], also known as chirps,

$$f\left(x\right) = x^H e^{2\pi i(m \log x / \log a)} \left(m \in \mathbb{Z}, a > 1\right) . \tag{6}$$

A characteristic feature of the above functions is that they show also some high frequency oscillations, thus suggesting us that somehow it should be possible to study fractals with high oscillating functions.

2.1 Self-similar Functions with Finite Energy

Our investigation will be restricted to the family of self-similar functions $f(x)$, such that $f\left(x\right) \in L_2\left(\mathbb{R}\right)$.

Definition 2. Let $f\left(x\right) : \mathbb{R} \to \mathbb{C}$, $f\left(x\right) \in L_2\left(\mathbb{R}\right)$ if $\left[\displaystyle\int_{-\infty}^{\infty} \left|f\left(x\right)\right|^2 dx\right]^{1/2} < \infty.$

Therefore some additional hypotheses should be done, on the self-similar function, so that it would belong to $L_2\left(\mathbb{R}\right)$. However it can be easily checked that

Theorem 1. *If $f(x)$ is self-similar then $f(x) \notin L_2(\mathbb{R})$.*

Proof. In fact, by taking into account definition 2, it is

$$\left[\int_{-\infty}^{\infty} |f(x)|^2 dx \right]^{1/2} \leq M < \infty$$

and

$$\left[\int_{-\infty}^{\infty} |f(\mu x)|^2 dx \right]^{1/2} \overset{(1)}{=} \left[\int_{-\infty}^{\infty} \left| \mu^H f(x) \right|^2 dx \right]^{1/2} = \mu^H \left[\int_{-\infty}^{\infty} |f(x)|^2 dx \right]^{1/2} \leq M < \infty$$

that is

$$\mu^H \left[\int_{-\infty}^{\infty} |f(x)|^2 dx \right]^{1/2} \leq \mu^H M \leq M < \infty \,,$$

so that we have

$$\mu^H \leq 1 \Rightarrow \begin{cases} 0 < \mu < 1 \,, \ 0 < H \\ \mu > 1 \quad , H < 0 \,. \end{cases}$$

Since it is, by definition, $H > 0$ it should be $0 < \mu < 1$. On the other hand, by a substitution in the integral variable, it is

$$\left[\int_{-\infty}^{\infty} |f(\mu x)|^2 dx \right]^{1/2} = \frac{1}{\mu} \left[\int_{-\infty}^{\infty} |f(x)|^2 dx \right]^{1/2} \leq M$$

and, by a comparison, there follows

$$\left[\int_{-\infty}^{\infty} |f(\mu x)|^2 dx \right]^{1/2} = \frac{1}{\mu} \left[\int_{-\infty}^{\infty} |f(x)|^2 dx \right]^{1/2} = \mu^H \left[\int_{-\infty}^{\infty} |f(x)|^2 dx \right]^{1/2} \leq M$$

that is

$$\frac{1}{\mu} = \mu^H \Rightarrow H = -1$$

which is in contradiction with the definition of H. ∎

Therefore, in general, there not exists a self-similar function which is defined on the entire $\mathbb{R}$ and belongs to $L_2(\mathbb{R})$. Instead, in a finite interval $[a, b] \subset \mathbb{R}$ it is

$$\left[\int_{a}^{b} |f(\mu x)|^2 dx \right]^{1/2} \overset{(1)}{=} \left[\int_{a}^{b} \left| \mu^H f(x) \right|^2 dx \right]^{1/2} = \mu^H \left[\int_{a}^{b} |f(x)|^2 dx \right]^{1/2} \leq M < \infty \,,$$

that is $H \log \mu \leq \log M$.

On the other hand by a substitution in the integral variable we have

$$\left[\int_a^b |f(\mu x)|^2 dx \right]^{1/2} = \frac{1}{\mu} \left[\int_{\mu a}^{\mu b} |f(x)|^2 dx \right]^{1/2}$$

so that,

$$
\begin{cases}
\mu^H \left[\int_a^b |f(x)|^2 dx \right]^{1/2} = \left[\int_a^b |f(\mu x)|^2 dx \right]^{1/2} > \frac{1}{\mu} \left[\int_a^b |f(x)|^2 dx \right]^{1/2} & , \mu > 1 \\[3ex]
\mu^H \left[\int_a^b |f(x)|^2 dx \right]^{1/2} = \left[\int_a^b |f(\mu x)|^2 dx \right]^{1/2} < \frac{1}{\mu} \left[\int_a^b |f(\mu x)|^2 dx \right]^{1/2} & , \mu < 1
\end{cases}
$$

or

$$
\begin{cases}
1 > \dfrac{1}{\mu^{H+1}} \, , \ \mu > 1 \\[3ex]
1 < \dfrac{1}{\mu^{H+1}} \, , \mu < 1 \, .
\end{cases}
\tag{7}
$$

There follows that, in order to have a self-similar function belonging to $L_2(\mathbb{R})$ the self-similar function should be at least defined on a compact support interval and the vertical factor μ has to fulfill the above condition (7) (see also [14]).

3   Harmonic Wavelet Basis

In this section, the main definitions and properties of harmonic wavelets (see also [4,6,7,8,9,10,18,19]) are given. The dilated and translated instances of the harmonic wavelets are

$$
\begin{cases}
\varphi_k^n(x) \overset{\text{def}}{=} 2^{n/2} \dfrac{e^{2\pi i(2^n x - k)} - 1}{2\pi i(2^n x - k)} \, , \\[3ex]
\psi_k^n(x) \overset{\text{def}}{=} 2^{n/2} \dfrac{e^{4\pi i(2^n x - k)} - e^{2\pi i(2^n x - k)}}{2\pi i(2^n x - k)} \, ,
\end{cases}
\tag{8}
$$

with $n \in \mathbb{N}$, $k \in \mathbb{Z}$. It is

$$\lim_{x \to 0} \varphi_k^n(x) = 2^{n/2} \delta_{k0} \, , \ \lim_{x \to 0} \psi_k^n(x) = 2^{n/2} \delta_{k0} \tag{9}$$

and

$$\lim_{x \to k/2^n} \varphi_k^n(x) = 2^{n/2} \delta_{k0} \, , \ \lim_{x \to k/2^n} \psi_k^n(x) = 2^{n/2} \delta_{k0}.$$

By a simple computation there follows from (8)

$$
\begin{cases}
\dfrac{1}{2}[\varphi_k^n(x) + \overline{\varphi}_k^n(x)] = 2^{n/2}\dfrac{\sin 2\pi(2^n x - k)}{2\pi(2^n x - k)} \\[3mm]
\dfrac{1}{2}\left[\psi_k^n(x) + \overline{\psi}_k^n(x)\right] = 2^{n/2}\left[\dfrac{\sin 2\pi(2^n x - k)}{2\pi(2^n x - k)} - 2\dfrac{\sin 4\pi(2^n x - k)}{4\pi(2^n x - k)}\right]
\end{cases}
$$

where the bar stands for complex conjugate. Thus the real part of harmonic wavelets

$$
\Re[\varphi_k^n(x)] \overset{\text{def}}{=} \Phi_k^n(x), \Re[\psi_k^n(x)] \overset{\text{def}}{=} \Psi_k^n(x) , \tag{10}
$$

are the so-called Shannon scaling and wavelet functions [5, 8]

$$
\begin{cases}
\Phi_k^n(x) \overset{\text{def}}{=} 2^{n/2}\dfrac{\sin 2\pi(2^n x - k)}{2\pi(2^n x - k)} \\[3mm]
\Psi_k^n(x) \overset{\text{def}}{=} 2^{n/2}\left[\dfrac{\sin 2\pi(2^n x - k)}{2\pi(2^n x - k)} - 2\dfrac{\sin 4\pi(2^n x - k)}{4\pi(2^n x - k)}\right] .
\end{cases} \tag{11}
$$

By using the Fourier transform:

$$
\widehat{\varphi}(\omega) = \mathcal{F}[\varphi(x)] \overset{\text{def}}{=} \frac{1}{2\pi}\int_{-\infty}^{\infty}\varphi(x)e^{-i\omega x}dx
$$

it can be easily shown that (see e.g. [5,7-9]):

$$
\begin{cases}
\widehat{\varphi}_k^n(\omega) = \dfrac{2^{-n/2}}{2\pi}e^{-i\omega k/2^n}\chi(2\pi + \omega/2^n) \\[3mm]
\widehat{\psi}_k^n(\omega) = \dfrac{2^{-n/2}}{2\pi}e^{-i\omega k/2^n}\chi(\omega/2^n)
\end{cases} \tag{12}
$$

where $\chi(\omega)$ is the box (characteristic) function

$$
\chi(\omega) \overset{\text{def}}{=} \begin{cases} 1, & 2\pi \le \omega \le 4\pi , \\ 0, & \text{elsewhere} . \end{cases} \tag{13}
$$

For the $L_2(\mathbb{R})$ space of functions, which is an Hilbert space, it can be defined the inner product, which fulfills the Parseval equality, as follows

$$
\langle f, g\rangle \overset{\text{def}}{=} \int_{-\infty}^{\infty} f(x)\overline{g(x)}dx = 2\pi\int_{-\infty}^{\infty}\widehat{f}(\omega)\overline{\widehat{g}(\omega)}d\omega = 2\pi\langle\widehat{f},\widehat{g}\rangle , \tag{14}
$$

for any $f(x) \in L_2(\mathbb{R})$ and $g(x) \in L_2(\mathbb{R})$.

Harmonic wavelets are orthonormal functions [4, 7, 8, 9, 10, 19], in the sense that

$$
\langle \psi_k^n(x), \psi_h^m(x)\rangle = \delta^{nm}\delta_{hk} ,
$$

where $\delta^{nm}(\delta_{hk})$ is the Kronecker symbol.

Analogously it can be shown that

$$
\begin{cases}
\langle \varphi_k^n (x), \varphi_h^m (x) \rangle = \delta^{nm} \delta_{kh}, \ \langle \overline{\varphi}_k^n (x), \overline{\varphi}_h^m (x) \rangle = \delta^{nm} \delta_{kh}, \ \langle \varphi_k^n (x), \overline{\varphi}_h^m (x) \rangle = 0, \\[2mm]
\langle \overline{\psi}_k^n (x), \overline{\psi}_h^m (x) \rangle = \delta^{nm} \delta_{kh}, \ \langle \psi_k^n (x), \overline{\psi}_h^m (x) \rangle = 0, \\[2mm]
\langle \varphi_k^n (x), \overline{\psi}_h^m (x) \rangle = 0, \ \langle \overline{\varphi}_k^n (x), \psi_h^m (x) \rangle = 0, \ .
\end{cases}
$$

Harmonic wavelets have many interesting characteristic features. A good localization in frequency, but mostly they are orthogonal to the Fourier basis, thus making possible the orthogonal decomposition of the space $L_1(\mathbb{R}) \oplus L_2(\mathbb{R})$, being $\oplus$ the direct sum of spaces. In other words, any signal which is a linear combination of a periodic function (which can be represented by a Fourier series) and a localized function can be decomposed into each part. This is a consequence of this fundamental theorem [8, 9, 10]:

Theorem 2. *Harmonic wavelets are orthogonal to the Fourier basis $\{e^{2\pi mix}\}$ with $m \in \mathbb{N}$:*

$$
\langle \varphi_k^n (x), e^{2\pi mix} \rangle = 0, \quad \langle \psi_k^n (x), e^{2\pi mix} \rangle = 0 \quad, \quad m > 2^n \tag{15}
$$

and

$$
\begin{cases}
\langle \varphi_k^n (x), e^{2\pi mix} \rangle = 2^{-n/2} e^{-2\pi kim/2^n}, m \le 2^n \\[2mm]
\langle \psi_k^n (x), e^{2\pi mix} \rangle = 2^{-n/2} e^{-2\pi kim/2^n}, m \le 2^{n+1} \ .
\end{cases}
$$

In particular, it is

$$
\langle \varphi_k^0 (x), e^{2\pi mix} \rangle =
\begin{cases}
0, \ m > 1 \\[2mm]
e^{-2\pi ki} = (-1)^k \ , \ m = 1 \ .
\end{cases}
$$

Analogously for the conjugate basis.

Proof: Let us consider first the product $\langle \varphi_k^n (x), e^{2\pi mix} \rangle$. The Fourier transform of the harmonic basis is:

$$
\mathcal{F}\left[e^{2\pi mix}\right] = \delta (\omega - 2m\pi)
$$

being $\delta (\omega)$ the Dirac delta. So that by performing the scalar product in the Fourier domain we have

$$
2\pi \langle \widehat{\varphi_k^n (x)}, \widehat{e^{2\pi mix}} \rangle = 2\pi \int_{-\infty}^{\infty} \frac{2^{-n/2}}{2\pi} e^{-i\omega k/2^n} \chi (2\pi + \omega/2^n) \delta (\omega - 2m\pi) \, d\omega
$$

$$
= 2^{-n/2} e^{-2\pi ikm/2^n} \chi (2\pi + 2m\pi/2^n) \ ,
$$

that is

$$2\pi\left\langle \widehat{\varphi_k^n(x)}, \widehat{e^{2\pi mix}} \right\rangle = 2^{-n/2}e^{-2\pi ikm/2^n}\chi\left(2\pi + 2m\pi/2^n\right)$$
$$= \begin{cases} 0, & m > 2^n \\ 2^{-n/2}e^{-2\pi ikm/2^n}, & m \leq 2^n . \end{cases}$$

From where, since

$$e^{\pi in} = \begin{cases} 1, & n = 2k, \quad k \in \mathbb{Z} \\ -1, & n = 2k+1, \quad k \in \mathbb{Z}, \end{cases} \tag{16}$$

we get

$$\left\langle \varphi_k^n(x), e^{2\pi mix} \right\rangle = \begin{cases} 0, & m > 2^n \\ 1, & m \leq 2^n \end{cases}$$

and, in particular,

$$\left\langle \varphi_k^0(x), e^{2\pi mix} \right\rangle = 0, \ m > 1 .$$

For the second product $\left\langle \psi_k^n(x), e^{2\pi mix} \right\rangle$ we have, according to (12),

$$\left\langle \psi_k^n(x), e^{2\pi mix} \right\rangle = 2\pi \int\limits_{-\infty}^{\infty} \frac{2^{-n/2}}{2\pi}e^{-i\omega k/2^n}\chi\left(\omega/2^n\right)\delta\left(\omega - 2m\pi\right)d\omega$$

from where

$$\left\langle \psi_k^n(x), e^{2\pi mix} \right\rangle = \begin{cases} 0, & m > 2^{n+1} \\ 1, & m \leq 2^{n+1} . \end{cases}$$ ∎

4   Harmonic Wavelet Reconstruction

Let $f(x) \in L_2(\mathbb{R})$, its reconstruction in terms of harmonic wavelets can be obtained by the formula (see e.g. [4, 7, 8, 9, 10, 19])

$$f(x) = \left[\sum_{k=-\infty}^{\infty} \alpha_k \varphi_k^0(x) + \sum_{n=0}^{\infty}\sum_{k=-\infty}^{\infty} \beta_k^n \psi_k^n(x) \right]$$
$$+ \left[\sum_{k=-\infty}^{\infty} \alpha_k^* \overline{\varphi}_k^0(x) + \sum_{n=0}^{\infty}\sum_{k=-\infty}^{\infty} \beta_k^{*n} \overline{\psi}_k^n(x) \right] \tag{17}$$

which involves (8) and the corresponding conjugate functions. The series at the r.h.s. represents the projection of $f(x)$ into the ∞-dimensional subspace of $L_2(\mathbb{R})$, done by the operator $\mathcal{W}_\infty^\infty$; so that formally we can write

$$f(x) = \mathcal{W}_\infty^\infty[f(x)]$$

and since $f(x) \in L_2(\mathbb{R})$ there follows its convergence.

Thanks to the orthonormality of the basis, the wavelet coefficients can be computed by

$$
\begin{cases}
\alpha_k = \langle f(x), \varphi_k^0(x) \rangle = \displaystyle\int_{-\infty}^{\infty} f(x) \overline{\varphi}_k^0(x)\, dx \\[2ex]
\alpha_k^* = \langle f(x), \overline{\varphi}_k^0(x) \rangle = \displaystyle\int_{-\infty}^{\infty} f(x) \varphi_k^0(x)\, dx \\[2ex]
\beta_k^n = \langle f(x), \psi_k^n(x) \rangle = \displaystyle\int_{-\infty}^{\infty} f(x) \overline{\psi}_k^n(x)\, dx \\[2ex]
\beta_k^{*n} = \left\langle f(x), \overline{\psi}_k^n(x) \right\rangle = \displaystyle\int_{-\infty}^{\infty} f(x) \psi_k^n(x)\, dx
\end{cases}
\tag{18}
$$

or, according to (14), with the equivalent corresponding equations in the Fourier domain [4, 7, 8, 9, 10],

$$
\begin{cases}
\alpha_k = 2\pi\langle \widehat{f(x)}, \widehat{\varphi_k^0(x)} \rangle = \displaystyle\int_{-\infty}^{\infty} \widehat{f}(\omega)\overline{\widehat{\varphi_k^0}(\omega)}\, d\omega = \displaystyle\int_0^{2\pi} \widehat{f}(\omega) e^{i\omega k}\, d\omega \\[2ex]
\alpha_k^* = 2\pi\langle \widehat{f(x)}, \widehat{\overline{\varphi}_k^0(x)} \rangle = \ldots = \displaystyle\int_0^{2\pi} \widehat{f}(\omega) e^{-i\omega k}\, d\omega \\[2ex]
\beta_k^n = 2\pi\langle \widehat{f(x)}, \widehat{\psi_k^n(x)} \rangle = \ldots = 2^{-n/2}\displaystyle\int_{2^{n+1}\pi}^{2^{n+2}\pi} \widehat{f}(\omega) e^{i\omega k/2^n}\, d\omega \\[2ex]
\beta_k^{*n} = \langle \widehat{f(x)}, \widehat{\overline{\psi}_k^n(x)} \rangle = \ldots = 2^{-n/2}\displaystyle\int_{2^{n+1}\pi}^{2^{n+2}\pi} \widehat{f}(\omega) e^{-i\omega k/2^n}\, d\omega \ ,
\end{cases}
\tag{19}
$$

being $\widehat{f(x)} = \overline{\widehat{f}(-\omega)}$.

The wavelet approximation, at the scale N, is obtained by fixing an upper limit in the series expansion (17), so that $N < \infty$, $M < \infty$. There follows that the function $f(x)$ is approximated by the projection $W_M^N[f(x)]$ defined as

$$
f(x) \cong W_M^N[f(x)] \overset{def}{=} \left[\sum_{k=0}^{M} \alpha_k \varphi_k^0(x) + \sum_{n=0}^{N} \sum_{k=-M}^{M} \beta_k^n \psi_k^n(x) \right]
$$
$$
+ \left[\sum_{k=0}^{M} \alpha_k^* \overline{\varphi}_k^0(x) + \sum_{n=0}^{N} \sum_{k=-M}^{M} \beta_k^{*n} \overline{\psi}_k^n(x) \right] .
\tag{20}
$$

As a consequence of (15),(17),(20), we have the following

Corollary 1. *The harmonic wavelet projection of the Fourier basis* $\left\{e^{2\pi mix}\right\}$, *with* $m \in \mathbb{N}$, *is identically zero:*

$$W_M^N\left[e^{2\pi mix}\right] = 0 \quad , \qquad \forall M, N \leq \infty . \tag{21}$$

5   Scale Invariance and Wavelet Coefficients

Let us compute the wavelet coefficients (16) for a self-similar function (3). It can be shown that (see also [22])

Theorem 3. *The scaling coefficients of a self-similar function (3) are not independent on the lower scale* $(n = 0)$ *scaling coefficients since it is*

$$\alpha_k^C = 2^{-C(H+1/2)}\alpha_k \quad , \qquad \forall k \in \mathbb{Z} \tag{22}$$

with C *defined by (2).*

Proof: From (16), $\alpha_k = \int\limits_{-\infty}^{\infty} f(x)\,\overline{\varphi}_k^0(x)\,dx$, by the substitution $x = 2^C X$ it is

$$\alpha_k = 2^C \int\limits_{-\infty}^{\infty} f\left(2^C X\right) \overline{\varphi}_k^0\left(2^C X\right) dX$$

and, by taking into account the definition (8),

$$\overline{\varphi}_k^0\left(2^C X\right) = 2^{-C/2}\overline{\varphi}_k^C(X) .$$

Thus, we have

$$\alpha_k = 2^C \int\limits_{-\infty}^{\infty} f\left(2^C X\right) 2^{-C/2}\overline{\varphi}_k^C(X)\,dX \overset{(AA)}{=} 2^{C/2} \int\limits_{-\infty}^{\infty} 2^{CH} f(X)\overline{\varphi}_k^C(X)\,dX$$

$$= 2^{C(H+1/2)} \int\limits_{-\infty}^{\infty} f(X)\,\overline{\varphi}_k^C(X)\,dX$$

that is

$$\alpha_k = 2^{C(H+1/2)}\alpha_k^C \tag{23}$$

so that (22) follows.   ∎

Conditions (22),(23) can be also used to define a self-similar function, as well as the following:

Corollary 2. *A localized function is self-similar if there exists an integer* n *which is solution of the equation*

$$n = \frac{1}{H+1/2} \log_2 \left| \frac{\alpha_k}{\alpha_k^n} \right| . \tag{24}$$

Equation (24) shows a relation between the values of the scaling coefficients and the scale parameter n. As can been seen by a direct computation, a simple solution of (24) is

$$\alpha_k^n = a^{-(H+1/2)n} g(k) \quad , \quad a > 1 .$$

Analogously, for the detail coefficients, it can be shown from (16) that

Theorem 4. *The wavelet coefficients of a localized self-similar function are not independent on the lower scale* $(n = 0)$ *wavelet coefficients, since it is*

$$\beta_k^{n+C} = 2^{-C(H+1/2)} \beta_k^n \quad , \quad \forall n \in \mathbb{N}, \ k \in \mathbb{Z} \tag{25}$$

with C *defined by* (2).

Proof: From $\beta_k^n = \displaystyle\int_{-\infty}^{\infty} f(x) \overline{\psi}_k^n(x)\, dx$, by the substitution $x = 2^C X$, it is

$$\beta_k^n = 2^C \int_{-\infty}^{\infty} f(2^C X) \overline{\psi}_k^n (2^C X)\, dX$$

and, by taking into account (8),

$$\overline{\psi}_k^n (2^C X) = 2^{-C/2} \overline{\psi}_k^{n+C} (X) ,$$

there follows

$$\beta_k^n = 2^C \int_{-\infty}^{\infty} f(2^C X)\, 2^{-C/2} \overline{\psi}_k^{n+C} (X)\, dX \overset{(AA)}{=} 2^{C/2} \int_{-\infty}^{\infty} 2^{CH} f(X) \overline{\psi}_k^{n+C} (X)\, dX$$

$$= 2^{C(H+1/2)} \int_{-\infty}^{\infty} f(X) \overline{\psi}_k^{n+C} (X)\, dX ,$$

so that

$$\beta_k^n = 2^{C(H+1/2)} \beta_k^{n+C} . \qquad \blacksquare$$

In other words for a self-similar function, characterized by a factor C, there are only ∞^{C+1} independent scaling-wavelet coefficients:

$$\{\alpha_k\}_{k=-\infty}^{\infty}, \{\beta_k^0\}_{k=-\infty}^{\infty}, \{\beta_k^1\}_{k=-\infty}^{\infty}, ..., \{\beta_k^{C-1}\}_{k=-\infty}^{\infty} . \tag{26}$$

According to this, the description of self-similar functions requires a number of coefficients which is less than those needed for regular functions.

As a consequence of Eq. (25), we also have that

Theorem 5. *The wavelet coefficients of a self-similar function (3) are self-similar functions of the form:*

$$\beta_k^n = F(2^n, k) \tag{27}$$

with $F(2^n, k)$ self-similar function of the scale

$$F(\mu 2^n, k) = \mu^S F(2^n, k) \ , \tag{28}$$

being

$$\mu = 2^C \ , \ S = -\left(H + \frac{1}{2}\right) \quad ,$$

so that

$$F(2^{n+C}, k) = 2^{-C(H+1/2)} F(2^n, k) \ . \tag{29}$$

Proof: From equation (25) let us write $\beta_k^n = F(a^n, k)$ with the base a to be determined. It is

$$F(a^{n+C}, k) = F(a^C a^n, k)$$

and, according to (25),

$$F(a^C a^n, k) = 2^{-C(H+1/2)} F(a^n, k) \ .$$

Let us assume that $F(a^n, k)$ is a self-similar function so that

$$F(\lambda a^n, k) = \lambda^S F(a^n, k) \ .$$

Comparing with the previous equation it is

$$a^C = \lambda \ , \ \lambda^S = 2^{-C(H+1/2)}$$

from where

$$\lambda = 2^C \ , \ S = -(H + 1/2) \ . \qquad \blacksquare$$

For example, according to (29), we can take

$$F(2^n, k) = 2^{-n(H+1/2)} k^q, q \in \mathbb{R} \tag{30}$$

which fulfills (29). The wavelet coefficients are

$$\beta_k^n = 2^{-n(H+1/2)} k^q, q \in \mathbb{R} \tag{31}$$

and the equation (25) becomes

$$\beta_k^{n+C} = 2^{-(n+C)(H+1/2)} k^q, q \in \mathbb{R} \quad .$$

More in general we can write

$$F(2^n, k) = 2^{-n(H+1/2)} g(k)$$

and

$$\beta_k^n = 2^{-n(H+1/2)} g(k) \ .$$

6   Wavelet Coefficients of the Weierstrass Function

This section deals with the harmonic wavelet reconstruction of a slightly modified version of the Weierstrass function (4). In fact, in order to give a good representation of these kind of functions and to study their properties by using wavelets one should localized them in space. Therefore we consider the following Weierstrass function

$$f(x) = \sum_{k=-\infty}^{\infty} a^{-kH} \cos a^k x \quad , \quad (a > 1, 0 \leq H \leq 1) \tag{32}$$

with a further condition to be limited in space, i.e. with compact support, as should be any physical signal. This implies that the sum will be taken up to finite fixed values.

The Fourier transform of (32) is

$$\mathcal{F}[f(x)] = \sum_{k=-\infty}^{\infty} a^{-kH} \mathcal{F}[\cos a^k x] = \sum_{k=-\infty}^{\infty} a^{-kH} \frac{1}{2} \left[\delta\left(\omega - a^k\right) + \delta\left(\omega + a^k\right) \right] ,$$

so that according to (19) it is

$$\alpha_k = \int_0^{2\pi} \hat{f}(\omega) e^{-i\omega k} d\omega = \int_0^{2\pi} \sum_{h=-\infty}^{\infty} a^{-hH} \frac{1}{2} \left[\delta\left(\omega - a^h\right) + \delta\left(\omega + a^h\right) \right] e^{-i\omega k} d\omega$$

$$= \sum_{h=-\infty}^{\infty} a^{-hH} \frac{1}{2} \left(e^{ia^h k} + e^{-ia^h k} \right) = \sum_{h=-\infty}^{\infty} a^{-hH} \cos\left(ka^h\right) ,$$

that is

$$\alpha_k = \sum_{h=-\infty}^{\infty} a^{-hH} \cos\left(ka^h\right) \overset{(32)}{=} f(k) .$$

Analogously we get

$$\begin{cases} \alpha_k^* = -\sum_{h=-\infty}^{\infty} a^{-hH} \cos(ka^h) \overset{(32)}{=} - f(k) \\[2mm] \beta_k^n = 2^{-n/2} \sum_{h=-\infty}^{\infty} a^{-hH} \cos\left(\frac{k}{2^n} a^h\right) \overset{(32)}{=} f\left(\frac{k}{2^n}\right) \\[2mm] \beta^{*n}_k = -2^{-n/2} \sum_{h=-\infty}^{\infty} a^{-hH} \cos\left(\frac{k}{2^n} a^h\right) \overset{(32)}{=} -f\left(\frac{k}{2^n}\right) , \end{cases} \tag{33}$$

and the reconstruction formula (17) for the Weierstrass function (32) gives

$$
f(x) = \sum_{k=-\infty}^{\infty} f(k) \left[\varphi_k^0(x) + \overline{\varphi}_k^0(x) \right] + \sum_{n=0}^{\infty} \sum_{k=-\infty}^{\infty} f\left(\frac{k}{2^n}\right) \left[\psi_k^n(x) + \overline{\psi}_k^n(x) \right]
$$

$$
\overset{(11)}{=} \sum_{k=-\infty}^{\infty} f(k) \Phi_k^0(x) + \sum_{n=0}^{\infty} \sum_{k=-\infty}^{\infty} f\left(\frac{k}{2^n}\right) \Psi_k^n(x)
$$

$$
= \sum_{k=-\infty}^{\infty} \left\{ \sum_{h=-\infty}^{\infty} a^{-hH} \cos\left(ka^h\right) \right\} \Phi_k^0(x)
$$

$$
+ \sum_{n=0}^{\infty} \sum_{k=-\infty}^{\infty} \left\{ \sum_{h=-\infty}^{\infty} a^{-hH} \cos\left(\frac{k}{2^n} a^h\right) \right\} \Psi_k^n(x) \ .
$$

Thus for the Weierstrass function the scaling parameter n increases the frequency of oscillations of the wavelet coefficients.

Let us see, now, if the self similarity conditions can be fulfilled by the function (32), and then by the wavelet coefficients (22) that is let us show if there exists a factor C such that condition (25) holds.

Theorem 6. *The Weirstrass function (32) is self-similar, and fulfills (25), if*

$$
a^h = 2^{h+C} \ .
$$

Proof: From definition (22), it is

$$
\beta_k^{n+C} = 2^{-(n+C)/2} \sum_{h=-\infty}^{\infty} a^{-hH} \cos\left(\frac{k}{2^{n+C}} a^h\right)
$$

$$
= 2^{-C/2} \left[2^{-n/2} \sum_{h=-\infty}^{\infty} a^{-hH} \cos\left(\frac{k}{2^n} \frac{a^h}{2^C}\right) \right] \ .
$$

By assuming $a^h = 2^{h+C}$ we have

$$
\beta_k^{n+C} = 2^{-C/2} \left[2^{-n/2} \sum_{h=-\infty}^{\infty} 2^{-(h+C)H} \cos\left(\frac{k}{2^n} \frac{2^{h+C}}{2^C}\right) \right]
$$

$$
= 2^{-C(h+1/2)} \left[2^{-n/2} \sum_{h=-\infty}^{\infty} 2^{-hH} \cos\left(\frac{k}{2^n} 2^h\right) \right] = 2^{-C(h+1/2)} \beta_k^n \ ,
$$

with

$$
\beta_k^n = 2^{-n/2} \sum_{h=-\infty}^{\infty} 2^{-(h+C)H} \cos\left(\frac{k}{2^n} 2^{h+C}\right)
$$

$$
= 2^{-n/2} \sum_{h=-\infty}^{\infty} 2^{-(h+C)H} \cos\left(\frac{k}{2^n} 2^h\right) \ .
$$

Therefore the scale invariance is fulfilled for the Weierstrass series (32) when the series is taken in the form

$$f(x) = \sum_{k=-\infty}^{\infty} 2^{-kH} \cos 2^k x \quad , \ (0 \leq H \leq 1) \ . \tag{34}$$

■

7   Denoising Fractal Signals

As application of the above theory let us consider a few examples of wavelet projection of signals made by two components: a 2π–periodic function and a fractal function (Weierstrass–like (34)). We show that, given the signal, where the two components are mixed together, it is possible to separate the periodic function from the other by a suitable projection.

Let us consider, for example, the artificial signal (Fig. 1,a)

$$f(x) = \frac{1}{30} \cos 4\pi x + \frac{1}{20} \sum_{k=-5}^{5} 2^{-0.4k} \cos 2^k x \tag{35}$$

where the amplitude of the periodic function is small comparing with the localized fractal pulse, and assume that the r.h.s. is unknown, so that we know only that there is a 2π–periodic component . The unknown functions, the periodic and the fractal pulse, are drawn in Fig. 1,b,d, respectively. After projection by harmonic wavelets, taking into account Eq. (21), it is

$$W_{50}^{10} \left[\frac{1}{30} \cos 4\pi x \right] = 0 \ ,$$

and we get (Fig. 1,c)

$$W_{50}^{10}[f(x)] = W_{50}^{10} \left[\frac{1}{20} \sum_{k=-5}^{5} 2^{-0.4k} \cos 2^k x \right] \cong \frac{1}{20} \sum_{k=-5}^{5} 2^{-0.4k} \cos 2^k x \ , \tag{36}$$

so that the periodic function follows from the difference (Fig. 1, e)

$$\frac{1}{30} \cos 4\pi x \cong f(x) - W_{50}^{10}[f(x)] \ .$$

As a second example, let us take the artificial signal (Fig. 2,a)

$$f(x) = 5 \sin 2\pi x + \frac{1}{10} \sum_{k=-5}^{5} 2^{-0.4k} \cos 2^k x \tag{37}$$

where the fractal noise is neglectable with respect to the periodic function $5 \sin 2\pi x$, so that signal looks like a smooth function (Fig. 2,a). Let us assume

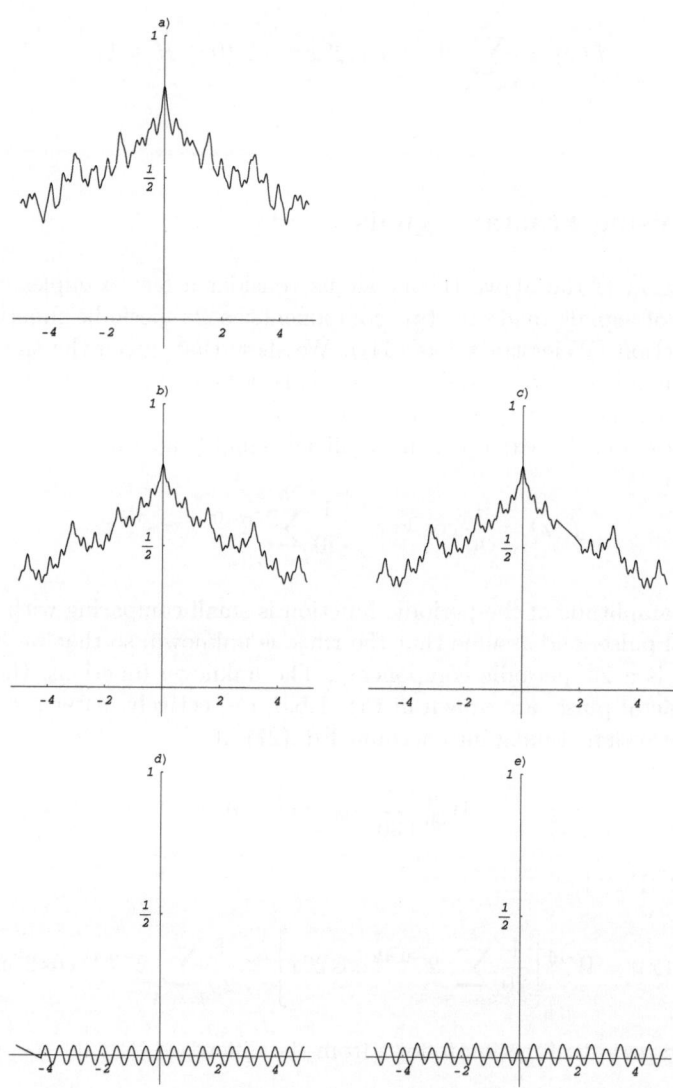

Fig. 1. Extraction of a periodic function from a noised signal (the amplitude of the periodic function is small): a) the given noised signal f, Eq. (35); b) the fractal component of the noised signal; c) the projection of the noised signal $W_{50}^{10}f$, Eq. (36), giving the approximation of the fractal component; d) the periodic component of the noised signal; e) the difference $f - W_{50}^{10}f$, giving the approximation of the periodic component

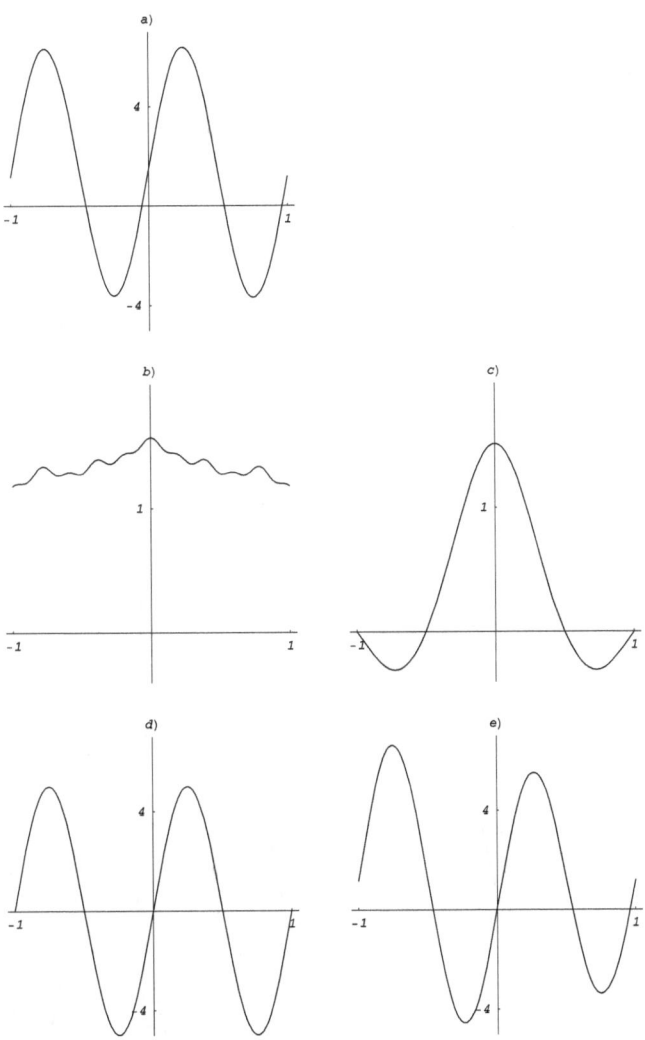

Fig. 2. Extraction of a fractal noise from a noised signal (the amplitude of the periodic function is large): a) the given noised signal f, Eq. (37); b) the fractal component of the noised signal; c) the projection of the noised signal $W_0^0 f$, giving a very rough approximation of the fractal component; d) the periodic component of the noised signal; e) the difference $f - W_0^0 f$, giving the coarse approximation of the periodic component

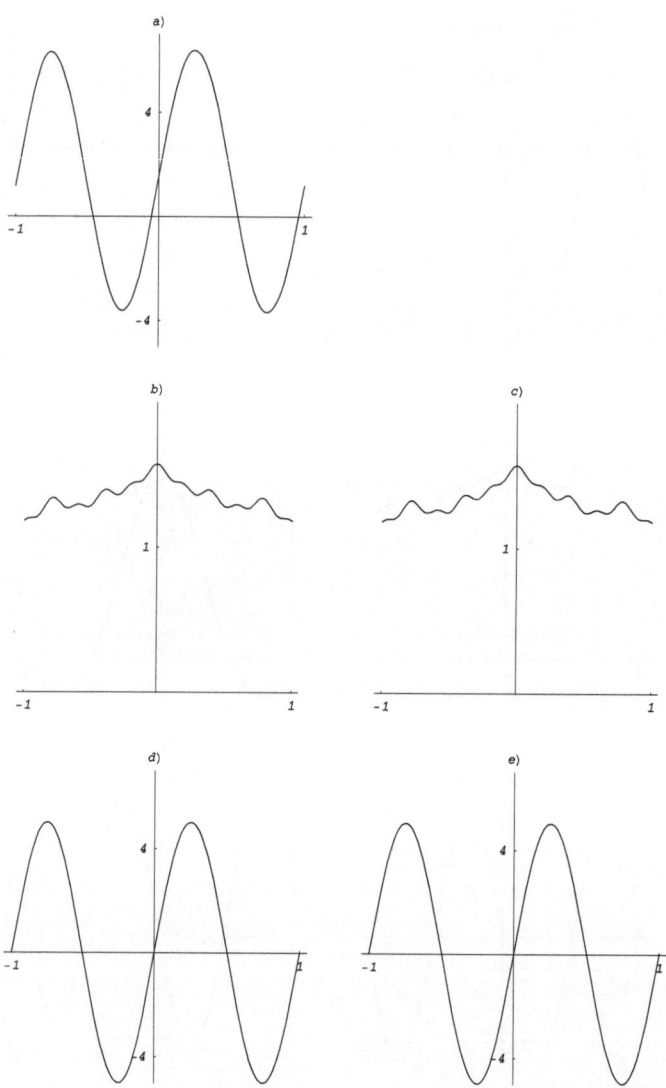

Fig. 3. Extraction of a fractal noise from a noised signal (the amplitude of the periodic function is large): a) the given noised signal f, Eq. (37); b) the fractal component of the noised signal; c) the projection of the noised signal $W_{50}^{10}f$, giving a good approximation of the fractal component; d) the periodic component of the noised signal; e) the difference $f - W_{50}^{10}f$, giving a good approximation of the periodic component

only that one component of the given signal (Fig. 2,a) is 2π–periodic. The unknown fractal and periodic components are drawn in Fig. 2,b,d, respectively. After projection by harmonic wavelets, taking into account Eq. (21), it is at the coarse level

$$W_0^0[5\sin 2\pi x] = 0 \ ,$$

and we get (Fig. 2,c)

$$W_0^0[f(x)] = W_0^0\left[\frac{1}{10}\sum_{k=-5}^{5} 2^{-0.4k}\cos 2^k x\right] \cong \frac{1}{10}\sum_{k=-5}^{5} 2^{-0.4k}\cos 2^k x \ .$$

The scale of approximation is very low and, as a consequence, the resulting periodic function from the difference (Fig. 2,e)

$$5\sin 2\pi x \cong f(x) - W_0^0[f(x)]$$

shows a quite large error of approximation. If we increase instead the scale of approximation (Fig. 3) it is still $W_{50}^{10}[5\sin 2\pi x] = 0$. We get (Fig. 3,c)

$$W_{50}^{10}[f(x)] = W_{50}^{10}\left[\frac{1}{10}\sum_{k=-5}^{5} 2^{-0.4k}\cos 2^k x\right] \cong \frac{1}{10}\sum_{k=-5}^{5} 2^{-0.4k}\cos 2^k x$$

with a better approximation. The periodic function follows from the difference (Fig. 3,e)

$$5\sin 2\pi x \cong f(x) - W_{50}^{10}[f(x)] \ .$$

8   Conclusion

In this paper harmonic wavelets were applied to the reconstruction of localized self-similar functions. The fundamental property of these wavelets of being orthonormal to the Fourier basis was used for the decomposition of a noised signal into a 2π-periodic and a localized fractal component. This method could be easily generalized to different periodicity components by a suitable definition of the harmonic wavelets (see also [9]).

References

1. Abry, P., Goncalves, P., Lévy-Véhel, J.: Lois d'éschelle, Fractales et ondelettes, Hermes (2002)
2. Borgnat, P., Flandrin, P.: On the chirp decomposition of Weierstrass-Mandelbrot functions, and their time-frequency interpretation. Applied and Computational Harmonic Analysis 15, 134–146 (2003)
3. Bunde, A., Havlin, S. (eds.): Fractals in Science. Springer, Berlin (1995)
4. Cattani, C.: Harmonic Wavelets towards Solution of Nonlinear PDE. Computers and Mathematics with Applications 50(8-9), 1191–1210 (2005)

5. Cattani, C.: Connection Coefficients of Shannon Wavelets. Mathematical Modelling and Analysis 11, 1–16 (2006)
6. Cattani, C., Rushchitsky, J.J.: Wavelet and Wave Analysis as applied to Materials with Micro or Nanostructure. Series on Advances in Mathematics for Applied Sciences, p. 74. World Scientific, Singapore (2007)
7. Cattani, C.: Wavelet extraction of a pulse from a periodic signal. In: Gervasi, O., Murgante, B., Laganà, A., Taniar, D., Mun, Y., Gavrilova, M.L. (eds.) ICCSA 2008, Part I. LNCS, vol. 5072, pp. 1202–1211. Springer, Heidelberg (2008)
8. Cattani, C.: Shannon Wavelets Theory. Mathematical Problems in Engineering 2008, Article ID 164808, 24 pages (2008)
9. Cattani, C.: Harmonic Wavelet Approximation of Random, Fractal and High Frequency Signals. To appear on Telecommunication Systems (2009)
10. Cattani, C.: Harmonic Wavelet Analysis of a Localized Fractal. International Journal of Engineering and Interdisciplinary Mathematics (to appear, 2009)
11. Chui, C.K.: An Introduction to Wavelets. Academic Press, New York (1992)
12. Daubechies, I.: Ten Lectures on wavelets. SIAM, Philadelphia (1992)
13. Dovgoshey, O., Martio, O., Ryazanov, V., Vuorinen, M.: The Cantor Function. Expositiones Mathematicae 24, 1–37 (2006)
14. Dutkay, D.E., Jorgensen, P.E.T.: Wavelets on Fractals. Rev. Mat. Iberoamericana 22(1), 131–180 (2006)
15. Falconer, K.: Fractal Geometry. John Wiley, New York (1977)
16. Feder, J.: Fractals. Pergamon, New York (1988)
17. Mallat, S.: A Wavelet tour of signal processing. Academic Press, London (1998)
18. Muniandy, S.V., Moroz, I.M.: Galerkin modelling of the Burgers equation using harmonic wavelets. Phys.Lett. A 235, 352–356 (1997)
19. Newland, D.E.: Harmonic wavelet analysis. Proc. R. Soc. Lond. A 443, 203–222 (1993)
20. Vicsek, T.: Fractal Growth Phenomenon. Word Scientific, Singapore (1989)
21. Weierstrass, K.: Über continuirliche Functionen eines reelles Arguments, die für keinen Werth des letzteren einen Bestimmten Differentialquotienten besitzen, König, Akad. der Wissenschaften, Berlin, July 18 (1872); Reprinted in: K. Weierstrass, Mathematische Werke II, pp. 71–74. Johnson, New York (1967)
22. Wornell, G.: Signal Processing with Fractals: A Wavelet-Based Approach. Prentice Hall, Englewood Cliffs (1996)

Wavelet Analysis of Spike Train
in the Fitzhugh Model

Carlo Cattani[1] and Massimo Scalia[2]

[1] diFarma, University of Salerno, Via Ponte Don Melillo, I-84084 Fisciano (SA)
ccattani@unisa.it
[2] Dept. of Mathematics, "G. Castelnuovo", University of Rome, "La Sapienza",
P.le A. Moro 2, I-00185 Roma
massimo.scalia@uniroma1.it

Abstract. This paper deals with the analysis of the wavelet coefficients for the nonlinear dynamical system which models the axons activity. A system with source made by a sequence of high pulses (spike train) is analyzed in dependence of the amplitude. The critical value of the amplitude, and a catastrophe are analyzed. The wavelet coefficients are computed and it is shown also that they are very sensitive to the local changes and can easily detect the spikes even on a nearly smooth function.

Keywords: Haar wavelets, Short Haar Wavelet Transform, Phase space, Catastrophe.

1 Introduction

This paper deals with the classical model of one excitatory neuron [1, 2, 8, 9, 10, 11, 12, 13, 14, 15, 19], with excitatory (inhibitory) synapses, in the simplified model given by Fitzhugh [1]. We will analyze the process of a spike train generation and the evolution in time will be described as a function of one parameter characterizing the dynamical system. The parametric dependence implies, at the initial time of neuron firing, either some stable (periodic) evolutions or unstable, both characterized by a bifurcation and critical time. However, in the numerical simulation, the critical point and the bifurcation as well, might be hidden by the evolution according to the amplitude of parameter. In order to easily single out the bifurcation a wavelet approach has been used. In fact, due to their localization wavelet coefficients can amplify local jumps and discontinuities. It should be noticed that there are different models of neurons (see e.g. [8] and references therein) such as the Integrate-and-fire, FitzHugh-Nagumo, Morris-Lecar and other which can be discussed and simulated, and the more general Hodgkin-Huxley model [12, 13, 14].

A simplified model for the neural activity, was proposed, in the early 60ties, by Fitzhugh [1] as a generalization of the Van der Pol equation thus showing the existence of a limit cycle and some periodicity. In this model, if we consider a simple system consisting of a neuron and a synapse, it is known that the

M.L. Gavrilova and C.J.K. Tan (Eds.): Trans. on Comput. Sci. VI, LNCS 5730, pp. 163–179, 2009.
© Springer-Verlag Berlin Heidelberg 2009

activity of stimulated axons can be detected by an abrupt change in the electrical potential. These pulse in a short time are called spikes or axons firing. The normal activity of neurons is usually characterized by a continuous process of axons firing, some of them can be related to some external action some of them (apparently) do not any external source. Thus being classified as the normal activity of axons. Therefore one of the main problems is to recognize among all generated spikes those which are caused by some external stimulations. Another problem follows from the pulse action of the firing process. In his system Fitzhugh suggest to use a kind of delta Dirac function. In other words, any changes in the electrical potential are localized in a short time interval and within this interval have a significant amplitude. It will be shown that the catastrophe (i.e. absence of uniqueness in the solution) cannot be detected by the phase portrait, in the sense that path in the phase plain is a continuous path with some small discontinuty in the tangent vectors. Instead by using the wavelet approach the loss of uniqueness can be single out from the high values of the wavelet coefficients.

In Sect. 2 some preliminary remarks about neuron models are given. Section 3 deals with the definitions about Haar wavelets and short Haar wavelet transform [3,4,5,6] are given. The qualitative and wavelet analysis of Fitzhugh system is given in section 4.

2   Preliminary Remarks on the General Model

The most successful and general model in neuroscience describes the neuron activity in terms of two conductances : G_{Na} sodium and G_K potassium and the activating gating particles $m(t)$, the inactivating particle for sodium $h(t)$ and the gating particles for potassium $n(t)$.

The general model HH (of Hodgkin-Huxley) [12,13,14] considers the neuron activity as an electrical circuit. Cells membrane store charges, electrochemical forces arise because of the imbalanced ion concentration in- and outside the cell. In the steady model, for the total membrane current, the capacitive current is linked with the ionic current:

$$C_m \frac{dV(t)}{dt} = I_{ionic}(t) \tag{1}$$

where the ionic current is based on three ionic currents: sodium, potassium (membrane dependent) and a leak conductance (independent of the membrane potential). The total ionic current is:

$$I_{ionic} = I_{Na} + I_K + I_{leak}. \tag{2}$$

Each ionic current depends on the membrane potential and it is expressed by a driving potential $V(t)$ and the reversal potential is given by the Nernst equation of the ionic species E_k. Neurons are connected by synapses and dendrites, and the number of links is constant in time. There are two kind of synapses for each neuron: excitatory and inhibitory synapses. Synaptic interactions are conductance-based, for neuron i we have:

$$I_{syn}(t) = \sum_j g_{ij}(t)(V_i - E_j), \tag{3}$$

where V_i is the membrane potential of neuron i, $g_{ji}(t)$ is the synaptic conductance of the synapse connecting neuron j to neuron i, and E_j is the reversal potential of that synapse. When a spike occurs in neuron j, the synaptic conductance g_{ji} is instantaneously incremented by a quantum value ($q_e = 6$ nS for excitatory and $q_i = 67$ nS for inhibitory synapses). Thus we have,

$$I_{ionic} = G_K(E_K - V) + G_{Na}(E_{Na} - V) + G_{leak}(E_{leak} - V) . \tag{4}$$

Individual currents can be either positive (outward) or negative (inward) depending on the membrane potential and the ion species reversal potential.

In the HH model there is a gating particle n function which is the probability that the gate for potassium conductance is active and two gates for sodium (m being the activation and h the inverse process). Functions G_* might have a complicated dependence on the gating values:

$$G_K = G_K(n\ , m\ , h)\ ,\ \ G_{Na} = G_{Na}(n\ , m\ , h)\ ,\ \ G_{leak} = G_{leak}(n\ , m\ , h) . \tag{5}$$

A fourth order model is the HH model which assumes:

$$G_K = \overline{G}_K n^4\ ,\ G_{Na} = \overline{G}_{Na} m^3 h\ , \tag{6}$$

so that

$$C_m \frac{dV(t)}{dt} = \overline{G}_K n^4(E_K - V) + \overline{G}_{Na} m^3 h(E_{Na} - V) + \overline{G}_{leak}(E_{leak} - V) . \tag{7}$$

The gating particle probabilities are nonlinearly depending on the potential so that the dynamics of neurons is described through a first-order nonlinear differential equations featuring two voltage time dependent terms (steady state activation and inactivation):

$$\begin{cases} \dfrac{dV}{dt} = \dfrac{1}{C_m}\Big\{\overline{G}_K n^4(E_K - V) + \overline{G}_{Na} m^3 h(E_{Na} - V)) \\[2mm] \qquad + \overline{G}_{leak}(E_{leak} - V) - I_{ext}(t)\Big\}, \\[4mm] \dfrac{dn}{dt} = \alpha_n(V)(1 - n) - \beta_n(V)n, \\[4mm] \dfrac{dm}{dt} = \alpha_m(V)(1 - m) - \beta_m(V)m, \\[4mm] \dfrac{dh}{dt} = \alpha_h(V)(1 - h) - \beta_h(V)h. \end{cases} \tag{8}$$

where

$$
\begin{cases}
\alpha_n(V) = \phi \cdot \dfrac{a_n(V - V_{\alpha_n})}{1 - e^{-(V-V_{\alpha_n})/K_{\alpha_n}}} & , \quad \beta_n(V) = \phi \cdot b_n e^{-(V-V_{\beta_n})/K_{\beta_n}} \\[3mm]
\alpha_m(V) = \phi \cdot \dfrac{a_m(V - V_{\alpha_m})}{1 - e^{-(V-V_{\alpha_m})/K_{\alpha_m}}} & , \quad \beta_m(V) = \phi \cdot b_m e^{-(V-V_{\beta_m})/K_{\beta_m}} \\[3mm]
\alpha_h(V) = \phi \cdot a_h e^{-(V-V_{\alpha_h})/K_{\alpha_h}} & , \quad \beta_h(V) = \phi \cdot \dfrac{b_h}{1 + e^{-(V-V_{\beta_h})/K_{\beta_h}}}
\end{cases}
$$

and, by assuming

$$ V(t) = u, \ C_m = 1, \ \mu F cm^{-2}, \ \overline{G}_{Na} = a_3, \ \overline{G}_K = a_1, \ \overline{G}_{leak} = a_5, \ E_{Na} = a_4 \ , $$

$$ E_K = a_2, \ E_{leak} = a_6 \ , a_n = a_7, \ V_{\alpha n} = a_8, \ K_{\alpha n} = a_9, \ b_n = a_{10}, \ V_{\beta n} = a_{11} \ , $$

$$ K_{\beta n} = a_{12}, \ a_m = a_{13}, \ V_{\alpha m} = a_{14}, \ K_{\alpha m} = a_{15}, \ b_m = a_{16}, \ V_{\beta m} = a_{17} \ , $$

$$ K_{\beta m} = a_{18}, \ a_h = a_{19}, \ V_{\alpha h} = a_{20}, \ K_{\alpha h} = a_{21}, \ b_h = a_{22}, \ V_{\beta h} = a_{23}, \ K_{\beta h} = a_{24} \ , $$

it can be written, in adimensional form, as

$$
\begin{cases}
\dfrac{du}{dt} = a_1 n^4 (a_2 - u) + a_3 m^3 h(a_4 - u) + a_5(a_6 - u) - I_{ext}(t) \ , \\[3mm]
\dfrac{dn}{dt} = \alpha_n(u)(1 - n) - \beta_n(u)n \ , \\[3mm]
\dfrac{dm}{dt} = \alpha_m(u)(1 - m) - \beta_m(u)m \ , \\[3mm]
\dfrac{dh}{dt} = \alpha_h(u)(1 - h) - \beta_h(u)h \ ,
\end{cases}
\tag{9}
$$

being

$$
\begin{cases}
\alpha_n(u) = \phi \cdot \dfrac{a_7(u - a_8)}{1 - e^{-(u-a_8)/a_9}} & , \quad \beta_n(u) = \phi \cdot a_{10} e^{-(u-a_{11})/a_{12}} \ , \\[3mm]
\alpha_m(u) = \phi \cdot \dfrac{a_{13}(u - a_{14})}{1 - e^{-(u-a_{14})/a_{15}}} & , \quad \beta_m(u) = \phi \cdot a_{16} e^{-(u-a_{17})/a_{18}} \ , \\[3mm]
\alpha_h(u) = \phi \cdot a_{19} e^{-(u-a_{20})/a_{21}} & , \quad \beta_h(u) = \phi \cdot \dfrac{a_{22}}{1 + e^{-(u-a_{23})/a_{24}}} \ .
\end{cases}
$$

2.1   Fitzhugh Model

Assuming only one gating activation, and a third order dependence the previous system can be simplified into the well known Fitzhugh model (or Fitzhugh-Nagumo) [1,2] $V \to x, \ n \to y \ , I_{ext} \to z$

$$\begin{cases} \dfrac{dx}{dt} = \dfrac{1}{\varepsilon}[x(1-x^2) - y] + \gamma z, \\[4mm] \dfrac{dy}{dt} = x - \beta y, \end{cases} \tag{10}$$

where ε is a small parameter $0 < \varepsilon \ll 1$, β is a fundamental parameter and $z = z(t)$ is a pulse function defined in a very short range interval (pulse function), having as limit case the delta Dirac. A single pulse source as a box function as been already considered in [7] in the following we will consider a spikes train. Parameter γ defines the amplitude of the pulse function.

This dynamical system depends [1, 2, 17, 18] on ε, β, γ in the sense that the evolution would be completely different, starting from some critical values. In the following we will consider the dependence on γ and show that the system undergoes to a catastrophe over a certain critical value of γ.

3   Short Haar Wavelet Transform

The *Haar scaling function* $\varphi(t)$ is the characteristic function on $[0, 1]$. By translation and dilation we have the family of functions defined (in $[0, 1]$)

$$\begin{cases} \varphi_k^n(t) \equiv 2^{n/2}\varphi(2^n t - k), & (0 \le n, \ 0 \le k \le 2^n - 1), \\[3mm] \varphi(2^n t - k) = \begin{cases} 1, t \in \Omega_k^n, \\ 0, t \notin \Omega_k^n. \end{cases} & \Omega_k^n \equiv \left[\dfrac{k}{2^n}, \dfrac{k+1}{2^n} \right), \end{cases} \tag{11}$$

The *Haar wavelet* family $\{\psi_k^n(t)\}$ is the orthonormal basis for the $L^2([0,1])$

functions:

$$\begin{cases} \psi_k^n(t) \equiv 2^{n/2}\psi(2^n t - k), & \|\psi_k^n(t)\|_{L^2} = 1, \\[3mm] \psi(2^n t - k) \equiv \begin{cases} -1, & t \in \left[\dfrac{k}{2^n}, \dfrac{k+1/2}{2^n} \right), \\[2mm] 1, & t \in \left[\dfrac{k+1/2}{2^n}, \dfrac{k+1}{2^n} \right), \\[2mm] 0, & \text{elsewhere}, \end{cases} & (0 \le n, \ 0 \le k \le 2^n - 1), \end{cases} \tag{12}$$

with $0 \le n$, $0 \le k \le 2^n - 1 \Longrightarrow \Omega_k^n \subseteq [0, 1]$. Let $\boldsymbol{Y} \equiv \{Y_i\}$, $(i = 0, \dots, 2^M - 1$, $2^M = N < \infty$, $M \in \mathbb{N})$, be a finite energy time-series and $t_i = i/(2^M - 1)$, the regular equispaced grid of *dyadic points*. Let the set $\boldsymbol{Y} = \{Y_i\}$ of N data be segmented into σ segments (in general) of different length. Each segment $\boldsymbol{Y}^s$, $s = 0, \dots, \sigma - 1$ is made of $p_s = 2^{m_s}$, $(\sum_s p_s = N)$, data:

$$\boldsymbol{Y} = \{Y_i\}_{i=0,\ldots,N-1} = \bigoplus_{s=0}^{\sigma-1}\{\boldsymbol{Y}^s\}\,, \qquad \boldsymbol{Y}^s \equiv \{Y_{sp_s},\, Y_{sp_s+1},\ldots,\, Y_{sp_s+p_s-1}\}\,,$$

being, in general, $p_s \neq p_r$. The short discrete Haar wavelet transform of $\boldsymbol{Y}$ is (see [3]) $\mathcal{W}^{p_s,\sigma}\boldsymbol{Y}$,

$$\begin{cases} \mathcal{W}^{p_s,\sigma} \equiv \displaystyle\bigoplus_{s=0}^{\sigma-1} \mathcal{W}_s^p\,, \quad \boldsymbol{Y} = \bigoplus_{s=0}^{\sigma-1} \boldsymbol{Y}^s\,, \\[3mm] \mathcal{W}^{p_s,\sigma}\boldsymbol{Y} \quad = \left(\displaystyle\bigoplus_{s=0}^{\sigma-1} \mathcal{W}^{p_s}\right)\boldsymbol{Y} = \left(\displaystyle\bigoplus_{s=0}^{\sigma-1} \mathcal{W}^{p_s}\boldsymbol{Y}^s\right)\,, \\[3mm] \mathcal{W}^{2^{m_s}}\boldsymbol{Y}^s \quad = \left\{\alpha_0^{0(s)},\, \beta_0^{0(s)},\, \beta_0^{1(s)},\, \beta_1^{1(s)},\ldots,\, \beta_{2^{m_s-1}-1}^{m_s-1(s)}\right\}\,. \end{cases}$$

with $2^{m_s} = p_s$, $\displaystyle\sum_{s=0}^{\sigma-1} p_s = N$. The *discrete Haar wavelet transform* is the operator $\mathcal{W}^N$ which maps the vector $\boldsymbol{Y}$ into the vector of the *wavelet coefficients* $\{\alpha,\, \beta_k^n\}$:

$$\mathcal{W}^N\boldsymbol{Y} = \{\alpha, \beta_0^0,\ldots,\beta_{2^{M-1}-1}^{M-1}\}\,, \qquad \boldsymbol{Y} = \{Y_0,\, Y_1,\, \ldots,\, Y_{N-1}\}\,. \tag{13}$$

The $N \times N$ matrix $\mathcal{W}^N$ can be computed by the recursive formula [4,5]

$$\mathcal{W}^N\boldsymbol{Y} \equiv \left[\prod_{k=1}^{M}\left((P_{2^k} \oplus I_{2^M-2^k})(H_{2^k} \oplus I_{2^M-2^k})\right)\right]\boldsymbol{Y}\,, \tag{14}$$

(being $\oplus$ the direct sum), which is based on the k-order identity matrix I_k, on the k-order permutation (shuffle) matrix P_k, which moves the odd (place) components of a vector $\boldsymbol{Y}$ into the first half positions and the even (place) components into the second half, and on the *lattice coefficients* k-order matrix H_k, which follows from a matrix factorization of the recursive equation coefficients:

$$H_2 = \begin{pmatrix} 1/\sqrt{2} & 1/\sqrt{2} \\ -1/\sqrt{2} & 1/\sqrt{2} \end{pmatrix}\,, \quad H_4 = H_2 \oplus H_2\,,\ldots\,.$$

For example, with $N = 4$, $M = 2$, assuming the empty set $I_0 \equiv \emptyset$ as the neutral term for the direct sum, it is

$$\begin{aligned} \mathcal{W}^4 &= \prod_{k=1,2} [(P_{2^k} \oplus I_{4-2^k})(H_{2^k} \oplus I_{4-2^k})] \\ &= [(P_2 \oplus I_2)(H_2 \oplus I_2)]_{k=1}\,[(P_4 \oplus I_0)(H_4 \oplus I_0)]_{k=2} \\ &= [(P_2 \oplus I_2)(H_2 \oplus I_2)]\,[P_4 H_4]\,. \end{aligned}$$

The wavelet coefficients of the discrete Haar transform have a simple interpretation in terms of finite differences. If we define the mean average value

$$\overline{Y}_{i,i+s} \equiv (s+1)^{-1} \sum_{k=i}^{i+s} Y_k$$

there follows, with easy computation [4], that

$$\begin{cases} \alpha & = \overline{Y}_{0,2^M-1} \,, \\ \beta_k^r & = 2^{(M-2-r)/2}\delta_{(M-1-r)h}\overline{Y}_{k+2^{M-2-r+2k},k+2^{M-2-r+2k}+M-r-2+2^{M-r-2}} \,, \\ \beta_k^{M-1} & = 2^{-1/2}\Delta_h Y_{2k} \,, \end{cases}$$

where $r = 0, \ldots, M-2$, $k = 0, \ldots, 2^{M-1}-1$, $h = 2^M$ and the forward and central (finite) difference formulas, as usual, are

$$\Delta_h Y_i \equiv (Y_{i+hN} - Y_i) \,, \qquad \delta_h Y_i \equiv (Y_{i+hN} - Y_{i-hN}) \,,$$

respectively. Therefore the wavelet coefficients β's, also called details coefficients, express (at least in the Haar wavelet approach) the finite differences, i.e. the first order approximate derivative. Thus being the more expedient tool for studying discontinuities, jumps and local behaviors like in the spikes firing.

4   Fitzhugh System with Solitary Spike

Let us study the nonlinear model (10) in the case where it depends on the single pulse (Fig. 1)

$$z(t) = \psi_0^4(t) = 2^2\psi(2^4 t) = 4 \times \begin{cases} -1 \,, t \in \left[0, \dfrac{1}{2^5}\right) \,, \\ 1 \,\,\,, t \in \left[\dfrac{1}{2^5}, \dfrac{1}{2^4}\right) \,, \\ 0 \,\,\,, t \notin \left[0, \dfrac{1}{2^4}\right) \,, \end{cases}$$

that is the jump function in the interval $[0, 2^{-4}]$. A pulse box-like has been studied in [7].

For the values of parameters we take $\varepsilon = 0.1$, $\beta = 1.1$ and as initial conditions it is assumed $x(0) = 0$, $y(0) = 0.2$.

The dynamics of equation (10) has been studied by analyzing the short wavelet transform of the time series obtained by a numerical computation of the solution of (10). By using the Runge-Kutta 4-th order method, with the accuracy 10^{-6}, we obtain in the interval $(0 < t \leq 8)$, four numerical solutions in correspondence with 4 values of the parameter: $\gamma = 5$, $\gamma = 31.86$, $\gamma = 31.87$, $\gamma = -15.7$. These sequences are discretized in $2^9 = 512$ time spots in order to obtain 512 values $\boldsymbol{Y} = \{Y_0, Y_1, \ldots, Y_{N-1}\}$, with $N = 512$ and $M = 9$. By using the short Haar wavelet transform, with $p = 4$, we have compared the wavelet coefficients of the

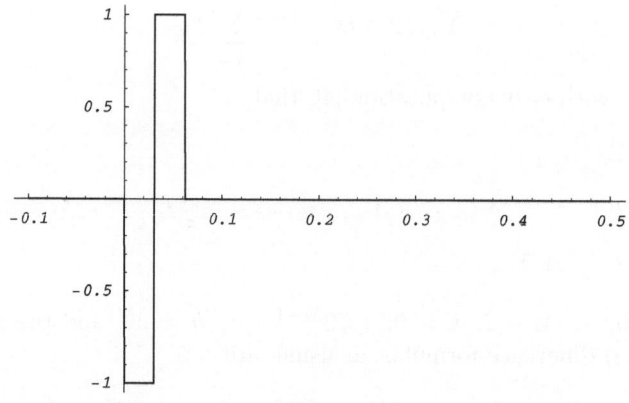

Fig. 1. Single Spike in the interval [0, 1]

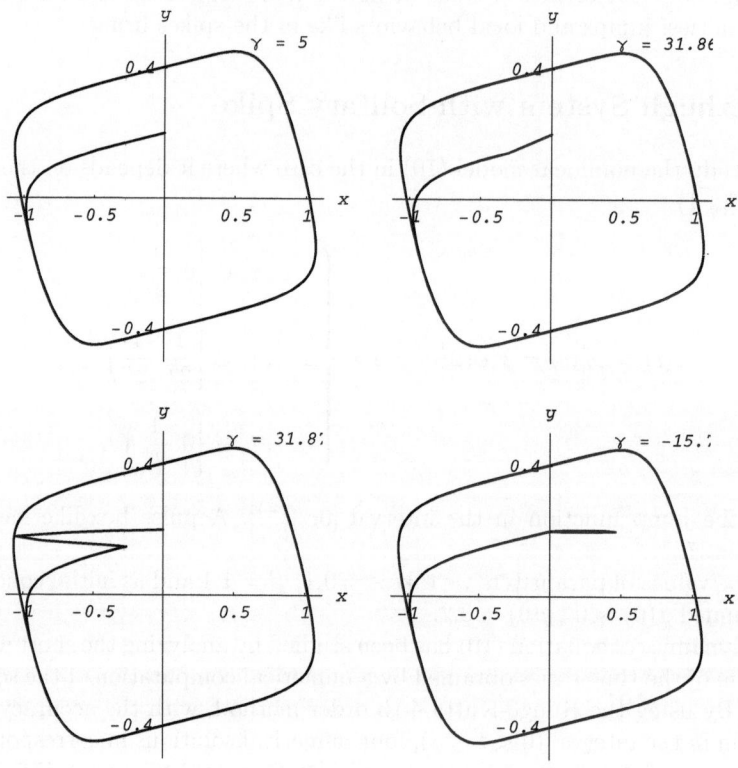

Fig. 2. Numerical solution in the phase space of system (10) with parameters $\varepsilon = 0.1, \beta = 1.1$, and initial conditions $x(0) = 0$, $y(0) = 0.2$, in correspondence of different values of $\gamma = 5$, $\gamma = 31.86$, $\gamma = 31.87$, $\gamma = -15.7$

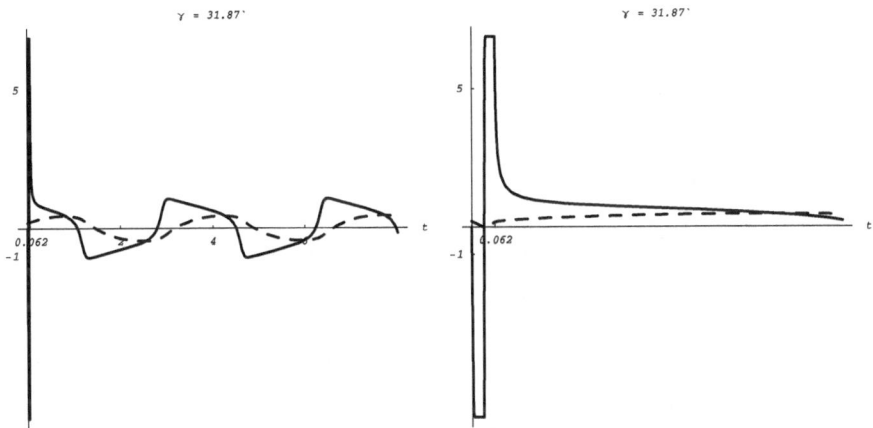

Fig. 3. Numerical solution (plain $x(t)$, dashed $y(t)$) of system (10) with parameters $\varepsilon = 0.1, \beta = 1.1$, (at the bifurcation) $\gamma = 31.87$ and initial conditions $x(0) = 0$, $y(0) = 0.2$, in the time interval $T = 8$ (left) and $T = 1$ (right)

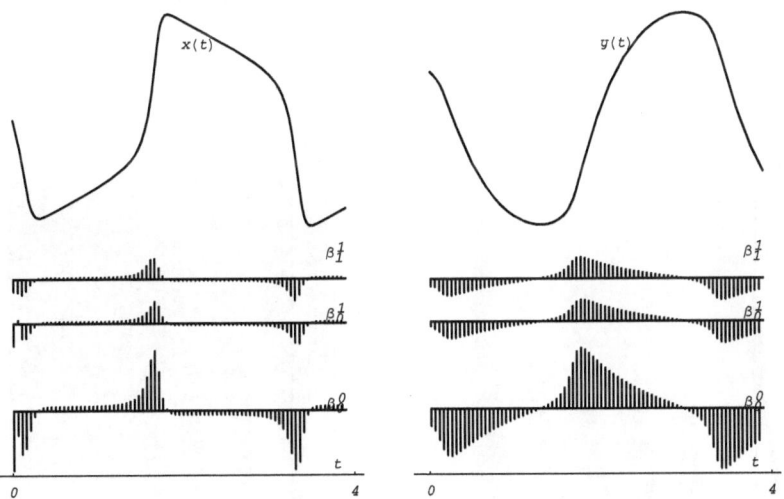

Fig. 4. Numerical solution and wavelet coefficients of 4-parameters of short Haar transform of the numerical solution $x(t)$ (left) and $y(t)$ (right) of system (10) with parameters $\varepsilon = 0.1, \beta = 1.1$, and initial conditions $x(0) = 0$, $y(0) = 0.2$, in correspondence of $\gamma = 5$

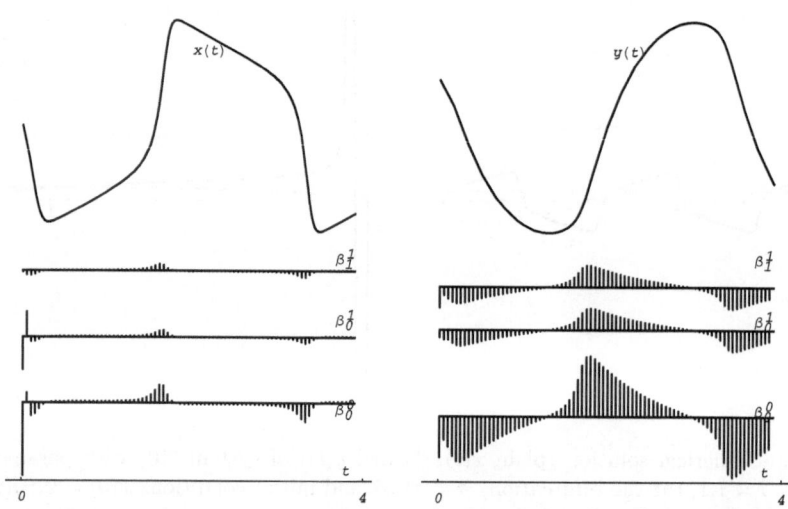

Fig. 5. Numerical solution and wavelet coefficient of 4-parameters of short Haar transform of the numerical solution $x(t)$ (left) and $y(t)$ (right) of system (10) with parameters $\varepsilon = 0.1, \beta = 1.1$, and initial conditions $x(0) = 0$, $y(0) = 0.2$, in correspondence of the bifurcation $\gamma = 31.87$

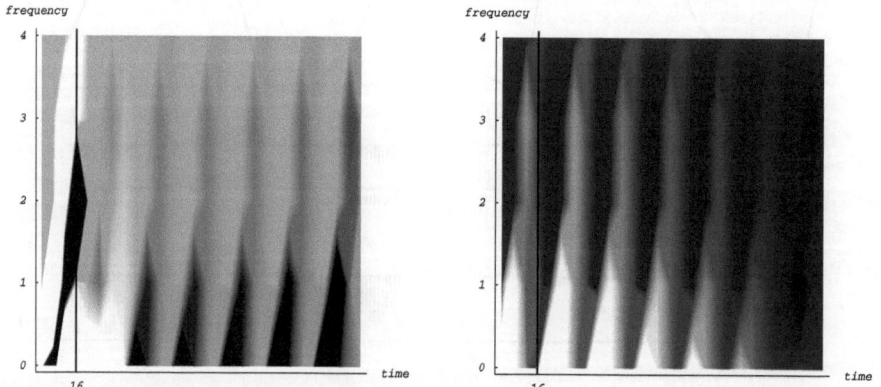

Fig. 6. Wavelet coefficients of the 16-parameters short Haar transform in the time-frequency plane for the numerical solution $x(t)$ (left) and $y(t)$ (right) of system (10) with parameters $\varepsilon = 0.1, \beta = 1.1$, and initial conditions $x(0) = 0$, $y(0) = 0.2$, in correspondence of the bifurcation $\gamma = 31.87$

4 time-series, near the bifurcation value of $\gamma = 31.87$ and the negative value $\gamma = -15.7$.

In correspondence of 4 different values of γ we have obtained 4 numerical solutions in the phase space as in Fig. 2. From a direct inspection it can be seen that we can identify 3 different kind of orbits, corresponding to 3 different physiological states:

1. *Periodic evolution*: when $-15.7 < \gamma < 31.86$, let say $\gamma = 5$ (Fig. 2 top) the orbit is closed, the motion is periodic. Orbits tend to the limit cycle. There exist some, periodically distributed, sharp jumps both for $x(t)$ and for $y(t)$. These jumps can be easily seen, in the numerical solution (Fig. 3) in the wavelet coefficients pictures (Fig. 4), but only for $x(t)$. However the jumps are not isolated, except at the initial time interval where the pulse is effective. In correspondence of $t^* = 0.062$ there is an isolated jump which can be thought as a sign of the existence of a singularity but its amplitude is quite small compared with the remaining jumps of the function (Fig. 4).

2. *Bifurcation*: when $\gamma = 31.87$, the uniqueness is going to be lost: in correspondence of the same value of t there are multiple values of the function $x(t)$ (see Fig. 2). The same can be seen from the first derivative which is singular so that the uniqueness of the solution is lost (Fig. 2 bottom, left) when $x^* \cong 0.9$, $y \cong 0.22$. The singularity can be seen from the high amplitude of the wavelet coefficient (nearby the initial time) see Fig. 5. The tangent vector (first derivative) of $x(t)$ is not unique. In correspondence of $t^* = 0.062$ there exists a cuspidal point. If we plot the wavelet coefficients in the frequency time-plane Fig. 6 it can be seen that the highest frequency is concentrated nearby $t^* = 0.062$ (mostly for $x(t)$).

3. *Chaotic behavior*: when $\gamma < -15.7$, the orbit in the phase plane is self crossing (Fig. 2 bottom, right). The uniqueness of motion is lost. The amplitude of the isolated jump, in correspondence of $t = 0.062$ it is well shown from the inspection of the wavelet coefficients (of $x(t)$). This amplitude is higher of the other jumps and isolated.

5   Fitzhugh System with Spike Train

Let us study the nonlinear model (10) in the case where it depends on the time varying pulse (see Fig. 7)

$$z(t) = \psi_0^4(t) + \psi_3^4(t) + \psi_6^4(t) \ .$$

For the values of parameters we take $\varepsilon = 0.1$, $\beta = 1.1$ and as initial conditions it is assumed $x(0) = 0$, $y(0) = 0.2$.

The dynamics of equation (10) has been studied by analyzing the short wavelet transform of the time series obtained by a numerical computation of the solution of (10). By using the Runge-Kutta 4-th order method, with the accuracy 10^{-6}, we obtain in the interval $(0 < t \leq 8)$, four numerical solutions in correspondence with 4 values of the parameter: $\gamma = 0.1$, $\gamma = 1$, $\gamma = 10$, $\gamma = 100$. These

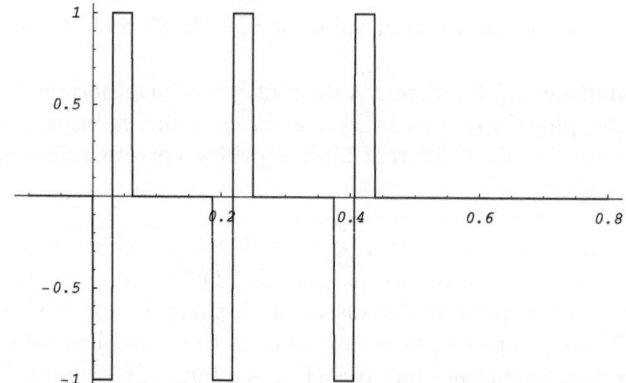

Fig. 7. Spike train in the interval [0, 1]

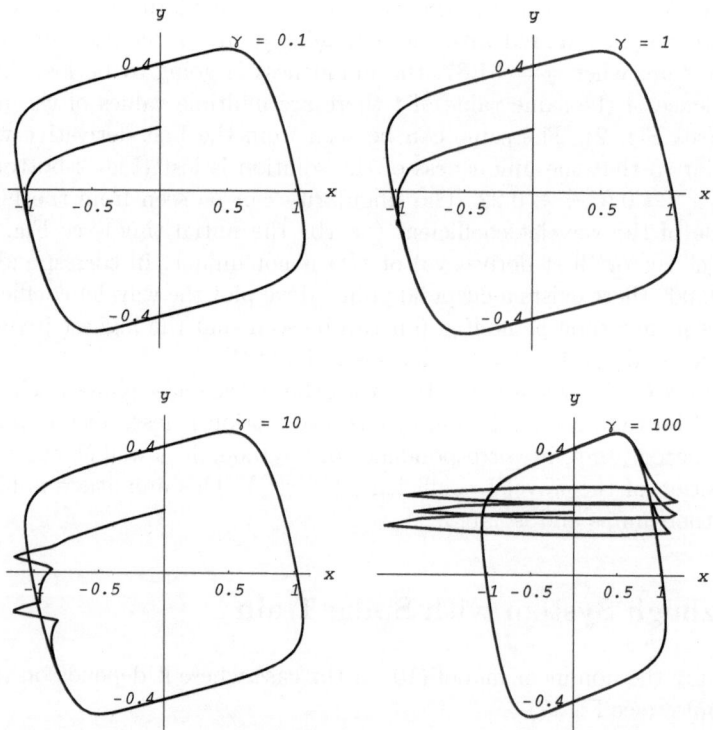

Fig. 8. Numerical solution in the phase space of system (10) with parameters $\varepsilon = 0.1, \beta = 1.1$, and initial conditions $x(0) = 0$, $y(0) = 0.2$, in correspondence of different values of $\gamma = 0.1$, $\gamma = 1$, $\gamma = 10$, $\gamma = 100$

sequences are discretized in $2^9 = 512$ time spots in order to obtain 512 values $\boldsymbol{Y} = \{Y_0, Y_1, \ldots, Y_{N-1}\}$, with $N = 512$ and $M = 9$. By using the short Haar wavelet transform, with $p = 4$, we have analysed the wavelet coefficients of the

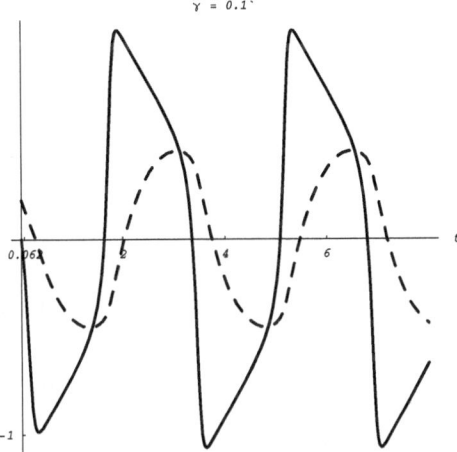

Fig. 9. Numerical solution (plain $x(t)$, dashed $y(t)$) of system (10) with parameters $\varepsilon = 0.1, \beta = 1.1$, $\gamma = 0.1$ and initial conditions $x(0) = 0$, $y(0) = 0.2$, in the time interval $T = 8$

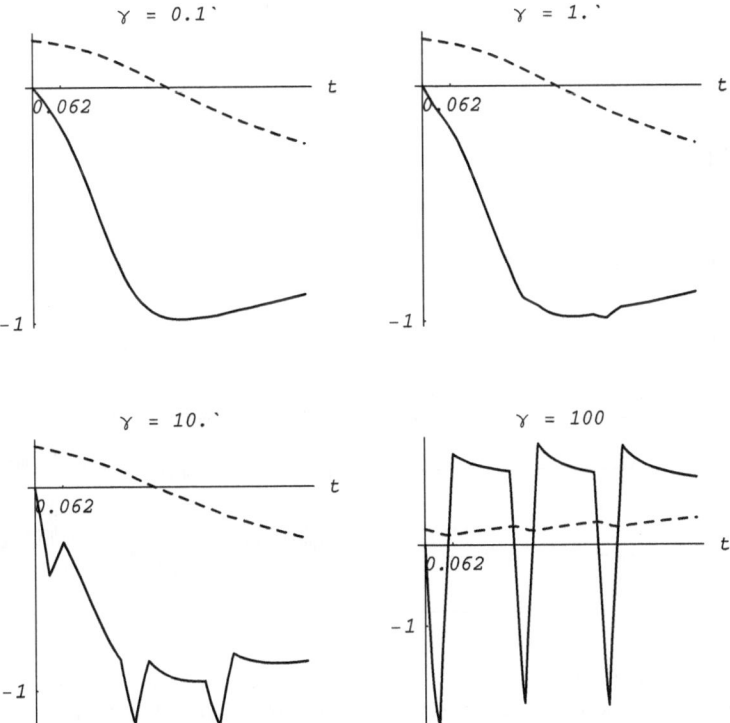

Fig. 10. Numerical solution (plain $x(t)$, dashed $y(t)$) of system (10) with parameters $\varepsilon = 0.1, \beta = 1.1$, and initial conditions $x(0) = 0$, $y(0) = 0.2$, in correspondence of different values of $\gamma = 0.1$, $\gamma = 1$, $\gamma = 10$, $\gamma = 100$, in the short time interval $T = 0.6$. The critical time is $t^* = 0.062$.

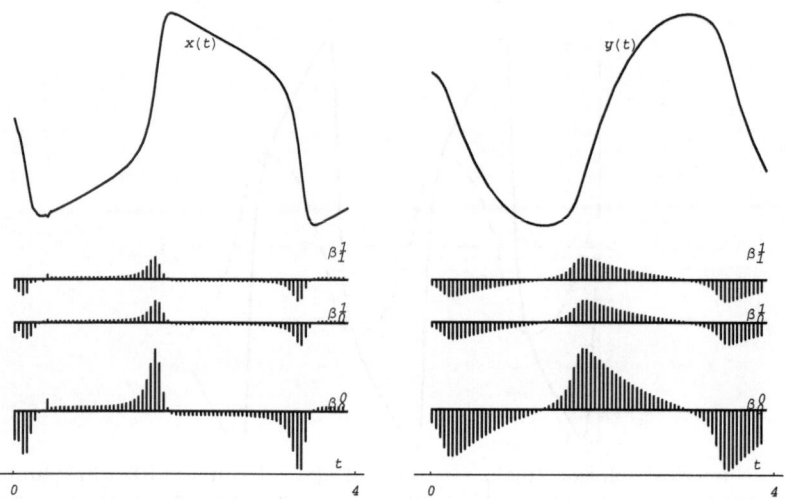

Fig. 11. Numerical solution and wavelet coefficients of 4-parameters of short Haar transform of the numerical solution $x(t)$ (left) and $y(t)$ (right) of system (10) with parameters $\varepsilon = 0.1, \beta = 1.1$, and initial conditions $x(0) = 0$, $y(0) = 0.2$, in correspondence of $\gamma = 0.01$

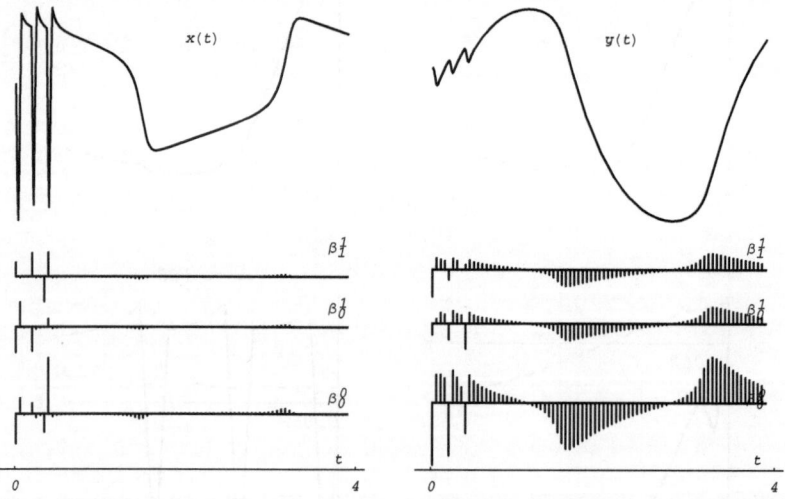

Fig. 12. Numerical solution and wavelet coefficient of 4-parameters of short Haar transform of the numerical solution $x(t)$ (left) and $y(t)$ (right) of system (10) with parameters $\varepsilon = 0.1, \beta = 1.1$, and initial conditions $x(0) = 0$, $y(0) = 0.2$, in correspondence of the chaotic $\gamma = 100$

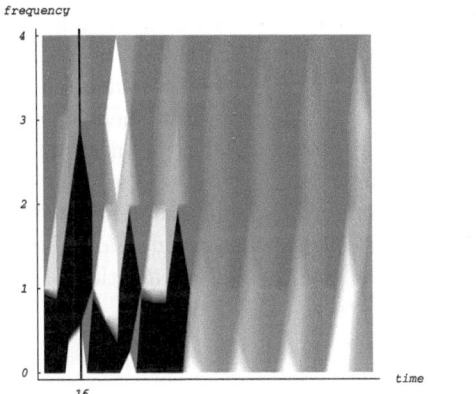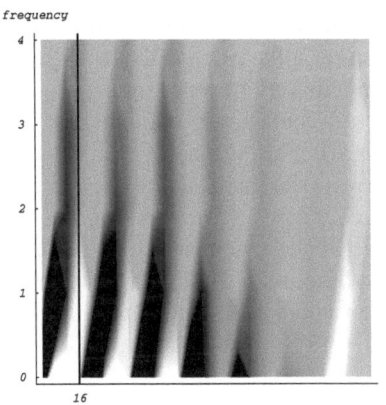

Fig. 13. Wavelet coefficients of the 16-parameters short Haar transform in the time-frequency plane for the numerical solution $x(t)$ (left) and $y(t)$ (right) of system (10) with parameters $\varepsilon = 0.1, \beta = 1.1$, and initial conditions $x(0) = 0$, $y(0) = 0.2$, in correspondence of the bifurcation $\gamma = 100$

4 time-series. In correspondence of 4 different values of γ we have obtained 4 numerical solutions in the phase space as in Fig. 8. and in time (Fig. 10). From a direct inspection it can be seen that we can identify 3 different kind of orbits, corresponding to 3 different physiological states:

1. *Periodic evolution*: when $\gamma \cong 0.1$, (Fig. 8 top, and Fig. 9) the orbit is closed, the motion is periodic. Orbits tend to the limit cycle. There exist some, periodically distributed, jumps both for $x(t)$ and for $y(t)$. These jumps can be easily seen in the wavelet coefficients pictures (Fig. 11). However the jumps are not isolated, except at the initial time interval where the pulse is effective. In correspondence of $t^* = 0.062$ there is an isolated jump which can be thought as a sign of the existence of a singularity but its amplitude is very small.
2. *Chaotic behaviour*: when $\gamma > 1$, the uniqueness is going to be lost: in correspondence of the same value of t there are multiple values of the function $x(t)$ (see Fig. 8). The same can be seen from the first derivative which is singular so that the uniqueness of the solution is lost. The singularity can be seen from the high amplitude of the wavelet coefficients (close to the initial time) see Fig. 12. If we plot the wavelet coefficients in the frequency time-plane Fig. 13 it can be seen that the highest frequency is concentrated nearby the initial time.

6   Conclusion

We have shown that by using the wavelet transform, even in the linear approximation of the short wavelet transform, it is possible to single out some featuring singularities. The Fitzhugh system with localized pulse has been studied and some critical values have been computed. In particular, when the amplitude γ of the single spike or a spike train gives instability to the system then it is possible to immediately observe high jumps in the wavelet coefficients. Only a small set of detail coefficients namely β_0^0, β_0^1, β_1^1, are able to give a sufficiently good information about the dynamical system, but also to emphasize the information hidden in the previous numerical approach. The large amplitude of a jump has been connected with the singularity of the dynamical system. In other words, since wavelet coefficients are sensitive to local jumps, when a singularity arise in the dynamical system, then it immediately reflects on a large variation of the wavelet coefficients, thus making possible to detect bifurcations.

References

[1] Fitzhugh, R.: Impulses and physiological states in theoretical models of nerve membrane. Biophys. Journal 1, 445–466 (1961)
[2] Nagumo, J., Arimoto, S., Yoshizawa, S.: An active pulse transmission line simulating nerve axon. Proc. Inst. Radio Eng. 50, 2061–2070 (1962)
[3] Cattani, C.: Haar Wavelet based Technique for Sharp Jumps Classification. Mathematical Computer Modelling 39, 255–279 (2004)
[4] Cattani, C.: Haar wavelets based technique in evolution problems. Proc. Estonian Acad. of Sciences, Phys. Math. 53(1), 45–63 (2004)
[5] Cattani, C.: Wavelet approach to Stability of Orbits Analysis. International Journal of Applied Mechanics 42(6), 136–142 (2006)
[6] Cattani, C., Rushchitsky, J.J.: Wavelet and Wave Analysis as applied to Materials with Micro or Nanostructure. Advances in Mathematics for Applied Sciences, p. 74. World Scientific, Singapore (2007)
[7] Cattani, C., Scalia, M.: Wavelet Analysis of pulses in the Fitzhugh model. In: Gervasi, O., Murgante, B., Laganà, A., Taniar, D., Mun, Y., Gavrilova, M.L. (eds.) ICCSA 2008, Part I. LNCS, vol. 5072, pp. 1191–1201. Springer, Heidelberg (2008)
[8] El Boustani, S., Pospischil, M., Rudolph-Lilith, M., Destexhe, A.: Activated cortical states: Experiments, analyses and models. Journal of Physiology, 99–109 (2007)
[9] Georgiev, N.V.: Identifying generalized FitzhughNagumo equation from a numerical solution of Hodgkin-Huxley model. Journal of Applied Mathematics, 397–407 (2003)
[10] Guckenheimer, J., Labouriau, I.S.: Bifurcation of the Hodgkin and Huxley equations: a new twist. Bull. Math. Biol., 937–952 (1993)
[11] Hassard, B.: Bifurcation of periodc solutions of the Hodgkin-Huxley model for the squid gain axon. J. Teor. Biol., 401–420 (1978)
[12] Hodgkin, A.L., Huxley, A.F.: Currents carried by sodium and potassium ions through the membrane of the giant axon of Loligo. J. Physiol. 116, 449–472 (1952)
[13] Hodgkin, A.L., Huxley, A.F.: The components of membrane conductance in the giant axon of Loligo. J. Physiol. 116, 473–496 (1952)
[14] Hodgkin, A.L., Huxley, A.F.: The dual effect of membrane potential on sodium conductance in the giant axon of Loligo. J. Physiol. 116, 497–506 (1952)
[15] Rinzel, J., Miller, R.N.: Numerical calculation of stable and unstable solutions to the Hodgkin-Huxley equations. Math Biosci., 27–59 (1980)

[16] Percival, D.B., Walden, A.T.: Wavelet Methods for Time Series Analysis. Cambridge University Press, Cambridge (2000)

[17] Toma, C.: An Extension of the Notion of Observability at Filtering and Sampling Devices. In: Proceedings of the International Symposium on Signals, Circuits and Systems Iasi SCS 2001, Romania, pp. 233–236 (2001)

[18] Toma, G.: Practical Test Functions Generated by Computer Algorithms. In: Gervasi, O., Gavrilova, M.L., Kumar, V., Laganá, A., Lee, H.P., Mun, Y., Taniar, D., Tan, C.J.K. (eds.) ICCSA 2005. LNCS, vol. 3482, pp. 576–584. Springer, Heidelberg (2005)

[19] Wang, J., Chen, L., Fei, X.: Analysis and control of the bifurcation of Hodgkin-Huxley model. Chaos, Solitons and Fractals, 247–256 (2007)

The Design and Implementation of
Secure Socket SCTP

Stefan Lindskog[1] and Anna Brunstrom[2]

[1] Centre for Quantifiable Quality of Service in Communication Systems
Norwegian University of Science and Technology
Trondheim, Norway
stefan.lindskog@q2s.ntnu.no

[2] Department of Computer Science
Karlstad University, Sweden
anna.brunstrom@kau.se

Abstract. This paper describes the design and implementation of secure socket SCTP (S²SCTP). S²SCTP is a new multi-layer, end-to-end
security solution for SCTP. It uses the AUTH protocol extension of
SCTP for integrity protection of both control and user messages; TLS is
the proposed solution for authentication and key agreement; Data confidentiality is provided through encryption and decryption at the socket
library layer. S²SCTP is designed to offer as much security differentiation support as possible using standardized solutions and mechanisms.
In the paper, S²SCTP is also compared to SCTP over IPsec and TLS
over SCTP in terms of packet protection, security differentiation, and
message complexity. The following main conclusions can be draw from
the comparison. S²SCTP compares favorably in terms of offered security differentiation and message overhead. Confidentiality protection of
SCTP control information is, however, only offered by SCTP over IPsec.

Keywords: SCTP, end-to-end security, protocol design, implementation, packet protection, security differentiation, message complexity.

1 Introduction

Stream control transmission protocol (SCTP) [23, 28] is a connection oriented
transport protocol standardized by the signaling transport (SIGTRAN) working group within Internet engineering task force (IETF). SCTP has evolved to
a mature, general purpose protocol on par with the traditional transport protocols user datagram protocol (UDP) and transmission control protocol (TCP).
Compared to UDP and TCP, SCTP provides many new and interesting transport features that make it a strong candidate for next generation communication
networks.

SCTP provides a reliable transport service, ensuring that data is transported
across a network without error and in sequence. An unordered delivery service is
also offered by the protocol. Prior to data transmission, an SCTP connection or

M.L. Gavrilova and C.J.K. Tan (Eds.): Trans. on Comput. Sci. VI, LNCS 5730, pp. 180–199, 2009.
© Springer-Verlag Berlin Heidelberg 2009

association, as it is called in SCTP, is created between the endpoints. An association is maintained until all data has been successfully transmitted. Associations in SCTP are initiated through a four-way handshake mechanism, which is aimed to protect against denial-of-service (DoS) attacks that try to block resources by sending a large number of forged setup requests. Furthermore, SCTP uses a flow and congestion control mechanism similar to TCP.

SCTP is message-oriented and supports framing of individual message boundaries based on the so-called chunk concept. Through the chunk concept, protocol extensions are easily implemented. In [25, 27, 30, 31], four such extensions that have reached a standard status are described.

Furthermore, SCTP provides advanced features for both multistreaming [13] and multihoming [7]. The idea with multistreaming is to decrease the impact of head-of-line (HoL) blocking. A stream in SCTP is a unidirectional channel within an association. Streams provide the ability to send separate sequences of ordered messages as independent flows. In particular, a packet loss in one stream does not inhibit the delivery of packets in other streams. Multihoming, on the other hand, provides the ability for a single SCTP endpoint to support multiple IP addresses. In SCTP, multihoming was introduced as a mean to provide network fault tolerance. By introducing multihoming, additional security attacks could, however, also be envisioned. In [26], unique and well-known security threats associated with SCTP and especially its multihoming facility are described.

Neither SCTP nor UDP or TCP provide mechanisms for end-to-end (E2E) security provision. For SCTP, E2E security is recommended to be implemented through the use of either IP security (IPsec) [12] or transport layer security (TLS) [4]. Both SCTP over IPsec [1] and TLS over SCTP [8] have, however, proven to introduce limitations and inefficiencies [32]. Two other solutions have therefore been proposed. These two are referred to as secure SCTP [32] and SCTP aware DTLS[1] [6, 29], respectively. However, even these two new solutions have their limitations [15].

The design and implementation of an alternative E2E security solution, referred to as secure socket SCTP (S^2SCTP), is described in this paper. During the design of S^2SCTP, the aim has been to provide an E2E security solution that is efficient and, at the same time, can offer a high degree of security differentiation based on standardized features. S^2SCTP provides E2E security through the new authenticated chunks (AUTH) extension [31] for integrity protection of messages; TLS for authentication and key agreement (AKA); and confidentiality protection of messages based on encryption and decryption at the socket library layer. The general idea behind S^2SCTP together with a qualitative comparison with the four above mentioned E2E security solutions are presented in [15]. In [14], an early version of the design of S^2SCTP is presented together with an analysis of the message complexity.

The remainder of the paper is organized as follows. In Sect. 2, the message concept in SCTP is described, as it is central to the design of S^2SCTP. The design and implementation of S^2SCTP is presented in Sect. 3. In Sect. 4, packet

---

[1] DTLS is an abbreviation for datagram TLS. The DTLS standard is described in [18].

protection and security differentiation support as well as message complexity are compared for S²SCTP, SCTP over IPsec, and TLS over SCTP. A brief summary of the security functionality provided by SCTP over IPsec and TLS over SCTP is therefore also included in this section. A discussion on implementing security solutions in a single- or multi-layer fashion together with future work on S²SCTP are provided in Sect. 5. Finally, concluding remarks are presented in Sect. 6.

2   SCTP Messages

As mentioned in the introduction, SCTP is message-oriented and supports framing of individual messages. Chunks are the concept used to frame messages within SCTP packets. Both control and user data are transferred as chunks. Furthermore, one or more messages may be bundled into the same SCTP packet. An example of an SCTP packet with two control chunks and three data chunks is illustrated in Fig. 1.

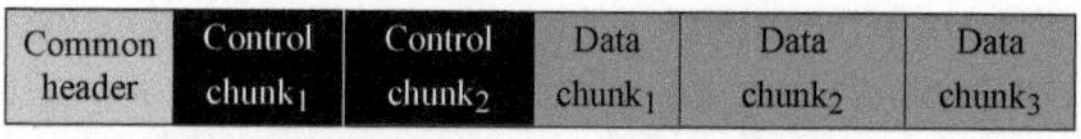

Common header	Control chunk$_1$	Control chunk$_2$	Data chunk$_1$	Data chunk$_2$	Data chunk$_3$

Fig. 1. An example of an SCTP packet with the mandatory SCTP common header (light grey), two control chunks (black), and three data chunks (dark grey)

As can be seen in Fig. 1, a common header is placed first in an SCTP packet. The common header contains a source and a destination port number, a 32-bit verification tag, and a 32-bit checksum. After the common header, the control chunks are included followed by as many data chunks as will fit in the packet. Each chunk has a type field, a flag field, and a length field. Most chunk types may also contain a value field. The content in the value field is dependent on the chunk type.

The number of chunks that can be carried within the same SCTP packet is limited by the path maximum transmission unit (PMTU). If a single message is larger than the PMTU minus the IP header and the common header, SCTP will fragment the message and send the fragments within multiple SCTP packets.

In the base SCTP protocol, 13 different chunk types are defined. Only one chunk type, i.e., DATA, is used for transferring user data. The other 12 chunk types carry control information. When initializing a SCTP association, INIT, INIT ACK, COOKIE ECHO, and COOKIE ACK are used. These are further described in Sect. 3. All standardized chunk types are furthermore presented in [14].

When designing SCTP a key feature was to make it extensible. The single most important feature for extensibility is the concept of chunks. By using the flexible chunk concept, four extensions have so far been standardized. These are partial reliability [25], padding chunk [30], authenticated chunks (referred to as the AUTH extension) [31], and dynamic address reconfiguration [27]. The extensions define new chunk types, new parameters for specific chunk types, and new error causes to be sent in ABORT and/or ERROR chunks. All in all, five additional chunk types have been defined in the four extensions. In S^{2}SCTP, the AUTH extension plays a major role and is more thoroughly described in the next section.

3   Design and Implementation of S^{2}SCTP

The design and implementation of S^{2}SCTP is described in detail in this section. An overview of the protocol is first given. Then, the procedure for establishing a secure association is described followed by a description of user data transfers. Next, S^{2}SCTP's socket application programming interface (API) is presented. Finally, a summary of our prototype implementation is provided.

3.1   Overview

When designing S^{2}SCTP, the goal has been to provide as much security differentiation as possible using features defined in the base SCTP protocol and in its standardized extensions. A key feature used in S^{2}SCTP is the concept of messages and chunks introduced in the previous section. With respect to the standardized extensions, only the AUTH extension is used. The structure of the involved components is illustrated in Fig. 2.

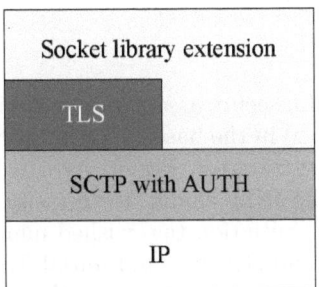

Fig. 2. The layered structure of S^{2}SCTP

As can be seen in the figure, S^{2}SCTP implements E2E security at multiple layers. It provides data integrity protection of control and data chunks, and data confidentiality protection of data chunks. Data integrity protection is provided through HMACs by using the AUTH extension, whereas data confidentiality is implemented through encryption and decryption at the socket library layer. Confidentiality protection is thus provided in user space. A new socket API, which extends the base SCTP socket API, has been defined. The API is further described in Sect. 3.4. In order to provide data integrity and confidentiality protection as described above, an AKA mechanism is needed. TLS is proposed to be used for this purpose.

3.2   Secure Association Establishment

A secure association in S^{2}SCTP is essentially established in two steps. This is illustrated in Fig. 3. In step 1, an SCTP association is established. In this step, key parameters for the AUTH extension are exchanged. In step 2, AKA is performed. After a successful completion of these two steps user data may be exchanged between the peers. A more detailed description of the two steps is provided below.

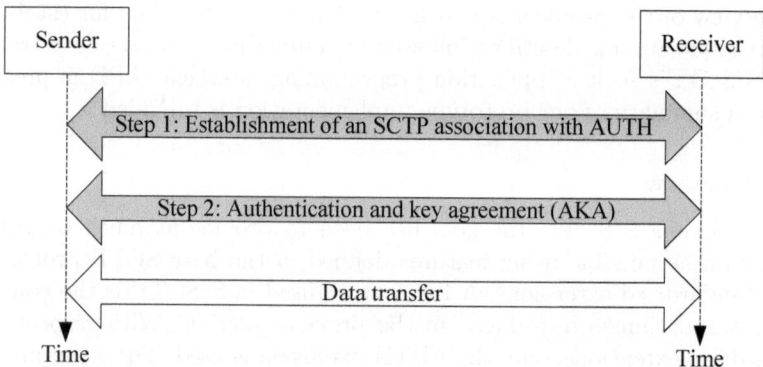

Fig. 3. An illustration of the two steps of a secure association establishment followed by the data transfer in S^{2}SCTP

Step 1: In the first step of a secure association establishment, a four-way hand-shake is performed as defined in the base SCTP protocol. The message exchange in such a handshake is illustrated in Fig. 4. As can be seen in the figure, the first chunk to be sent is the INIT chunk. INIT contains information related to the association, such as an initiation tag, wished number of outbound streams, maximum number of inbound streams, and initial TSN. The initiation tag is a random number that will later be used as a verification tag in the SCTP common header. The INIT chunk may also contain pre-defined optional parameters, such as additional IP version 4 (IPv4) addresses, IP version 6 (IPv6) addresses, and hostnames.

The second chunk exchanged is the INIT ACK chunk, which contains similar information to the INIT chunk except the mandatory state cookie parameter. The state cookie parameter is used as a defense against resource exhaustion attacks. The content of the cookie is implementation specific, but some guidelines are given in [23]. A signature for authentication when the state cookie comes back must at least be included.

When the peer receives the INIT ACK, a COOKIE ECHO chunk is replied. The COOKIE ECHO contains a copy of the received state cookie. Upon reception of a COOKIE ECHO chunk, the included state cookie is verified. If the cookie is valid, a COOKIE ACK is sent. This is the final chunk sent in the four-way

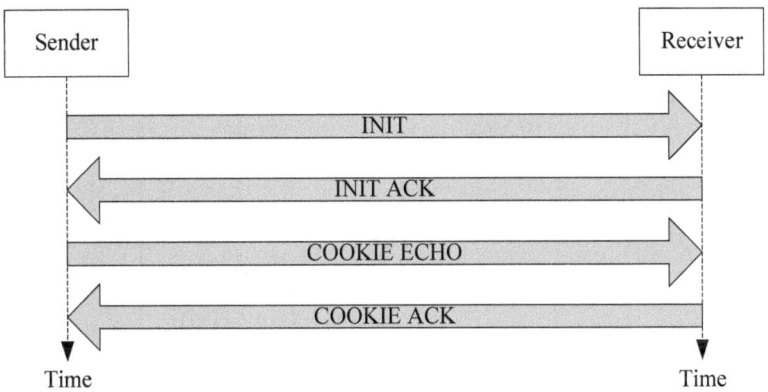

Fig. 4. Step 1 of the secure association establishment in S^2SCTP. This message flow is in SCTP referred to as the four-way handshake.

handshake. If, on the other hand, the state cookie is invalid, an ABORT chunk is replied. The COOKIE ACK chunk is an example of a chunk type that only contains the three mandatory fields, i.e., type, flag, and length, as described in Sect. 2. Note that both the COOKIE ECHO and the COOKIE ACK could be bundled with other chunks, e.g., DATA and SACK, in an SCTP packet.

When the AUTH extension is used, three additional parameters are exchanged in the INIT and/or INIT ACK chunks: a random parameter (RANDOM), a chunk list parameter (CHUNKS), and a requested HMAC algorithm parameter (HMAC-ALGO). RANDOM contains an arbitrary random number that is included once in the INIT or INIT ACK chunk. CHUNKS contains a list of chunk types that are required to be protected by an AUTH chunk. INIT, INIT ACK, SHUTDOWN COMPLETE, and AUTH are not allowed to be included in this list. HMAC-ALGO, finally, contains a list of supported HMAC algorithms in sender preference order. Two algorithms have so far been defined: secure hash algorithm number 1 (SHA-1) and SHA-256 [16]. These two produce a HMAC of 160 and 256 bits, respectively.

The actual integrity protection is provided through the AUTH chunk. The fields in an AUTH chunk type are illustrated in Fig. 5. Except the basic chunk fields, i.e., type, flag, and length, four additional fields are included. The shared key identifier field is used to notify the receiver which key to use. The HMAC identifier field specifies the HMAC algorithm that has been used to produce the following HMAC. The third field is the HMAC itself. The last field is a padding field, which is used to align the length of the chunk to a multiple of 32 bits. This implies that the length of the padding field could be 0, 8, 16, or 24 bits. For both SHA-1 and SHA-256, the length of the padding field will be 0, since the total length of an AUTH chunk with SHA-1 will be 224 bits and an AUTH chunk with SHA-256 will be 320 bits. Both 224 and 320 are multiples of 32.

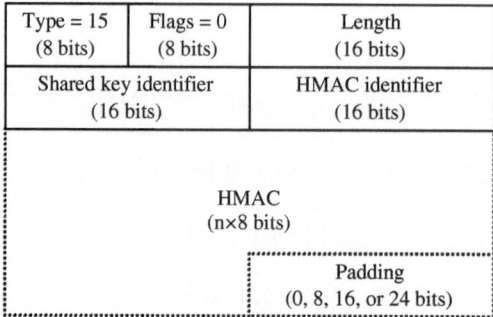

Fig. 5. Fields in an AUTH chunk

Step 2: Once an SCTP association is established, the next step is to perform AKA. In S^{2}SCTP, TLS is the preferred AKA mechanism. The choice of TLS is simply motivated by the fact that it is a working, standardized, and highly flexible solution. In TLS, authentication is achieved through X.509 certificates that are signed using either RSA[2] or the digital signature standard (DSS). We recommend that mutual authentication is used, which implies that both peers must send its certificate to the other side. For key exchange, three different algorithms could be used, i.e., RSA, fixed Diffie-Hellman (DH), or ephemeral DH (DHE). Authentication and cryptographic algorithm negotiation are performed by the handshake protocol defined in the TLS protocol, and the message flow in step 2 of S^{2}SCTP is according to the standard [3].

After the TLS handshake, the next task is to establish the cryptographic keys. This is performed as follows. A pre-master secret is first exchanged. Next, a master secret is generated by each peer. From the master secret, a set of keys are derived. One key is used for authentication and another key for encryption and/or decryption.

An alternative to TLS is to use a pre-shared key (PSK) approach. Pre-shared means that the peers agree in advance on a shared, secret (master) key that is later used (either directly or indirectly) for authentication and/or encryption and decryption. The main problem with PSK is to maintain fresh keys. PSK is therefore not a viable solution in large networks.

As mentioned in the introduction, SCTP supports the concept of multistreaming in order to decrease the impact of HoL blocking. Each established stream has a stream identifier that identifies to which stream a message belongs. The number of streams within an association is negotiated in the first two messages, i.e., INIT and INIT ACK, sent in step 1 of the secure association establishment. In S^{2}SCTP, streams 0 in both directions are reserved for the TLS handshake procedure described above.

---

[2] RSA is an abbreviation for Rivest, Shamir, and Adleman, who are the inventors of the algorithm.

3.3   User Data Transfer

After successfully finishing steps 1 and 2, user data may be transferred. User data are in SCTP transferred as DATA chunks. The fields in a DATA chunk type is illustrated in Fig. 6.

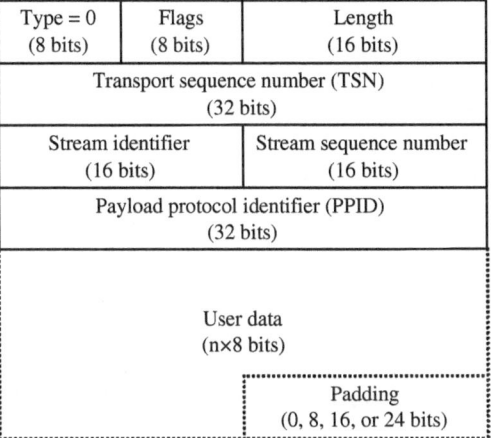

Fig. 6. Fields in a DATA chunk

For the DATA chunk type, three flags have been defined. These are the lower three bits in the flag field and are referred to as U, B, and E, respectively. The U bit is set if data carried in the chunk is unordered. The B bit is set to indicate that the chunk is the first part of a fragmented message, while the E bit is set to denote that this is the last part of a fragmented message. If an entire message fits into a single chunk, both the B and the E bit is set. When transferring data pieces in the middle of a message, neither B nor E is set.

The TSN field is a sequence number for the association, while the stream sequence number is a sequence number for the stream. The stream identifier indicates which stream the chunk belongs to. The payload protocol identifier (PPID) field is a user supplied value. This field is ignored by SCTP, but is an important field in S²SCTP. In S²SCTP, the PPID value is divided into two halves with 16 bits each. The first 16 bits are used to carry a confidentiality key identifier and the last 16 bits are used to specify which confidentiality algorithm that has been used to encrypt the message. Note the similarity with the AUTH chunk.

The remaining two fields in the DATA chunk type is the user data field and the padding field. The user data field contains the actual user message, and the padding field is used to align data to 32-bit boundaries. When data confidentiality protection is needed, encryption is applied to user data using symmetric block ciphers. Encryption of data is furthermore performed at the application level and on a per-message basis.

3.4 Socket API Extension

A new socket API is provided for S^{2}SCTP, which extends the current SCTP socket API described in [24]. An early version of the SCTP socket API is presented in [22]. In the most recent version of the S^{2}SCTP API, the following eight functions are defined.

- s2sctp_createAuthKey()
- s2sctp_setActiveAuthKey()
- s2sctp_deleteAuthKey()
- s2sctp_createConfKey()
- s2sctp_deleteConfKey()
- s2sctp_createSecureAssoc()
- s2sctp_send()
- s2sctp_receive()

The first three functions, i.e., s2sctp_createAuthKey(), s2sctp_setActive-AuthKey(), and s2sctp_deleteAuthKey() are all used to manage AUTH keys. New keys are created using the s2sctp_createAuthKey() function. To refer to an existing key, a key identifier is used. Multiple keys, all with different key identifiers, can be created but only one key can be active in a given association at the same time. A key is activated with s2sctp_setActiveAuthKey(), and a key that will not be used anymore should be deleted using the s2sctp_delete-AuthKey() function.

The functions s2sctp_createConfKey() and s2sctp_deleteConfKey() are used to create and delete confidentiality keys. Such keys are used for encryption and/or decryption of messages. Multiple confidentiality keys can be created and managed simultaneously. To distinguish between different confidentiality keys, key and algorithm identifiers are used.

The s2sctp_createSecureAssoc() function establishes an S^{2}SCTP association as described in Sect. 3.2. This function must therefore be called before any data can be transferred.

The s2sctp_send() function is used to send messages. The parameters are the same as in the sctp_send() function provided in [24] with one exception—two extra parameters containing a key identifier and a confidentiality algorithm identifier have been added. The key identifier specifies which confidentiality key and the confidentiality algorithm identifier which confidentiality algorithm to use for the message.

The s2sctp_receive() function is used to receive and, if needed, decrypt messages at the receiver side. The decryption algorithm is, however, only applied on data in chunks with a non-zero key identifier.

Note that an earlier version of the S^{2}SCTP socket API was outlined in [14]. Here, only four API functions were described. The new functions that have been added are the key management functions described above.

3.5   Prototype Implementation

The S²SCTP protocol is currently being implemented in the FreeBSD 7.0 operating system [5] using the C/C++ programming language. In the current release of S²SCTP, TLS-based AKA is not yet fully finalized. Instead, AKA is performed using PSK. PSK works fine in small and static networks, such as our current experimental environment, but is not a recommended solution in large, dynamic networks. However, our focus in the development of the prototype has been on the implementation of the data integrity and confidentiality protection functionality in S²SCTP. The motivation for this is that these two functionalities are the most critical components with respect to data transfer performance, since they will be used much more frequently than the AKA mechanism.

Data Integrity. Data integrity functionality is implemented using the AUTH extension. In the implementation of AUTH in FreeBSD 7.0, two different HMAC algorithms are supported: SHA-1 and SHA-256, which both use a 128-bit key. The selection of which chunk types to authenticate is specified through a list provided by each side in the INIT or INIT ACK chunk. The preferred HMAC algorithms are also specified as a list in order of preference and are also included in these two chunks. In the list of HMAC algorithms, SHA-1 must always be a member. This implies that if SHA-256 is the preferred HMAC algorithm, it should be included first in the list.

An important feature is that the provided integrity protection can be applied in an asymmetric way. For example, the sender may request that all control chunks sent from the receiver must be protected with an AUTH chunk; while the receiver may only request that the DATA chunks are protected using AUTH. Furthermore, the sender may recommend that the receiver uses SHA-256 when producing the AUTH, while the receiver is satisfied with the default SHA-1 algorithm. This type of asymmetry is fully supported in the current implementation.

Data Confidentiality. Data confidentiality protection is implemented at the socket library layer through the use of the OpenSSL crypto library [17]. Currently, the basic data encryption standard (DES) [21] algorithm and advanced encryption standard (AES) [2] algorithm are supported. DES and AES are both standardized block ciphers. DES has a static block and key length of 64 bits, whereas AES uses a block length of 128 bits and a key length of 128, 192, or 256 bits. In S²SCTP, all these AES key lengths are supported. Furthermore, both algorithms may operate in a number of different modes. We have, however, limited the current prototype to only support cipher block chaining (CBC) mode for all algorithm configurations. Support for other block ciphers, such as Triple DES (3DES) [21], Blowfish [19], and Twofish [20], and other modes of operation, such as electronic code book (ECB) and cipher Feedback (CFB) mode, can easily be provided.

When block ciphers are used, special treatment of the last block is needed. This is due to the fact that such algorithms operate on fixed size blocks, which implies that some form of padding is necessary of the last block. In S^2SCTP, padding of the last block is added and the length of padding is stored in the last byte of the block. After that, encryption is performed. Similarly, when the last block is decrypted by the receiver, the padding is removed based on the padding length indicator stored in the last byte of the block. The added padding causes an overhead, which is directly dependent on the block length. The longer block length, the more overhead is added. The overhead caused by an encryption algorithm for a given message can be expressed as follows:

$$PAD = BL - ML \ mod \ BL \tag{1}$$

where PAD denotes the length of padding, and BL and ML is the block and message lengths, respectively. Remember also that padding is added by SCTP to align data to 32-bit boundaries, see Fig. 6. When DES and AES are used as confidentiality algorithms as in S^2SCTP, padding at the SCTP layer is not needed since data provided from the application layer will always be 32-bit aligned. The overhead calculation caused by the data confidentiality protection mechanism in (1) is thus overly pessimistic.

In S^2SCTP, the PPID field in a DATA chunk is used to carry a key and a confidentiality algorithm identifier. A zero value key identifier denotes that the message is unencrypted. Unencrypted messages can thus be sent by either using the new s2sctp_send() described in the previous subsection with a key identifier of 0 or using one of the send functions defined in the SCTP socket API. A summary of the reserved confidentiality algorithm identifiers in S^2SCTP is given in Table 1.

Table 1. Reserved confidentiality algorithm identifiers in S^2SCTP

Algorithm	Algorithm identifier
NULL	0x0000
AES-128[a]	0x0001
AES-192	0x0002
AES-256	0x0003
DES	0x0004

[a] AES-128 denotes that the AES algorithm with a 128-bit key is used. Similarly, AES-192 and AES-256 denotes that AES with a key length of 192 and 256 bits are used, respectively.

By sending the key identifier in each DATA chunk, a new key can potentiality be used for each message. In addition, the confidentiality algorithm can also easily be changed at run-time. Thanks to the usage of the PPID field, a high degree of security differentiation can thus be applied on user data sent by an application.

In the current implementation, each peer keeps track of two separate lists of confidentiality key contexts. Both lists are indexed by the key and algorithm identifiers. For each key context, an algorithm, a key, and an initialization vector (IV) [19] of a certain length dependent on the algorithm are stored. For AES-128, a key and IV of 128 bits are stored. Similarly, a key and IV of 64 bits are stored when DES is used. The algorithm identifier and the key remain the same until they are either deleted using `s2sctp_deleteConfKey()` or replaced with a new key using `s2sctp_createConfKey()`. The IV, on the other hand, is constantly updated as a side-effect of an encryption/decryption operation when the CBC mode of operation is used.

4   Comparison of Security Solutions

A comparison of S²SCTP with the two standardized E2E security solutions for SCTP is provided in this section. An overview of the standardized E2E solutions is therefore first given. Then, packet protection and security differentiation are compared, followed by a comparison of the message complexity when transferring user data.

4.1   Standardized E2E Security Solutions

The two standardized E2E security solutions for SCTP—SCTP over IPsec and TLS over SCTP—are briefly summarized below.

SCTP over IPsec. The use of IPsec to secure E2E SCTP traffic is described in [1]. IPsec operates on the network layer and was designed to be independent of the transport protocol. The current version of IPsec has evolved to a fairly complex protocol. IPsec could in fact be seen as a number of protocols. When only integrity protection is needed the authentication header (AH) protocol [10] is used. If both integrity and confidentiality protection is needed, the encapsulating security payload (ESP) protocol [11] is used instead. To be able to dynamically establish security associations (SAs), a protocol referred to as the Internet key exchange (IKE) protocol [9] might be used.

The major problem with this solution, as described in [1], is the dynamic creation of SAs using IKE for multihomed hosts. When two multihomed hosts with n and m IP addresses, respectively, are establishing an SCTP association, $2 * m * n$ SAs must also be established. The creation of SAs is known to be an expensive operation and must be negotiated separately. When SCTP is used over IPsec, all messages belonging to the same SCTP association are also protected in the same way. Hence, no security differentiation may be applied within a session at run-time. This is mainly due to the fact that IPsec is neither aware of SCTP messages nor of SCTP streams. Securing all data transferred between two hosts may produce unnecessary computational burden at the endpoints.

TLS over SCTP. Another solution to implement E2E security is through the use of TLS over SCTP as described in [8]. TLS is a byte-oriented protocol that requires an underlying reliable and in-sequence delivery service, which is offered by SCTP as long as it does not use the unordered delivery service or the extension for partially reliable transport. Application data transfers are made over a TLS connection. Such a connection must be bidirectional. TLS provides AKA, data integrity through hashed message authentication codes (HMACs), and/or data confidentiality through encryption and decryption algorithms.

The most serious disadvantage with TLS over SCTP is that SCTP control chunks exchanged between two peers are completely unprotected. This is caused by the fact that TLS operates on a layer above SCTP and is thus not aware of the control chunks. Due to the requirement of TLS on reliable ordered delivery, the SCTP unordered delivery service as well as the extension for partially reliable transport is not at all supported. Note also that a bidirectional TLS connection is established using two unidirectional SCTP streams. Hence, a single unidirectional SCTP stream cannot be protected by TLS.

4.2   Packet Protection and Security Differentiation

When protecting an SCTP packet, additional data is included. Depending on the chosen E2E security solution data is added at different places. Figure 7 depicts the IP version 4 (IPv4) datagrams that result when transferring the SCTP packet introduced in Fig. 1 in five different protection scenarios. In (a), no protection at all is applied. In scenario (b) and (c), IPsec is used with the AH and ESP protocol, respectively. Scenario (d) illustrates when TLS is used, and, finally, in (e) S²SCTP is used.

As is illustrated in Fig. 7, IPsec provides two different protection modes. If data integrity protection is sufficient, an AH is added to each IP datagram. If both data integrity and confidentiality protection is needed, an ESP header is used instead. When the choice of protection type has been made, all SCTP packets are protected in the same way, i.e., an all-or-nothing approach is applied to all SCTP chunks for both user and control data.

When TLS is used to protect data transferred over SCTP, only user data is protected as can be seen in Fig. 7. This is due to the fact that the data protection algorithms are applied to data above the SCTP layer, which implies that SCTP control data is transferred unprotected. For user data, security differentiation could be conducted on a per-stream basis.

In S²SCTP, both user and control chunks could be integrity protected using the AUTH extension. The selection of which chunk types to integrity protect is decided on a per-association basis. In the example given in Fig. 7, it is assumed that all illustrated control chunks and data chunks are integrity protected. Data confidentiality protection in S²SCTP is provided through encryption and decryption implemented at the socket library layer. Encryption is thus only applied on user data and could selectively be applied on a per-message basis.

Comparing the protocols, SCTP over IPsec provides the strongest protection and S²SCTP the most flexible security differentiation. In addition to the

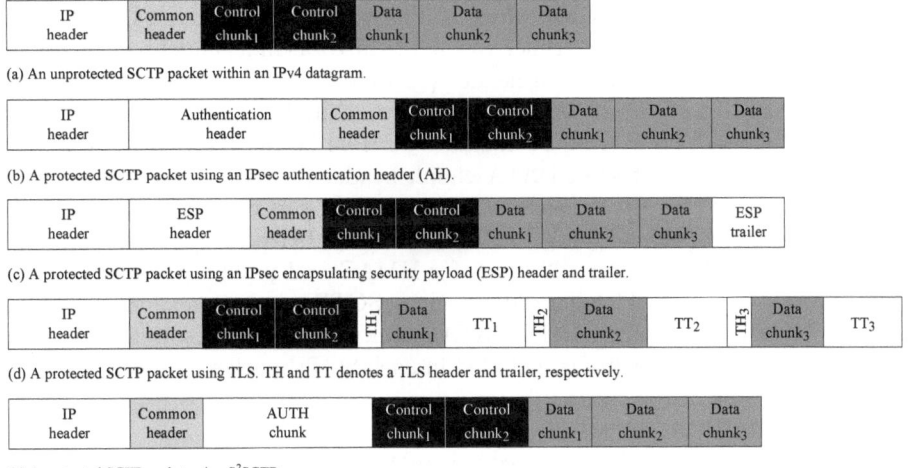

(a) An unprotected SCTP packet within an IPv4 datagram.

(b) A protected SCTP packet using an IPsec authentication header (AH).

(c) A protected SCTP packet using an IPsec encapsulating security payload (ESP) header and trailer.

(d) A protected SCTP packet using TLS. TH and TT denotes a TLS header and trailer, respectively.

(e) A protected SCTP packet using S²SCTP.

Fig. 7. This figure illustrates the extra overhead added when using (b) the IPsec AH protocol, (c) the IPsec ESP protocol, (d) TLS, and (e) S^{2}SCTP to protect the SCTP packet in (a)

type of security differentiation described above, all three E2E security solutions compared in this paper also allow security differentiation through algorithm selection, i.e., selection of AKA mechanism, HMAC algorithm, and encryption algorithm.

4.3 Message Complexity

A comparison of the message complexity when transferring user data using SCTP over IPsec, TLS over SCTP, and S^{2}SCTP is emphasized in this section. We consider cases when user messages are unbundled as well as bundled. For data integrity and confidentiality protection, the SHA-1 and AES-128 algorithm in CBC mode are assumed. Since the focus is on the message overhead introduced when transferring user data, the secure association establishment is not considered. This event is also infrequent in comparison to data transfers.

Unbundled Data Transfers. Sending unbundled user messages means that each SCTP packet only carries a single user message. Such transfers will therefore result in a higher message overhead compared to a bundled transfer, simply due to the fact that some protocol headers must be duplicated in each SCTP packet. Unbundled data transfers are typically used to achieve as low latency as possible, which may be a requirement in signaling and multimedia applications.

A summary of the message complexity for the three different investigated E2E security solutions when transferring user messages unbundled is provided in Table 2. The figures in the table include all higher layer protocol headers starting with the IP layer and a user message of the specified size. IPv4 is assumed to be used here. Hence, when SCTP over IPsec is used, an IPv4 header (20 bytes), an IPsec header/trailer (32 bytes in total independent of if AH or ESP is used), an SCTP common header (12 bytes), and a DATA chunk header (16 bytes), and as much of the user message that fits is included in each SCTP packet. Similarly, TLS over SCTP includes an IPv4 header (20 bytes), an SCTP common header (12 bytes), a DATA chunk header (16 bytes), and as much of the user message that fits in each SCTP packet. In this case, the user message contains both the payload data, the TLS header (5 bytes for type, version, and length) and the TLS trailer (23 bytes for HMAC and padding). Finally, when S^2SCTP is used, an IPv4 header (20 bytes), an SCTP common header (12 bytes), an AUTH chunk (28 bytes), and a DATA chunk header (16 bytes), and as much of the user message that fits is included in each SCTP packet.

Table 2. Message complexity expressed in bytes for different sized user messages when transferred unbundled using IPv4. The PMTU is assumed to be 1500 bytes.

E2E security solution	Message size in bytes				
	128	256	1024	4096	16384
SCTP over IPsec	208	336	1104	4336	17344
TLS over SCTP	204	332	1100	4268	16988
S^2SCTP	204	332	1100	4324	17296

As can be seen in the Table 2, S^2SCTP and TLS over SCTP produce the same amount of traffic for user messages that fit in a single SCTP packet. When message fragmentation is needed, i.e., when transferring user messages that do not fit in a single SCTP packet, TLS causes slightly less traffic compared to S^2SCTP. From the table it is also evident that SCTP over IPsec results in the highest amount of network traffic although the difference with S^2SCTP is small.

In Fig. 8, the message overhead in percent added for unbundled data transfers by the different solutions are depicted. The message overheads for all solutions are high when transferring small user messages, but rather small when transferring large messages.

Bundled Data Transfers. When bundled data transfers are used, multiple user messages could potentially be carried by a single SCTP packet. Bundling is, however, only used when multiple chunks are ready to be sent and fit completely in a single SCTP packet. If that is not the case, each message is sent in one or more SCTP packets depending on the size of the data. By using bundling, the message overhead is reduced. Bundled data transfers could therefore be used with advantage by, e.g., notification and data log services. In such services, messages exchanged are typically small, i.e., a few hundred bytes. Slightly increased latency is also typically acceptable in such services.

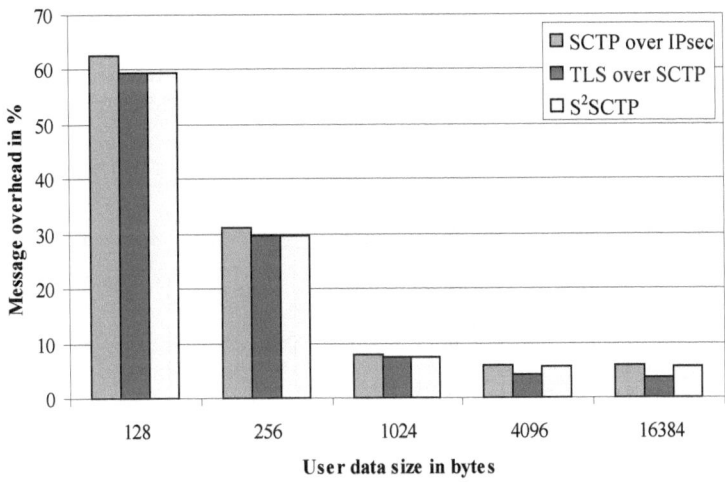

Fig. 8. Message overhead for different sized unbundled user messages

The message complexity for the three studied E2E security solutions with different amount of message bundling is depicted in Table 3. All messages are assumed to be of equal length, i.e., 256 bytes in this case. Such an assumption is fair when considering notification as well as data log services. When bundling is used, TLS causes the highest message complexity. This is due to the extra protocol overhead added by TLS for each message, i.e., the TLS header (5 bytes) and the TLS trailer (23 bytes) as described above. When TLS is used, five bundled user messages will therefore not fit into a single SCTP packet. Instead, the four first messages are sent in one SCTP packet and the fifth is sent in a separate one, assuming that no further messages arrive to the send queue before the second packet is sent. When comparing S²SCTP and SCTP over IPsec, the message complexity is again similar.

Table 3. Message complexity expressed in bytes for different amount of message bundling using IPv4. Each user message is 256 bytes and the PMTU is assumed to be 1500 bytes.

E2E security solution	Number of messages				
	1	2	3	4	5
SCTP over IPsec	336	608	880	1152	1424
TLS over SCTP	332	632	932	1232	1564[a]
S²SCTP	332	604	876	1148	1420

[a]For this case, two SCTP packets are needed to transfer these five user messages, since the total data size is greater than the PMTU.

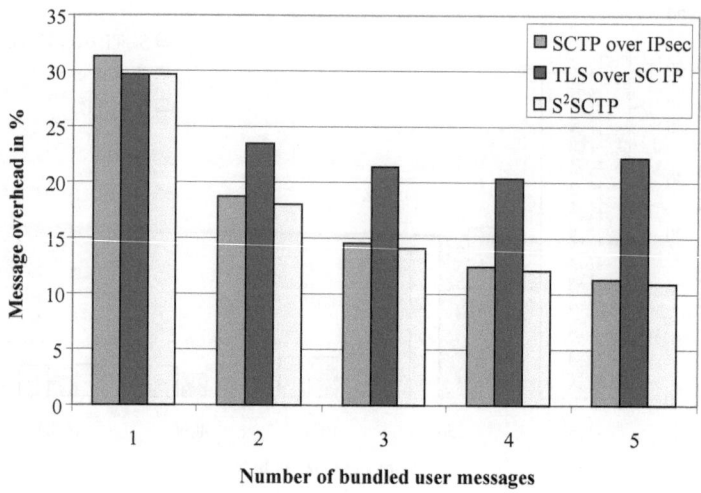

Fig. 9. Message overhead when user messages are bundled

Figure 9 illustrates the high overhead introduced when TLS over SCTP is used to transfer bundled user messages. In the worst case illustrated in Fig. 9, i.e., five bundled messages, the overhead is more than twice as big as compared to S^2SCTP and SCTP over IPsec. This implies that TLS is not a preferable E2E security solution when transferring small messages that can be bundled together in SCTP packets. Finally, the difference between S^2SCTP and SCTP over IPsec with respect to message overhead is almost negligible.

5   Discussion and Future Work

Security may be implemented at various communication layers. In wireless networks, security functionalities are often implemented at the link layer in order to achieve a similar level of protection as an equivalent wired network. IPsec and TLS provide security at the network and transport layer, respectively. In other cases, security is implemented at the application layer. Furthermore, in the wireless community physical layer security has recently become a hot research topic. Hence, security is mainly implemented at a single layer, and very few existing solutions make use of security functionalities at different layers. This implies that similar functionalities may be applied at multiple layers without knowing about each other. The result is increased computational and network overhead, and an unclear improvement, if any, of security.

It is today unclear where in the communication stack security functionalities are most efficiently provided. Certification verification and key establishment procedures used for AKA seem impossible to perform in the kernel due to unpredictable blocking times [6]. In practice, this implies that AKA needs to be implemented above the transport layer. Regarding data integrity protection through HMACs and confidentiality protection through encryption/decryption it is more unclear where such functionalities are most efficiently implemented. We believe that a better understanding of where in the communication stack different security functionalities are most efficiently implemented—with respect to both security and performance—will result in more suitable E2E security solutions for the next generation communication networks. Furthermore, by allowing an increased flexibility between the layers, security configurations can be applied based on need. In S^2SCTP, we aim to provide an efficient and flexible security solution by combining security mechanisms available at multiple layers.

The next step in the development of S^2SCTP is to perform a quantitative performance evaluation of the data integrity and confidentiality protection mechanisms implemented in the prototype. Then, AKA support through TLS will be added. When that is finished, a comparison with existing solutions will be conducted. We are especially interested in a comparison with TLS over SCTP, since that solution is the one that is most similar to S^2SCTP with respect to security differentiation. By comparing S^2SCTP with TLS over SCTP, we hope to discover new findings regarding the efficiency of providing security functionality at either different layers as in S^2SCTP or at the same layer as in TLS over SCTP.

6   Concluding Remarks

The design and implementation of S^2SCTP has been presented in this paper. The main focus has been on mechanisms for data integrity and confidentiality protection. Data integrity protection is provided through the new AUTH extension for SCTP, whereas data confidentiality protection is implemented at the socket library layer. Furthermore, AKA is proposed to be performed using certificate based methods provided by the TLS protocol.

S^2SCTP is also compared to SCTP over IPsec and TLS over SCTP in terms of packet protection, security differentiation, and message complexity. In conclusion, SCTP over IPsec offers the lowest degree of security differentiation, but the highest level of security. TLS over SCTP produces the least communication overhead for large messages. S^2SCTP, finally, provides the finest degree of security differentiation and produces the least communication overhead when messages are bundled.

Acknowledgment. The work at the Norwegian University of Science and Technology is financially supported by the research council of Norway. The work at Karlstad University is supported by grants from the Knowledge Foundations of Sweden with TietoEnator and Ericsson as industrial partners.

References

1. Bellovin, S., Ioannidis, J., Keromytis, A., Stewart, R.: RFC 3554: On the use of stream control transmission protocol (SCTP) with IPsec (July 2003)
2. Daemen, J., Rijmen, V.: The Design of Rijndael: AES—The Advanced Encryption Standard. Springer, Heidelberg (2002)

3. Dierks, T., Allen, C.: RFC 2246: The TLS protocol version 1.0 (January 1999)
4. Dierks, T., Rescorla, E.: RFC 4346: The transport layer security (TLS) protocol version 1.1 (April 2006)
5. FreeBSD homepage, http://www.freebsd.org/ (visited December 9, 2008)
6. Hohendorf, C., Rathgeb, E.P., Unurkhaan, E., Tüxen, M.: Secure end-to-end transport over SCTP. Journal of Computers 2(4), 31–40 (2007)
7. Iyengar, J.R., Amer, P.D., Stewart, R.: Concurrent multipath transfer using SCTP multihoming over independent end-to-end paths. IEEE/ACM Transactions on Networking 14(5), 951–964 (2006)
8. Jungmair, A., Rescorla, E., Tuexen, M.: RFC 3436: Transport layer security over stream control transmission protocol (December 2002)
9. Kaufman, C.: RFC 4306: Internet key exchange (IKEv2) protocol (December 2005)
10. Kent, S.: RFC 4302: IP authentication header (December 2005)
11. Kent, S.: RFC 4303: IP encapsulating security payload (ESP) (December 2005)
12. Kent, S., Seo, K.: RFC 4301: Security architecture for the Internet protocol (December 2005)
13. Ladha, S., Amer, P.D.: Improving file transfers using SCTP multistreaming. In: Proceedings of the 2004 IEEE International Conference on Performance, Computing, and Communication (IPCCC 2004), Phoenix, AZ, USA, April 15–17, pp. 513–522 (2004)
14. Lindskog, S., Brunstrom, A.: The design and message complexity of secure socket SCTP. In: Gervasi, O., Murgante, B., Laganà, A., Taniar, D., Mun, Y., Gavrilova, M.L. (eds.) ICCSA 2008, Part II. LNCS, vol. 5073, pp. 484–499. Springer, Heidelberg (2008)
15. Lindskog, S., Brunstrom, A.: An end-to-end security solution for SCTP. In: Proceedings of the Third International Conference on Availability, Reliability and Security (ARES 2008), Barcelona, Spain, March 4–7, 2008, vol. 7, pp. 526–531 (2008)
16. National Institute of Standards and Technology (NIST). Secure hash standard, August 1 (2002), http://csrc.nist.gov/publications/fips/fips180-2/fips180-2.pdf
17. OpenSSL homepage, http://www.openssl.org/ (visited December 9, 2008)
18. Rescorla, E., Modadugu, N.: RFC 4347: Datagram transport layer security (April 2006)
19. Schneier, B.: Applied Cryptography: Protocols, Algorithms, and Source Code in C, 2nd edn. John Wiley & Sons, New York (1996)
20. Schneier, B., Kelsey, J., Whiting, D., Wagner, D., Hall, C., Ferguson, N.: The Twofish Encryption Algorithm: A 128-Bit Block Cipher, 2nd edn. John Wiley & Sons, New York (1999)
21. Stallings, W.: Cryptography and Network Security: Principles and Practice, 4th edn. Prentice-Hall, Upper Saddle River (2006)
22. Stevens, W.R., Fenner, B., Rudoff, A.M.: UNIX Networking Programming: The Sockets Networking API, 3rd edn., vol. 1. Addison-Wesley, Boston (2004)
23. Stewart, R.: RFC 4960: Stream control transmission protocol (September 2007)
24. Stewart, R., Poon, K., Tuexen, M., Yasevich, V., Lei, P.: Sockets API extensions for stream control transmission protocol (SCTP), draft-ietf-tsvwg-sctpsocket-19.txt (work in progress) (Expires: August 20, 2009)
25. Stewart, R., Ramalho, M., Xie, Q., Tuexen, M., Conrad, P.: RFC 3578: Stream control transmission protocol (SCTP) partial reliability extension (May 2004)
26. Stewart, R., Tuexen, M., Camarillo, G.: RFC 5062: Security attacks found against the stream control transmission protocol (SCTP) and current countermeasures (September 2007)

27. Stewart, R., Xie, Q., Tuexen, M., Maruyama, S., Kozuka, M.: RFC 5061: Stream control transmission protocol (SCTP) dynamic address reconfiguration (September 2007)
28. Stewart, R.R., Xie, Q.: Stream Control Transmission Protocol (SCTP): A Reference Guide. Addison-Wesley, Boston (2002)
29. Tuexen, M., Seggelmann, R., Rescorla, E.: Datagram transport layer security for stream control transmission protocol, draft-ietf-tsvwg-dtls-for-sctp-01.txt (work in progress) (Expires: January 9, 2010)
30. Tuexen, M., Stewart, R., Lei, P.: RFC 4820: Padding chunk and parameter for the stream control transmission protocol (SCTP) (March 2007)
31. Tuexen, M., Stewart, R., Lei, P., Rescorla, E.: RFC 4895: Authenticated chunks for stream control transmission protocol (SCTP) (August 2007)
32. Unurkhaan, E., Rathgeb, E.P., Jungmair, A.: Secure SCTP: A versatile secure transport protocol. Telecommunication Systems 27(2-4), 273–296 (2004)

A General-Purpose Geosimulation Infrastructure for Spatial Decision Support

Ivan Blecic, Arnaldo Cecchini, and Giuseppe A. Trunfio

Laboratory of Analysis and Models for Planning (LAMP)
Department of Architecture and Planning - University of Sassari,
Palazzo Pou Salit, Piazza Duomo 6, 07041 Alghero, Italy
{ivan,cecchini,trunfio}@uniss.it

Abstract. In this paper we present the general-purpose simulation infrastructure MAGI, with features and computational strategies particularly relevant for strongly geo-spatially oriented simulations. Its main characteristics are *(1)* a comprehensive approach to geosimulation modelling, with a flexible underlying meta-model formally generalising a variety of types of models, both from the cellular automata and from the agent-based family of models, *(2)* tight interoperability between GIS and the modelling environment, *(3)* computationally efficiency and *(4)* user-friendliness. Both raster and vector representation of simulated entities are allowed and managed with efficiency, which is obtained through the integration of a geometry engine implementing a core set of operations on spatial data through robust geometric algorithms, and an efficient spatial indexing strategy for moving agents. We furthermore present three test-case applications to discuss its efficiency, to present a standard operational modelling workflow within the simulation environment and to briefly illustrate its look-and-feel.

Keywords: Geosimulation, software, multi-agent systems, cellular automata, GIS, urban modelling, open source.

1 Introduction

Geosimulation is a simulation technique for modelling phenomena taking place in geographical environments through the so called bottom-up approaches [1-9], such as cellular automata (CA) or multi-agent systems (MAS). Typically, a geosimulation model uses geo-spatial datasets (i.e. entities in space such as buildings, infrastructures, terrain, and so on) to represent the environment and the "active" entities (cells, agents) interacting locally. The use of advanced characteristics of artificial entities (like autonomy, pro-activity, ability to perceive the space, mobility, etc.) combined with explicit and faithful representations of the geographical space make this family of models an effective technique for simulating complex geographical systems.

Indeed, it has been recognised that such approaches may be of great potential for verifying and evaluating hypotheses about how real geographical complex systems operate, and are therefore often considered as key computer-based tools for territorial analysis and decision support, especially in the field of spatial planning [5].

M.L. Gavrilova and C.J.K. Tan (Eds.): Trans. on Comput. Sci. VI, LNCS 5730, pp. 200–218, 2009.
© Springer-Verlag Berlin Heidelberg 2009

The entering of the geosimulation modelling in its age of maturity [1-8], the growing computational capabilities of nowadays computers, the consolidation of general-purpose as well as specialised GIS-based geo-analysis tools together with their availability as open-source libraries and applications, in our view offer the necessary critical mass to foster the development of robust and scalable general-purpose modelling tools. Tools going beyond the toy-like, demonstrative, proof-of-concept status, thus allowing operational modelling and simulation of real-world geographical systems.

As such, these tools should encompass a series of requirements. In particular, they should (1) be grounded on a sufficiently generalised and flexible meta-model, allowing the implementation of a variety of types of geosimulation models; (2) allow a tight interoperability with GIS where the spatial datasets are stored and are pre- or post-processed and analysed; (3) be computationally efficient; and (4) user-friendly.

MAGI (Multi-agent Geosimulation Infrastructure) was developed as a general-purpose integrated infrastructure for geosimulation complying with these requirements. The objective of this paper is to present the main characteristics and features of MAGI relevant for strongly geo-spatially oriented simulations.

2   The Meta-model

The underlying meta-model of MAGI is an attempt to formally generalise a variety of types of models, both from the cellular automata as well as from the agent-based family of models. Therefore, MAGI should foremost be seen as a software application wrapping around this meta-model, without making compromises to the simplicity and elegance of the modelling practice, while offering rich modelling, simulation and experimentation environment.

The meta-model in MAGI is basically a set of objects and agents organised in layers, where each layer can host one specific type of entity. Formally, the environment Env is defined as:

$$Env = \langle \mathcal{P}_G, \mathcal{F}_G, \mathcal{L} \rangle \tag{1}$$

where $\mathcal{P}_G$ is the set of *global parameters*; the set $\mathcal{F}_G$ groups together a series of *global functions* f_G; and $\mathcal{L}$ is the set of layers. In particular, each layer is defined as:

$$L = \langle \mathcal{P}_L, \mathcal{F}_L, \mathcal{A} \rangle \tag{2}$$

where $\mathcal{P}_L$ is the set of *layer parameter*; the set $\mathcal{F}_L$ collects a series of *layer functions* f_L; and $\mathcal{A}$ is the set of model entities (i.e. either agents or simpler objects like static entities or CA cells).

Global and layer parameters can be used in models to affect the behaviour of entities populating the environment. This permits forms of global steering of the model evolution, which has been recognised as important for simulation of phenomena not completely reducible to local interactions (i.e. factors outside the local spatial context can be accounted for in the entities' behaviour).

Moreover, in the MAGI meta-model, parameters may also be a way to store aggregate information from a layer, to be used as such in other layers. For example, thanks to the global and layer-level functions, which are scheduled for execution during the simulation, the local evolution of a phenomenon evolving on a layer L_i may depend on some global indicators of the phenomena evolving on L_j.

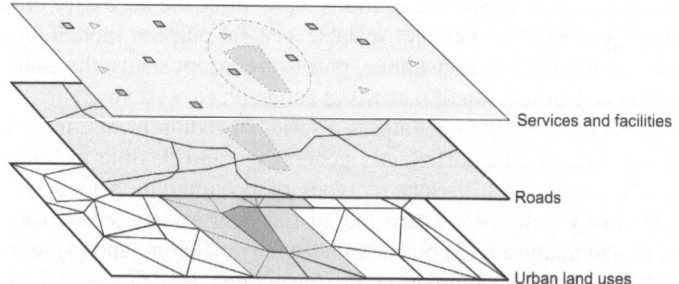

Fig. 1. Example of a model organised in layers

The multi-layer approach presents many advantages in spatial modelling, such as:

- simpler coupling of many spatial models of dynamic phenomena evolving in the same geographical space (e.g. regional and urban models requiring different spatial and temporal resolution);
- easier integration among several different models (evolving on different layers) especially if developed by distinct field experts;
- more direct correspondence of GIS data to the layer-based structure of the model, which simplifies the acquisition of model inputs and the exportation of results.

It is important to point out that the meta-model in discussion includes the possibility of having auxiliary layers containing only static and "inactive" entities. This may prove very useful for urban modelling, for example to account for the effects on other layers of urban entities such as roads, services, facilities and infrastructures (see Fig. 1). In the hereby presented meta-model, the modelling of such interactions is possible because (as we shall see below) every single entity belonging to one layer can interact with entities belonging to other layers, including therefore also auxiliary layers with only static entities.

2.1 The Entity (MAGI Agent)

Layers are populated by agents of different types τ. In particular, an agent a_τ of type τ is defined as:

$$a_\tau := \langle s, g, C \rangle \tag{3}$$

Where s is the internal data structure describing agent's state and history. During the simulation, s can change as a result of interactions with other agents, between the agent and its environment, and due to its behavioural specification. The data structure g defines the agent's geo-spatial attributes. It includes both a vector geometrical shape (like point, line, polygon, circle), representative of the agent, as well as its positioning (i.e. location and rotation). Clearly, during the simulation g can be updated (for example the agent can change its position). Finally, C is the current *spatial context* of the agent a_τ, represented by a set of references to objects and agents being relevant to a_τ, and therefore potentially subject to its actions or requests for action. Clearly, during the simulation, every object in C can be dynamically updated by the agent as a result of its perception activity.

An agent type τ is defined as:

$$\tau := \langle S_\tau, G_\tau, \Sigma_\tau, \Theta_\tau, \delta_\tau, \gamma_\tau \rangle \tag{4}$$

where:

- S_τ is the set of possible agent's internal data structures s;
- G_τ is the set of admissible geometrical shapes for agents of type τ;
- Σ_τ is the set of possible actions defining behavioural specification (actions can modify internal states and geo-spatial attributes of agents);
- Θ_τ is the set of *perception functions* which can be used by agents to perceive their environment. A function $\theta_i \in \Theta_\tau$ maps the environment to the agent's context C. The perception activity is of particular importance for agents in a geo-spatial environment, since it is one of the main components of their spatial cognitive capabilities.
- δ_τ is the *decision function*. Based on its current state and context, this function is used by agents at every turn in order to decide which operation to execute. An operation may be one of agent's own actions and/or requests to other agents to execute their actions.
- γ_τ is the *agreement function*. This function is used by agents to decide, based on their state and context, whether to agree on the execution of an action from Σ_τ, when requested to do so by other agents.

The above definition of agents allow for special case of agents with purely reactive behaviour. As a consequence of this, the agreement on actions asked to reactive agents is decided through a specific agreement function directly belonging to the environment.

2.2 Cellular Automata as a Special Case

The simplest type of "agents" available through this meta-model are *cells* which can be used for modelling the environment as a CA. Such cells are endowed with a shape and dynamic multiple neighbourhoods (i.e. the agent's *spatial context*). Moreover, MAGI makes available a rich set of functions for building regular lattices consisting of rectangular or hexagonal cells with standard or custom neighbourhoods.

Given the possibility of multi-layer structure of models, we thus have a so called multi-layered model in which the automata are organised in different layers [10].

Even if many CA models adopt a strictly local neighbourhood, spatial interactions in geographical phenomena can often take place over greater distances. For this reason, accordingly with the notion of the proximal space deriving from the research in "cellular geography" [11], in CAs developed in MAGI every cell can have a different neighbourhood defined by relations of proximity of spatial geographical entities, where proximity can be both a topological relation or a generic functional influence.

In other words, similarly to the spatial context of an agent defined above, the neighbourhood of a CA cell can be defined as a generic set of agents-cells whose states can influence the variation of the cell's state. It is important to note that neighbourhoods can be non-stationary, since they can vary in size and shape during the system's evolution, thanks to the reiteration of the spatial queries attached to the cell. In general, depending on the model requirements, various queries can be associated to each cell, so that different possible neighbourhoods, even belonging to different layers, correspond respectively to different rules of the transition function.

3   Overview of MAGI Architecture

From architectural point of view, MAGI is composed of three components: *(i)* a modelling and simulation environment, *(ii)* a GIS interoperability bridge, and *(iii)* a class library developed in C++ language. The latter can be used for developing model plug-ins for MAGI simulation environment as well as independent stand-alone models.

3.1   Modelling and Simulation Environment

The modelling environment has two working modes, the *design mode* and the *simulation mode*. As shown in Figure 2, the design mode allows the user to define and fully specify the model with all its input-output constraints, and its domain and preconditions of application. The simulation mode is used to execute the model for a specific scenario. In general, a scenario is defined by a set of parameters and by a set of geo-spatial layers, with their initial and boundary conditions.

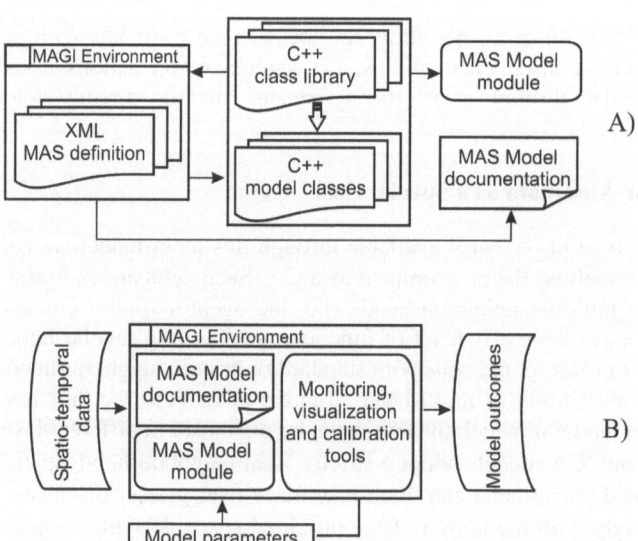

Fig. 2. A scheme of MAGI: A) Modelling phase; B) Simulation phase

Internally, a model structure in MAGI is an XML document describing the organisation of the model in layers, model parameters, properties of elementary entities, and so on. Such XML model description is used by the framework to generate a skeleton of the C++ source code of the model, according to the MAGI C++ class library (see Figure 2). Subsequently, the generated source code can be imported into a standard C++ development environment for further coding and debugging.

In addition, the simulation module contains specific user interfaces for invoking the simulation plug-in and for configuring the preferences and initial conditions relevant for the simulation, among which the values of global and layer parameters, the number of time steps, and the output options.

The execution of the simulation engine produces output scenarios in XML format, together with several synthetic statistics of the results of simulation (like the evolution of global and layer parameters over simulation steps).

3.2 GIS Interoperability Bridge

Recent research in the field of geosimulation focused mainly on techniques to improve models of spatial processes, to propose new conceptualisations of spatial entities and their mutual relationships, to apply simulation models to real-world problems and to develop new software tools for supporting the process of modelling. This latter research direction has produced modelling and simulation frameworks – like Swarm, Repast, OBEUS [5,6], among others (see [5] for a review) – supporting various forms of integration of geo-spatial data. However, the interoperability with GIS offered by existing tools is still limited.

The integration of simulation engines *into* existing GIS systems has been proposed elsewhere (e.g. loose coupling based on the Remote Procedure Call paradigm or other proprietary protocols) [4,12,13]. However, the coupling of simulation engines with proprietary GIS can hardly provide the necessary modelling flexibility and a satisfactory computational efficiency. For this reason, MAGI has been specifically designed to allow an easy and effective importation and exportation of geo-spatial datasets from and to standard GIS formats. In particular, the user interface of MAGI offers functionalities for: *(i)* creating entities (i.e. agents) from features of vector data files; *(ii)* mapping the feature attributes directly into agents' states and objects' properties; *(iii)* importing raster files as cellular spaces where cells' attributes can correspond to pixels samples, *(iv)* powerful importing capabilities from common GIS formats, and *(v)* the integration of specific computational strategies for efficiency and robustness.

3.3 Class Library

As mentioned above, MAGI is endowed with a C++ class library for supporting the development of geosimulation models according to the meta-model described in section 2. Clearly, the MAGI class library can also be used by software developers as a stand-alone library for deriving and implementing models, independently from the MAGI environment.

The main advantage of providing a class library lies in the greater simplicity of model-designing phase in comparison to a development from scratch. Indeed, at the beginning, the modeller needs only to specify the model components, and is relieved from a number of implementation details and complications. In addition, the use of the XML standard for outlining the model structure permits an easy merging of different sub-models, simplifying and promoting the collaboration among different experts. Another important implication of the use of XML consists in high readability of the model structure, which is reputed important for avoiding black-box effects.

The MAGI class library implements and adopts several computational strategies and offers specific algorithms which are utterly relevant for the computational efficiency of geosimulations. Given their relevance, we present few of them in the remaining of the present section, namely the algorithms for indexing of agents and the embedded geometry engine.

Indexing of moving agents. MAGI class library implements a simple but efficient data structure known as R-Tree [14] which is widely adopted with many variations (e.g. [15,16]) for indexing spatial data in GIS and CAD applications.

Briefly, an R-tree (the "R" stands for "Rectangle") is a tree-like data structure where: *(i)* each leaf node contains a variable number of pointers to spatial objects (e.g. buildings, pedestrians, cars) together with their *minimum bounding rectangle* (MBR); *(ii)* every non-leaf node contains a variable number of pointers to child nodes together with the bounding box of all their MBRs (see Figure 3). In practice, an R-tree corresponds to a variable number of MBRs hierarchically nested but with the possibility of spatial overlapping (minimising such overlapping is of crucial importance for the efficiency of an R-tree query). The insertion procedure uses MBRs to ensure that two objects close enough in the space are also placed in the same leaf node. If the maximum number of entries for a leaf node is reached during the insertion, then that leaf node is split into two nodes according to some heuristics, and such a split can propagate along the tree. The clear advantage of this kind of spatial indexing structure emerges when the searching algorithms use bounding boxes to decide whether or not to search inside a child node. In fact, in such a searching strategy, most of the tree's nodes are never visited thanks to a computationally cheap geometric tests on rectangles. In addition, in MAGI the R-tree is stored in the main memory and this further increases the efficiency of the application.

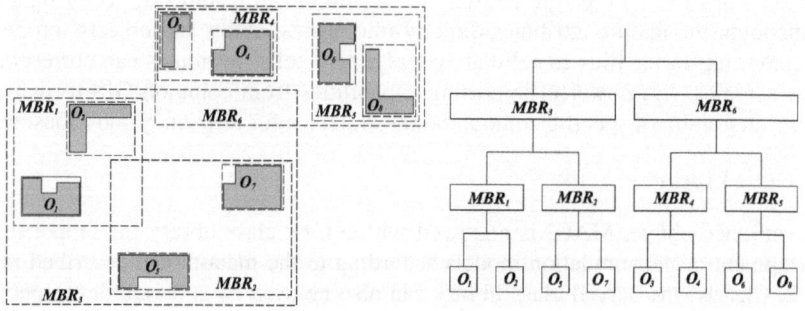

Fig. 3. R-tree spatial indexing scheme

The use of such an indexing strategy provides clear gains in terms of computational efficiency and scalability. One very common use is for simulating visual perception of moving agents, which is the example more extensively discussed in the test cases section of this paper (see below section 4.1). There are also other types of spatial perception that can take advantage of the spatial indexing structure available in MAGI. For example, in assessing the physical admissibility of agent's movement, the agent itself (or a specific environment function) must check if the desired trajectory intersects objects or other agents.

It is worth mentioning that the algorithms of spatial perception in MAGI consider all agents with their actual bi-dimensional shapes (e.g. circles, segments, polygons), so agents may reduce or block the view to each other. Therefore, agents' shapes must be indexed in the R-tree as any other entity in the environment. However, since agents can move, the update operations are necessary for maintaining the coherence and efficiency of the spatial indexing structure. Unfortunately, since the change of object position in a standard R-tree corresponds to one remove and one insert operation (potentially with the need to split or merge nodes), that would become significantly expensive for simulating models with many thousands of moving agents. Indeed, one of the reasons of reduced efficiency of geosimulation models developed *within* a GIS environment is related to the use of therein available indexing structures which were conceived mainly for static spatial data and therefore exhibit poor update performances.

Since the problem of efficiently indexing the actual position of moving objects arises in many applications (e.g. real-time traffic simulation and monitoring, transportation managements, mobile computing) [17], it has been object of great attention in recent years. To reduce the number of update operations, many existing approaches (e.g. [18,19]) use a function to describe movements of objects and then update the data structure only when the parameters of that function change (e.g. when the speed or the direction of a car or a pedestrian change). However, frequently there is no sufficiently good function available to describe some specific movements. For this reason, other alternatives to the standard indexing techniques have been proposed for accelerate updating operations. One of such indexing scheme, which fits well to the characteristics of moving agents, has been implemented in MAGI. It is based on processing updates in the R-tree using a bottom-up approach instead of the standard top-down procedure [20,21,22]. Specifically, this approach processes an update from the leaf node of the old entry to be updated, and tries to insert the new entry into the same leaf node or into its sibling node. For accessing the leaf node, a secondary index (i.e. a direct link) is maintained for all objects. The particular scheme implemented in MAGI is the so called LUR-tree (Leaf-prior-Update R-tree) [21]. For further details on the visual perception problem and LUR-tree implementation in MAGI see [23].

Geometry engine. Given the vector representation of objects associated to entities in the environment, both spatial perception and spatial reasoning require the use of a geometry engine. Such engine must provide: *(i)* the capability of creating and using geometrical objects; and *(ii)* suitable computational geometry algorithms for handling these objects, such as computation of spatial relationships and properties, shape combinations, overlay, buffer operations, among others.

For this purpose, the GEOS (Geometry Engine - Open Source) class library [24] has been integrated into MAGI, together with a simple wrapper interface. GEOS provides a complete model for specifying 2-D linear geometry and implements a set of operations on spatial data using robust geometry algorithms. The main reasons for integrating GEOS in a geosimulation class library are:

- *Robusteness*. Geometry algorithms are susceptible to problems of robustness, i.e. an inexact behaviour due to the round-off errors and the numerical errors propagation. In GEOS, the fundamental geometric operations (e.g. line intersections), on which most of other operations are based, are implemented using robust algorithms. Also, the binary predicate algorithm, which is particularly important for the process of agents' spatial perception, is fully robust. Many other algorithms in GEOS minimise the problems of robustness and those known not to be fully robust (i.e. spatial overlay and buffer algorithms) are rarely used in agent modelling and work correctly in the majority of cases.
- *Computational requirements*. As geosimulation models may deal with thousands of agents, the computational complexity of every geometric operation used by agents can greatly affect the overall requirements of computational resources. Unfortunately, robust geometric algorithms often present poor computational performances. GEOS uses specific techniques to produce good performance when operating on common types of input data.
- *Rich set of available features and spatial operations*. In terms of the spatial model, geometric objects and method definitions, GEOS complies well with the OpenGIS Simple Features Specification [30]. In particular, it provides a set of Boolean predicates for computing common spatial relationships (e.g. *disjoint, intersects, crosses, within, contains, overlaps, touches, equals*, etc.) as well as common algorithms for computing intersections, distances, areas and many others.

Thanks to specific interfaces made available in MAGI, it becomes easy to access GEOS functionalities and to use them for the development of complex spatial reasoning algorithms.

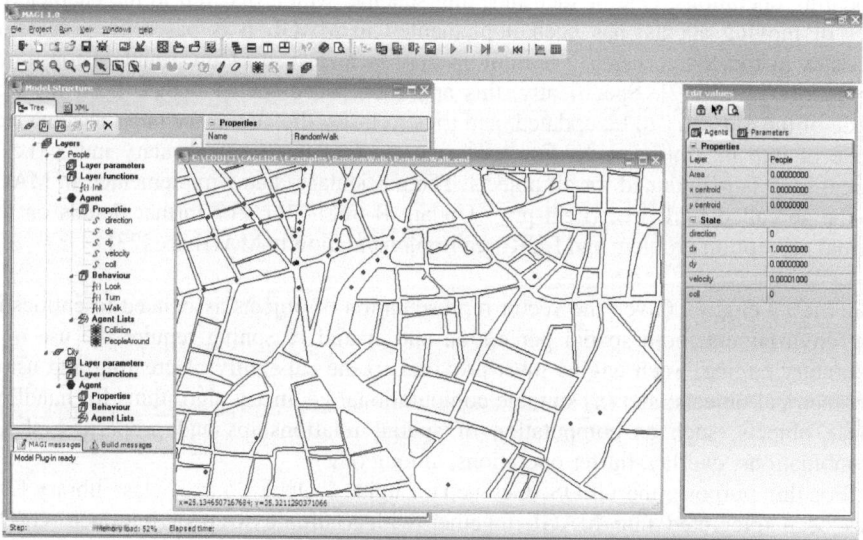

Fig. 4. A screen capture of MAGI

4 Test Cases

MAGI has been extensively used for implementing different MAS and CA models. In this section, we discuss three test cases to better show the functioning and to point out the advantages of MAGI. The first one is a multi-agent model of visual perception of moving agents; the second is a CA simulation of urban growth [3,4], and the third is a CA land-use model [25,26,27].

4.1 Visual Perception of Moving Agents

As mentioned before, the spatial perception of agents represents one of the main components of their spatial cognitive capabilities in geosimulation models. An agent must be able to perceive objects and other surrounding agents to build an internal representation of its spatial context, and to use it in her decision-making process. In some cases, spatial perception implies not only detecting presence/absence of entities (i.e. agents in the language of our meta-model), but also estimating their geometric characteristics (e.g. the distance from them, the relative distances among different objects, objects' areas or shape types).

Of particular importance for many models (for example those involving pedestrian movements or traffic models) is the ability of *visual perception*, which is based on the determination of the visibility of entities lying within agent's field of vision. Visual perception represents a typical computation-intensive task that can dominate the computational effort in a simulation. Thus, when the objective is to produce fast simulations on standard computers involving thousands of agents, some approximate treatment is usually adopted (e.g. see [28]).

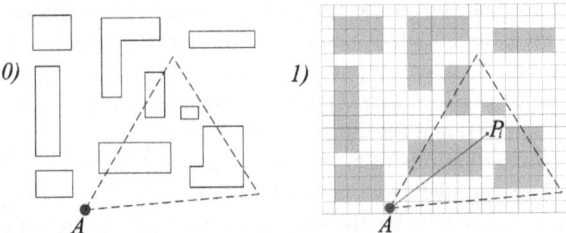

Fig. 5. Visual perception in most raster contexts. First, the problem in 0) is represented in the raster space 1). Then, in 1) for each cell P_i in the agent's field of vision, the line of sight AP_i is computed; if the line AP_i intercepts an object, then the cell P_i is not visible.

In most geo-spatial simulation models, agents are constrained to move to adjacent sites, often represented by the vertices of a graph [10] or by the cells of a grid (i.e. in a raster environment) [28]. In such a setting, the way in which the environment is structured determines the way in which spatial indexing is performed, which on its turn influences the way in which the model of agents' spatial perception must be implemented. For example, in a raster space, an agent's perception of immediately surrounding objects could simply be based on an exploration of immediately neighbouring cells. For farther entities, the visual perception can be based on the computation of the line of sight between the agent's position and the target cell, for every cell in agent's field of vision (see Fig. 5) [28].

As mentioned before (see subsection 2.1), in MAGI all agents are associated to a geometrical shape with a vector representation. This however does not eliminate the possibility to use raster layers as cellular spaces. Also, agents may in the line of principle move to any position (although using some specific types of agents in MAGI, the action *move* must be evaluated by the environment, and its successful completion is not guaranteed in general).

The possibility to use vector shapes allows more realistic simulations, offers greater flexibility, and a better interoperability with GIS applications. In models based entirely on cellular spaces, the dimension of cells determines both the accuracy of the representation of complex shapes (e.g. buildings) and the spatial resolution of agents' movements. However, such a rigidity is not present in models using vector shapes for entities, be they static or moving agents. Furthermore, the accuracy of spatial perception based on vector representation can take significant advantages from the existing algorithms of computational geometry. Another manifest benefit offered by the possibility to treat with vector entities lies in a greater simplicity of use of –, and interoperability with most spatial data (e.g. demographic data, land-uses, buildings shapes, etc.) managed by GIS software, without the need of their previous rasterisation. In fact, the geo-spatial features can directly be mapped to agents of the model, and vice versa.

However, to these advantages could correspond greater computational costs of algorithms managing agents' movements and spatial perception. In many situations this may not be of crucial importance (for example when the model involves only few agents or when its purpose is to analyse a complex system without the need for real-time simulation). Nevertheless, the containment of computational requirements is still a relevant factor, as that allows the use of more detailed models (e.g. more evolved agents) with given computational resources. In order to obtain such objectives of efficiency, the process of agents' spatial perception in MAGI is grounded on the two features mentioned above: a specific purpose-oriented spatial indexing technique for moving objects, and a fast and robust geometry engine.

We have undertaken various tests in order to assess the efficiency of different implementations of moving agents capable of visual perception. In Figure 6 we report the results of a scalability benchmark with respect to the number of agents used. The test model used is a simple simulation of agents walking in an urban environment with two increasing degrees of capability of spatial perception. In the case A, agents can avoid static obstacles (e.g. buildings) ignoring the presence of other agents in their proximity. In the case B, agents can also perceive other agents. At each step of the simulation, every agent makes a single spatial movement following a direction chosen to avoid collisions.

The experiment was conducted on a standard 1.8 Ghz PC, with a fixed number of 200 simulation steps, starting from scenarios populated with different number of agents, from 1000 to 15.000.

Clearly, the case A does not produce significant computational problems as the number of agents increases. Indeed, the model scales linearly even in naïve implementations (i.e. without any spatial indexing technique), although the computational time is about 90% greater than with the implementation using R-tree.

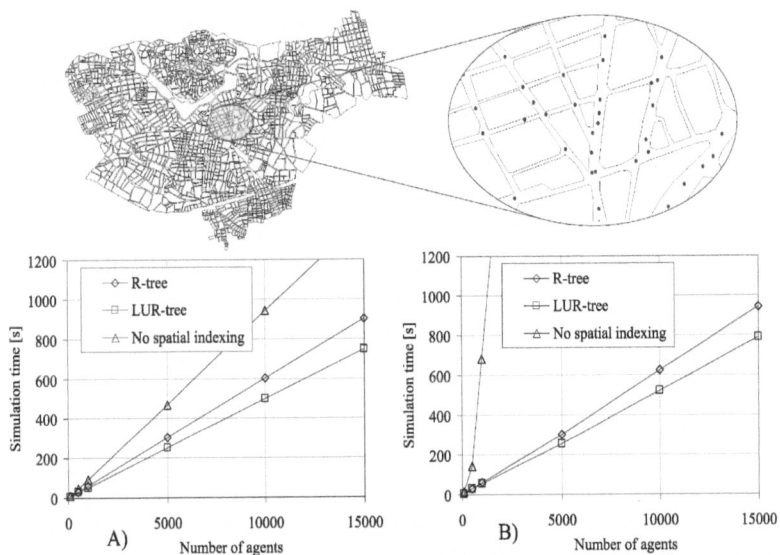

Fig. 6. Results of the test on walking agents. In A) agents walk avoiding collisions but ignoring other agents; in B) agents walk avoiding collisions and perceiving the presence of other agents in their field of vision.

As expected, in the case *B* where agents need to perceive the presence and the position of other agents, the naïve approach leads to the impossibility to simulate systems with thousands of agents within reasonable time, since the computational complexity of perception is $O(n^2)$, where n is the number of the agents. On the contrary, the computational time, obtained through the use of standard algorithms available in MAGI, exhibits a linear dependence on the number of agents.

In both cases the use of the LUR-tree approach leads to a computational time of about 16% lower than that of the standard R-tree.

On the whole, these tests show that the advantages deriving from vector-based geosimulation models using thousands of agents can effectively be put in action with MAGI, at affordable computational costs.

4.2 Urban Growth CA Model

An operational assessment of the effectiveness of MAGI, as simulation environment tightly interoperable with GIS, has been undertaken through the development a SLEUTH-like urban growth model.

SLEUTH is one of the most popular CA-based simulation models of urban growth and of the related evolution of land uses as a diffusion process over time [3, 4]. The name comes from the GIS data layers required as inputs to the model: *Slope*, *Land use*, *Exclusion layer*, *Urban*, *Transportation*, and *Hillshade*.

The CA urban growth model (UGM) in SLEUTH has been widely applied [29]. The UGM uses square cells, with neighbourhoods of eight cells. The cell states represent several static characteristics, corresponding to the above mentioned GIS data layers (the layer *Hillshade* is used only for representation purposes and the *Land use* layer is not used by the UGM), as well as dynamic characteristics of the urbanisation. The transition function depends on five integer parameters defined on the interval from 1 to 100. These parameters are:

- *Dispersion*, which determines the overall dispersal of urbanisation;
- *Breed*, which expresses the likelihood of a new isolated urban cell to initiate an independent growth cycle;
- *Spread*, controlling the diffusion by contagion from existing urbanised cells;
- *Slope Resistance*, influencing the likelihood of urbanisation on steep slopes;
- *Road Gravity*, which regulates the generation of new urbanisation towards and along roads.

The UGM transition function is composed of four probabilistic rules: *(i) Spontaneous Growth*, which consists of random spontaneous urbanisation of a fixed number of cells at each time step; *(ii) New Spreading Centres*, determining whether any of the new, spontaneously urbanised cells become a new urban spreading centre; *(iii) Edge Growth*, defining the component of growth which stems from existing spreading centres; *(iv) Road-Influenced Growth*, accounting for the urbanisation related to the presence of transportation infrastructure and therefore of the resulting better accessibility (for details see [29]).

Since the main purpose of our implementation of SLEUTH UGM in MAGI is the assessment of effectiveness of the software infrastructure, the model described here does not include details such as the so called parameter self-modification [3].

The SLEUTH UGM represents an interesting test case for the evaluation of the characteristics of MAGI, mainly because:

1. it is a parameter-dependent CA model, thus fitting well with the characteristics of the MAGI library that allows the definition of a set $\mathcal{P}_L$ of layer parameters;
2. it is a "constrained" CA, since in the *Spontaneous Growth* rule the number of cells changing their state from *non urban* to *urban* is fixed, and this can be handled in a natural way using a specific layer-level function in the set $\mathcal{F}_L$;
3. the transition function determines the interaction among each cell and its nearest neighbouring cells (for the *New Spreading Centres* and *Edge Growth* rules), as in a standard CA, but also the interaction with cells representing roads within a radius considerably larger than the size of single cells; this can be easily implemented defining two distinct neighbourhoods of cell, as contemplated by the meta-model described above.

In particular, the simulation model under discussion is composed in MAGI of one layer including the five layer-level parameters mentioned above. Each cell is characterised by a state composed of the following five scalar properties::

- *Slope*, representing the terrain slope in percent;
- *Exclusion*, defining the level of resistance to urbanisation (e.g. open water bodies take the value 100, while areas available for urban development take the value 0);
- *Urban*, which is the only dynamic property and represents the urbanisation state (can take the two values: *urban* and *non urban*);

- *Road*, derived from a road network GIS layer, and is also of binary type (*road* or *non-road*) indicating whether the cell contains a road or not.

In order to implement the UGM transition function in MAGI (i.e. the four above mentioned rules of the transition function), two distinct neighbourhoods are associated to each cell. This is obtained by defining the set Θ of selection queries as:

$$\Theta = \{\theta_M, \theta_R\} \tag{5}$$

where θ_M gives the standard Moore neighbourhood while θ_R returns the nearest cell, located within a predefined radius, having $Road \neq 0$. Both selection queries are executed only at the first time step of the simulation. While the query θ_M is provided by the library, in order to implement the query θ_R in MAGI, the modeller have to develop an easy override of the two virtual functions executeQuery and includeEntity, as shown in Fig. 7. In particular, in executeQuery a filter shape representing a circle centered in the cell's centroid is used for invoking the library function nearestNeighbourhoodSearch, which on its turn performs an efficient preliminary search in the R-tree based on the MBRs of the involved shapes. The latter function also invokes the includeEntity function for refining the result through the check of the property *Road* of the candidate cells as well as its spatial relationship with the filter shape (i.e. using the function contains provided by GEOS). This approach in construction CA neighbourhoods, being not dependent on the regularity of the space tessellation, presents the advantage of a more flexible modelling approach without sacrificing the computational efficiency. For example, the extension towards a model based on irregular cells would not require the rewriting of the functions in Fig. 7.

The raster layers used for the production of the input scenario file have been elaborated from several raster files available on the SLEUTH web site [29]. Even if not strictly necessary, the visualisation capabilities of MAGI are useful in the preprocessing phase for a preliminary assessment of qualitative characteristics of spatial data which can influence some modelling-related decisions (such as the extension of the area under study or the size of cells).

The SLEUTH UGM has been used in MAGI with many different input scenarios and, on the whole, the test case has definitely highlighted the effectiveness of the framework in making this family of well-established CA models more accessible to standard GIS users.

4.3 Constrained Land-Use Model

Another test case model developed and experimented in MAGI is a constrained CA for the simulation of the local land-use dynamics [25,26,27].

This model is "constrained" in the sense that at each time step the aggregate level of demand for every land use is fixed by an exogenous constraint. This constraint is defined on a macro-level by external a-spatial demographic, economic and environmental models.

```
void NearestRoadCells::executeQuery()
{
    auto_ptr<Geometry> filter(world.shapeFactory->createCircle(
        ownerEntity->getXCentroid(),
        ownerEntity->getYCentroid(),
        world.road_gravity_radius));

    nearestNeighbourhoodSearch(1, filter);
}


bool NearestRoadCells::includeEntity(Entity &candidate,
                                     Geometry* filter)
{
    try
    {
        Cell &c = dynamic_cast<Cell&>(candidate);
        if ( c.Road!=0   &&
             filter->contains(ownerEntity->getShape()) )
        return true;
    }
    catch (const bad_cast& e)
    {
        cerr << e.what() << endl;
    }
    return false;
}
```

Fig. 7. The C++ functions required by the MAGI library for implementing the θ_R query in the SLEUTH-like urban growth model

Our model is a simplified version of the one discussed in [25,26,27] since it has been devised for MAGI testing purposes. However, its extension to a more detailed version is plainly straightforward.

In particular, the only dynamic property of the elementary automaton (i.e. cell) state represents the land use, and has been limited to the set $\{residential, rural, industrial\}$. Other properties of cells represent static land-use characteristics. These are: cell *accessibility* $A \in [0,1]$, *suitability* $S_j \in [0,1]$ for different land uses (taking into account features like slope, terrain aspect and the presence of specific infrastructures), a Boolean value Z_j for each land use, defining the exclusion of the j-th land use (for example due to zoning regulations or physical constraints). The state of the cell also includes a *transition potential* P_j for each land use j, expressing the level of propensity of the land to acquire the j-th type of use.

The cell neighbourhood is defined as the circular region of a given radius around the cell.

The first phase of the transition function consists of the computation of the transition potentials for every cell, according to the equation:

$$P_j = \nu A_j S_j Z_j N_j \tag{6}$$

where ν is a random perturbation factor and N_j is the so called *neighbourhood effect*. The latter represents the sum of all the attractive and repulsive effects of land uses and land covers within the neighbourhood, on the j-th land use which the current cell may assume. Since, in general, more distant cells in the neighbourhood have smaller influence, in our MAGI version of the model, the factor N_j is computed as:

$$N_j = I_k + \sum_{c \in V} \frac{2\, a_{ij}}{1 + e^{\frac{\gamma\,(d-s)}{d_{ij}}}} \tag{7}$$

where the summation is extended to all cells of the neighbourhood V (which does not include the owner cell itself). In Eq. 7 the current land use of the cell $c \in V$ is denoted by i; s is the cell size; d is the distance between the cell c and the current cell; the parameter a_{ij} represents the maximum influence (positive or negative) of the land use i on the use j; d_{ij} is the distance in correspondence of which the contribution to the neighbouring effect is a fraction α of the maximum value (currently $\alpha = 0.05$) and $\gamma = \ln((1 - \alpha)/\alpha)$. The positive term I_k, where k denotes the current land use of the cell, accounts for the effect of the cell on itself (zero-distance effect) and represents an inertia effect due to the costs of transformation from one land use to another.

The second phase of the transition function takes place on a non-local basis and consists of transforming each cell into the state with the highest potential, given the constraint of the overall number of cells in each state imposed by the exogenous trend for that iteration.

As for the SLEUTH UGM model discussed in section 4.2, this land-use model represents another significant test case for the purpose of evaluating MAGI. In fact, the presence of a non-standard neighbourhoods and of a parameterised transition function including also non-local transformations, are both characteristics which can naturally be treated by the MAGI library.

In the model developed for MAGI, a single layer of square cells has been used for the automaton. However, MAGI library could also allow irregular cells be used for this model without the need of changing the model source code.

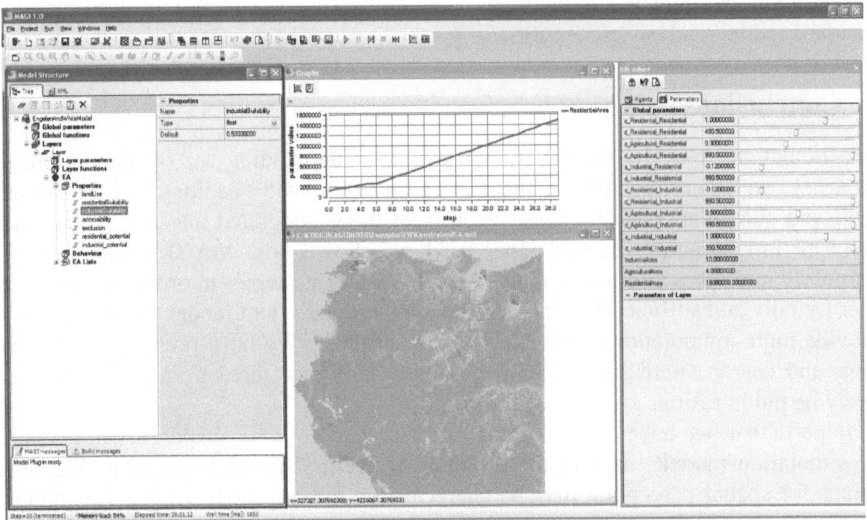

Fig. 8. A screenshot of MAGI running the constrained CA for the simulation of the land-use functions

The neighbourhood has been defined through a selection query executed only at the first simulation time step, which returns the cells located within a predefined radius of 1000 m, excluding the owner cell. To this purpose, as for the model discussed in section 4.2, specific overrides for the library functions executeQuery and includeEntity, have been developed.

A set of layer parameters hold the values of I_k, a_{ij} and d_{ij} of Eq. 7 and, once the XML model definition has been loaded in MAGI, their value can be easily set by the user and stored in a scenario file.

At the first time step of the simulation, a layer function loading trends of land-use demand for all the subsequent time steps is executed.

The local component of the transition function (i.e. the standard transition function executed on the cell level) computes the transition potential P_j according to Eqs. 6 and 7, while a layer function implements the transition algorithm. In the latter, all cells are ranked by their highest potential, and cell transitions begin with the highest ranked cell and proceed downwards until a sufficient number of cells have been attributed to every land use.

The model has been applied on a region in the western part of Sicily (Italy) using a grid of 135×135 square cells of size 400x400 meters. The initial scenario, corresponding to the situation in 1970, has been produced efficiently in MAGI where some input raster layers have been associated to cells' properties and incorporated into the XML scenario file, together with the values of layer parameters.

Since the model under discussion is probabilistic, a series of 30-step simulations (each simulating 30 years of system's evolution) has been performed and the output scenarios have been used for the production of a probability map through a specific function in MAGI. In Fig. 8, the final outcome of the simulation is depicted.

This test case also showed that the architecture and the features of MAGI can make CA-based modelling and simulation as accessible to users as other advanced but more standard spatial analysis procedures.

5   Conclusions

Geosimulations may be computationally quite a demanding set of simulation approaches, and therefore require the use of an articulated "ecology" of techniques, tools and computational strategies which have to be integrated into a well organised software infrastructure. In this paper we have shown how MAGI implements and purposely integrates a series of approaches, tools and strategies in order not only to be user friendly and sufficiently general to host a large variety of model types, but also to provide high computational efficiency and robustness. We have presented few situations and cases, where such techniques and strategies offered by MAGI can effectively be put in action.

In particular, we have discussed one of the crucial cognitive capability of agents in geosimulation models, namely the problem of agents' spatial perception. The algorithms for spatial perception may be characterised by very high computational costs, especially in models with vector representation for entities. That notwithstanding, in MAGI agents can operate in a vector space, but can still sense the environment rather efficiently thanks to the use of a robust and efficient geometric engine together with a specifically implemented spatial indexing technique.

We have also briefly presented two test drives of MAGI on two widely know CA models, which we hope have shown its effectiveness and user-friendliness not only for the modelling and simulation tasks, but also the usefulness of its GIS interoperability bridge for all the ancillary, often time-consuming activities of pre-processing of spatial data, scenario initialisation, and analysis of results of simulation.

Taking into consideration the current use experiences of MAGI – notwithstanding they have still been quite limited in number – there exist preliminary clues that different aspects of MAGI are fairly competitive with other similar simulation frameworks (like Swarm, Repast and OBEUS) with respect to characteristics such as user-friedliness, interoperability with GIS, and computational efficiency due to the adoption of optimized libraries, specifically implemented computational approaches and the use of a programming language (i.e. C++) generally exhibiting better performance.

From the point of view of system's extension, in the future we plan to focus on the development of a support tool for calibrating and validating geosimulation models, as well as on few other implementational aspects of MAGI, such as the development of a version using thread-pooling techniques, capable of taking greater advantage of multicore processors.

The architectural characteristics of the integrated environment of MAGI delivers satisfactory computational power without sacrificing the generality of the base meta-model. This is a particularly relevant point given the fact that MAGI was designed as a general-purpose modelling and simulation tool for geosimulations.

References

1. Cecchini, A.: Urban modelling by means of cellular automata: generalised urban automata with the help on-line (AUGH) model. Environment and Planning B, 721–732 (1996)
2. Batty, M.: Urban systems as cellular automata. Environment and Planning B 24, 159–164 (1997)
3. Clarke, K., Hoppen, S., Gaydos, L.: A self-modifying cellular automaton model of historical urbanization in the San Francisco Bay Area. Environ. Plan B 24, 247–261 (1997)
4. Clarke, K., Gaydos, L.: Loose-coupling a cellular automaton model and gis: long-term urban growth predictions for San Francisco and Baltimore. Int. J. Geog. Inf. Sci. 12, 699–714 (1998)
5. Benenson, I., Torrens, P.M.: Geosimulation: object-based modeling of urban phenomena. Comput. Environ. Urban Syst. 28, 1–8 (2004)
6. Torrens, P., Benenson, I.: Geographic automata systems. Int. J. of Geogr. Inf. Sci. 19, 385–412 (2005)
7. Castle, C., Crooks, A.: Principles and concepts of agent-based modelling for developing geospatial simulations. Working Paper 110, Centre for Advanced Spatial Analysis, University College London, London (2006)
8. Cecchini, A., Trunfio, G.A.: Supporting urban planning with CAGE: a software environment to simulate complex systems. Adv. Complex Syst. 10-2, 309–325 (2007)
9. Ferber, J.: Multi-agent systems: an introduction to distributed artificial intelligence. Addison-Wesley, Reading (1999)
10. Bandini, S., Manzoni, S., Vizzari, G.: Toward a platform for multi-layered multi-agent situated system (MMASS)-based simulations: focusing on field diffusion. Appl. Artif. Intell. 20, 327–351 (2006)

11. Tobler, W.: Cellular geography. In: Gale, S., Olsson, G. (eds.) Philosophy in Geography, pp. 379–386. Reidel, Dordrecht (1979)
12. Wu, F.: GIS-based simulation as an exploratory analysis for space-time processes. J. Geogr. Syst. 1, 199–218 (1999)
13. Wagner, D.: Cellular automata and geographic information systems. Environ. Plan. B 24, 219–234 (1997)
14. Guttman, A.: R-trees: A dynamic index structure for spatial searching. In: SIGMOD Conf., pp. 47–57 (1984)
15. Beckmann, N., Kriegel, H.P., Schneider, R., Seeger, B.: The R*-tree: An efficient and robust access method for points and rectangles. In: SIGMOD Conf., pp. 322–331 (1990)
16. Roussopoulos, N., Leifker, D.: Direct spatial search on pictorial databases using packed R-trees. In: SIGMOD Conf., pp. 17–31 (1985)
17. Wolfson, O., Xu, B., Chamberlain, S., Jiang, L.: Moving objects databases: Issues and solutions. In: 10th Int. Conf. Sci. Stat. Database Manag., pp. 111–122. IEEE Computer Society, Los Alamitos (1998)
18. Kollios, G., Gunopulos, D., Tsotras, V.J.: On indexing mobile objects. In: Proceedings of the Eighteenth ACM SIGACT-SIGMOD-SIGART Symp. Princip. Database Syst., pp. 261–272 (1999)
19. Tayeb, J., Ulusoy, Ä.O., Wolfson, O.A.: Quadtree-based dynamic attribute indexing method. Comput. J. 41, 185–200 (1998)
20. Kwon, D., Lee, S., Lee, S.: Indexing the current positions of moving objects using the lazy update R-tree. In: Proceedings of the Third Int. Conf. Mob. Data Manag., pp. 113–120 (2002)
21. Lee, M.L., Hsu, W., Jensen, C.S., et al.: Supporting frequent updates in r-trees: A bottom-up approach. In: VLDB, pp. 608–619 (2003)
22. Kwon, D., Lee, S., Lee, S.: Efficient update method for indexing locations of moving objects. J. Inf. Sci. Eng. 21, 643–658 (2005)
23. Blecic, I., Cecchini, A., Trunfio, G.A.: A Software Infrastructure for Multi-agent Geosimulation Applications. In: Gervasi, O., Murgante, B., Laganà, A., Taniar, D., Mun, Y., Gavrilova, M.L. (eds.) ICCSA 2008, Part I. LNCS, vol. 5072, pp. 375–388. Springer, Heidelberg (2008)
24. GEOS: Geometry Engine - Open Source (2008), http://geos.refractions.net/
25. White, R., Engelen, G.: Cellular automata and fractal urban form: A cellular modeling approach to the evolution of urban land-use patterns. Environment and Planning, 1175–1199 (1993)
26. White, R., Engelen, G., Uljee, I.: The use of constrained cellular automata for high-resolution modelling of urban land use dynamics. Environment and Planning B: Planning and Design 24, 323–343 (1997)
27. White, R., Engelen, G.: High-resolution integrated modelling of the spatial dynamics of urban and regional systems. Computers, Environment and Urban Systems 24, 383–400 (2000)
28. Moulin, B., Chaker, W., Perron, J., et al.: MAGS project: Multi-agent geosimulation and crowd simulation. In: Kuhn, W., Worboys, M.F., Timpf, S. (eds.) COSIT 2003. LNCS, vol. 2825, pp. 151–168. Springer, Heidelberg (2003)
29. Project Gigalopolis, NCGIA (2003), http://www.ncgia.ucsb.edu/projects/gig/
30. Open GIS Consortium, OpenGIS simple features specification for SQL revision 1.1(2007), http://www.opengis.org/techno/specs/99-049.pdf

Exploratory Spatial Analysis of Illegal Oil Discharges Detected off Canada's Pacific Coast

Norma Serra-Sogas[1], Patrick O'Hara[2], Rosaline Canessa[3], Stefania Bertazzon[4], and Marina Gavrilova[5]

[1] Department of Geography, University of Victoria
normas@uvic.ca
[2] Environment Canada – Canadian Wildlife Service,
Institute of Ocean Sciences, Sidney, BC
Patrick.OHara@dfo-mpo.gc.ca
[3] Department of Geography, University of Victoria
rosaline@uvic.ca
[4] Department of Geography, University of Calgary
bertazzs@ucalgary.ca
[5] Department of Computer Science, University of Calgary
marina@cpsc.ucalgary.ca

Abstract. In order to identify a model that best predicts spatial patterns it is necessary to first explore the spatial properties of the data that will be included in a predictive model. Exploratory analyses help determine whether or not important statistical assumptions are met, and potentially lead to the definition of spatial patterns that might exist in the data. Here, we present results from exploratory analyses based on data describing illegal oil spills detected by the National Aerial Surveillance Program (NASP) in Canada's Pacific Region, and marine vessel traffic, the possible source of these oil discharges. We identify and describe spatial properties of the oil spills, surveillance flights and marine traffic, to ultimately identify the most suitable predictive model to map areas where these events are more likely to occur.

Keywords: Illegal Oil Pollution, Spatial Autocorrelation, Moran's I, Geary's c, LISA, Canada's Pacific Region.

1 Introduction

The International Convention for the Prevention of Pollution from Ships and its Protocol of 1978 (MARPOL 73/78) specifies that oily wastes generated from routine vessel operations must be disposed of by approved on-board incinerators or at port-side facilities, and that any discharge at sea must contain less than 15 parts per million of hydrocarbon residue [1]. However, a small proportion of vessel operators are non-compliant and illegally discharge oil mixtures while en route. These activities are probably the largest contributor to what is commonly referred to as chronic oil pollution [2].

In Canada, the National Aerial Surveillance Program (NASP), operated by Transport Canada, is the principal surveillance mechanism for monitoring and enforcing

M.L. Gavrilova and C.J.K. Tan (Eds.): Trans. on Comput. Sci. VI, LNCS 5730, pp. 219–233, 2009.

ship compliance with MARPOL [3]. In the Pacific Region, patrol flights are primarily conducted based on the distribution of marine traffic (e.g., one of the programs objectives is to fly over as many commercial vessels as possible), and limited by weather conditions [3], and, until now, by the aircraft's limited flight range. In 2008, NASP program performance has been improved with the incorporation of a larger aircraft (a Dash 8) with a new and sophisticated array of remote sensors [4].

The complexity and extent of Canada's Pacific Coast continue to be a challenge for pollution surveillance. Focusing surveillance resources in areas where illegal oil discharges are more likely to occur can save Transport Canada NASP program tens of thousands of dollars. Generally, it is expected to find a higher concentration of oil spills in those areas with a higher density of marine traffic. However, it is believed that offshore areas where marine traffic is more dispersed or not confined to specific routes, such as Traffic Separation Schemes, might be more vulnerable to intentional discharges of oil as vessel operators perceive a little risk of being identified by patrol aircrafts or other vessels [5]. Therefore, it is paramount to incorporate the complete distribution of marine vessels within Canada's Pacific region in our analysis to extend our understanding of oil pollution to those areas that do not have the same degree of surveillance effort.

The principal aim of this study is to identify a model that best predicts spatial patterns of illegal oil discharges based on marine commercial vessel type and movement patterns, while controlling for the spatial distribution of surveillance flights. In order to identify a valid statistical model it is necessary to first explore the spatial properties (i.e., spatial dependency and spatial heterogeneity) of our data to assess whether or not important statistical assumptions are appropriately addressed with our data [6].

In this paper, we focus on exploring the presence of spatial dependence or spatial autocorrelation in oil spill counts (the dependent variable), surveillance flights counts, and vessel movement counts for different vessel type (the independent variables). Spatial autocorrelation statistics measure the degree to which observations are related to values of the same variable at different spatial locations [7][8]. The presence of positive (or negative) spatial autocorrelation can be defined as "a nuisance in applying conventional statistical methodology to spatial data" [9, pp. 265]. Standard statistical methods are utilized with the assumption that data collected for a set of observations are independent and uniformly distributed. However, "the assumption of independence cannot be sustained by spatial data" [10, pp. 330]. As stated by Tobler [11, pp.3] "the first law of geography: everything is related to everything, but near things are more related than distance things".

In particular, in spatial regression models the problem resides in spatial dependence in the regression residuals. A basic assumption of the regression model is that the error terms are independent (or spatially independent) and constant over the study area [8]. However, if the error terms are positively (or negatively) autocorrelated, these assumptions are violated, and thus invalidating the confidence intervals for the regression coefficients and the corresponding assessment of the significance of the regression model. [8][10].

Spatial autocorrelation is also defined as a descriptive index, measuring the way a set of observations are related and distributed in space [7]. Global measures of spatial autocorrelation estimate a single measure of the relation between observations (e.g., positive autocorrelation is estimated when high values are near to other high values);

whereas local measures of spatial autocorrelation provide a value at each location, considering the relationship between both its neighboring sites and the entire data set [12]. Other significant differences between global and local measures of spatial auto-correlation are: global measures assume stationarity, emphasize similarities over space and are non-mappable whereas local measures assume non-stationarity, identify variation and are mappable [12] [13].

In the remainder of the paper, we present: in section 2, an overview of the data used for this study; in section 3, a description of the methods for exploring the degree of spatial autocorrelation present in the distribution of detected oil spills, surveillance flights and shipping densities for different vessel type computing two of the most common measures global measures of spatial autocorrelation, Moran's I and Geary's c, as well as the local form of Moran's I, the local Moran's I_i, and G_i^*; in section 4, we present a description of the results; and finally, in section 5, a discussion of possible suitable predictive models based on the presented results and propose research question/hypothesis to further explore the relationships between oil spills and vessel movements.

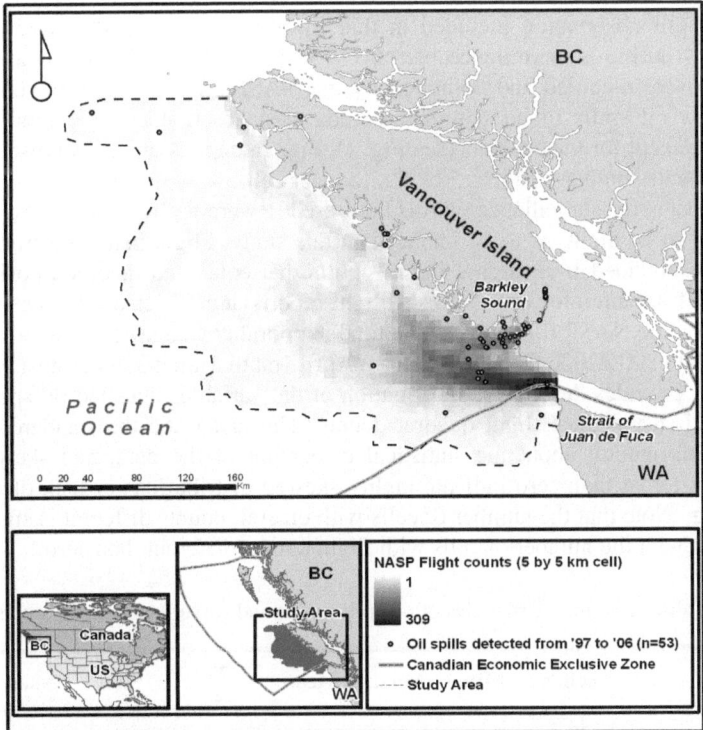

Fig. 1. Study area represented as the maximum bounding area covered by the NASP program off the West Coast of Vancouver Island

2 Data

2.1 Study Area

The larger and more heterogeneous the study region, the more difficult it is to measure the true spatial relationships among variables [12]. A possible solution is to reduce the complexity of the original study area, i.e. the entire Canada's West Coast, by selecting a more homogeneous sub-region. The maximum bounding area covered by the NASP flights off the west coast of Vancouver Island was chosen to test for presence of spatial autocorrelation in our variables. Fig. 1 shows the location and size of the study area which coincides with the spatial distribution of the surveillance intensity represented as number of surveillance flights per a 5 by 5 kilometer cell. This same regular grid cell of 5 by 5 kilometers (except in the edges with the coastline where cells are usually smaller) that extends only as far as the surveillance flights, was used to aggregate the rest of the variables; oil spill counts and vessel movement counts. For simplicity, in this case study we are assuming that all cells are of the same size.

2.2 Illegal Oil Discharges and Surveillance Flights

Each oil spill observation included in this study was visually identified by NASP flight crew during a surveillance patrol. Oil spill position (using GPS) and time of detection were recorded and compiled by the NASP crew in monthly summary reports. Only oil spills found within the study area described above (total of 53 oil spills) were kept for the analysis (see Fig. 1). Fig. 2a depicts the spatial distribution of oil spill counts summarized in a 5 by 5 kilometer cell.

Flight paths for surveillance patrols before 2001 were archived as hard copy maps, which had to be digitized to be included in this study. Flight path data from 2001 to 2006 were obtained from digital flight reports that contained recorded position, date and time of the aircraft every second. Flight reports and detected oil discharges were provided by the NASP flight crew. The total temporal coverage of this data set is from 1997/1998 to 2005/2006 fiscal years (from April 1rst to March 31rst of next year).

Table 1 provides descriptive information of the variables detected oil spill quadrat counts and surveillance flight quadrat counts. The mean and the standard deviation provide information about the statistical dispersion of the data; and skewness and kurtosis different than cero indicate highly skewed distributions or lack of normality in the data. Note that the number of cells with oil spill counts different is significantly lower than with the number of cells with flight counts different than cero.

Table 1. Summary statistics of oil spill counts and surveillance flight counts

	# cells*	Min.	Max.	Mean	St. dev.	Sum	Skew	Kurt.
Oil spill counts	44 (2,741)	0	3	0.02	0.162	53	9.85	113.6
Flight counts	2151 (2,741)	0	309	13.75	27.12	37,681	4.21	25.6

* The number of cells in parenthesis is the total number of cells in the study area, and the top number is the total number of cells with counts different than 0.

2.3 Description of Shipping Densities and Maritime Routes in British Columbia

Shipping densities are relatively high off the Canadian West Coast, concentrated in areas to the west and along Vancouver Island and The Juan de Fuca Strait. Bulk cargo and container vessels (carrier vessels and oil tankers Fig. 2b and 2c) generally follow one of the three Great Circle Routes (defined in this paper as the shortest geographical distances from and to principal Asian ports), or the Prince Rupert and Alaska routes (i.e., principal routes to and from Prince Rupert, BC, and Alaskan ports). All of these routes converge north and northwest of Vancouver Island, and this combined route passes along the west coast of Vancouver Island en route to and from the entrance to the Juan de Fuca Strait, or to continue south along the US West Coast.

Loaded crude oil tankers coming from Alaska to the Juan de Fuca Strait, also known as the Trans Canada Pipeline System, are advised to remain seaward of the Tanker Exclusion Zone (TEZ) situated 60 nm off the coast. Oil taker vessel operators respect the TEZ despite being a voluntary measure.

Tug vessels (Fig. 2d) and fishing vessels (Fig. 2e) have less specific routing patterns. Tugboats usually are smaller and slower vessels that choose to navigate closer and along the coastline for safety reasons. Fishing vessel patterns can be explained by the distribution of fishing resources and closeness to important fishing communities. Larger fishing vessels fishing in Alaska but based in Seattle, Washington, navigate along the Alaska route.

Finally, cruise ships (Fig. 2f) follow clearly defined routes that are distinct from other vessel types. The Gulf of Alaska route is one of the main and most popular cruising routes to and from Alaska, where large cruise ships sail northbound or southbound along the west coast of Vancouver Island, between the port of Vancouver and Seward, Alaska.

Table 2. Summary statistics of shipping traffic estimates

	# cells	Min.	Max.	Mean	St. dev.	Sum	Skew	Kurt.
Carrier counts	2464 (2,741)	0	4,614	134.23	290.19	367,927	8.1	104.9
Oil Tanker counts	2000 (2,741)	0	862	19.07	54.15	52,277	8.26	92.05
Tug counts	1810 (2,741)	0	1,235	20.71	73.14	56,768	7.2	72.55
Fishing vessel counts	2239 (2,741)	0	403	21.12	38.36	57,901	4.05	22.42
Cruise ships counts	1280 (2,741)	0	171	5.13	13.66	14,054	4.9	33.77

* The number of cells in parenthesis is the total number of cells in the study area, and the top number is the total number of cells with counts different than 0.

Shipping traffic minimum densities (i.e., number of vessel movements per cell) were estimated for each different vessel type included in this analysis, including: carriers, including all bulk, break bulk, cargo and ro-ro vessels (i.e., car cargo) (Fig. 2a), oil tankers (Fig. 2b), tugboats (Fig. 2c), fishing vessels (Fig. 2d) and cruise ships (Fig. 2e) (for more detailed information see [14]).

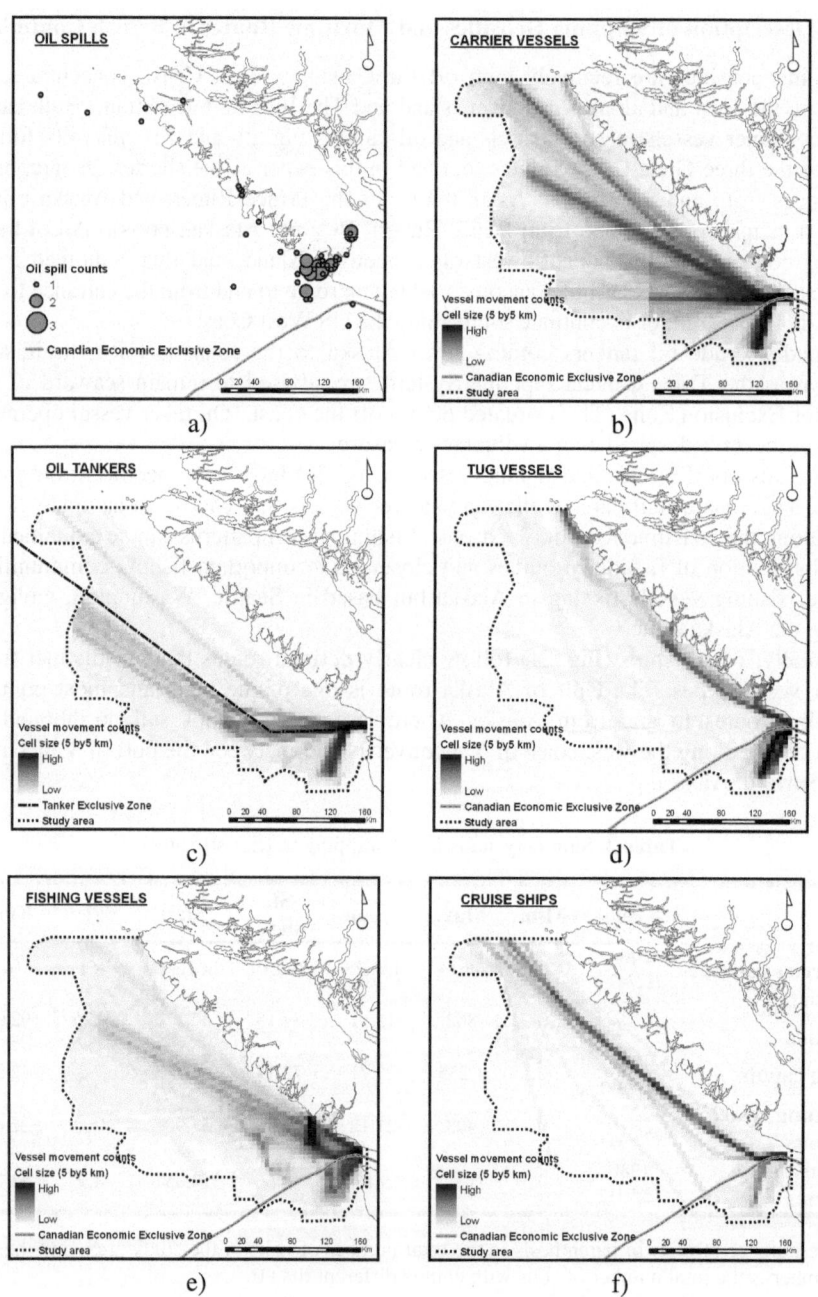

Fig. 2. Spatial distribution of (a) oil spill counts (proportional symbols were used to improve visualization), (b) carrier vessel movement counts, (c) oil tanker movement counts, (d) tugboat movement counts, (e) fishing vessel movement counts, and (f) cruise ship movement counts. Ship densities were summarized by 5*5km grid cell.

Table 2 presents summary statistics of shipping traffic estimates for each vessel type. Note the differences in maximum number of vessel movements among the different vessel types; the number of carriers is significantly higher compared with the rest of vessel types, especially if compared with the number of cruise ships. This is due to the low activity of cruise chips concentrated almost exclusively in the summer. Skewness and kurtosis values different than cero depict highly skewed distributions or lack of normality for all variables.

Marine traffic data for 2003 was provided by the Marine Communication Traffic Services in Vancouver, BC. It is important to note that shipping density patterns in BC have remained relatively constant during the last 10 years, thus assuming shipping densities in 2003 to be representative of the overall pattern [15].

3 Methods

3.1 Global Measures of Spatial Autocorrelation

Global measures of spatial autocorrelation examine the nature and extent of the dependence within model variables and produce a single value for the entire data set [12]. The two most common global measures are Moran's I and Geary's c [8].

Moran's I is calculated from

$$I = \frac{n}{\sum_{i=1}^{n}(y_i - \bar{y})^2} \frac{\sum_{i=1}^{n}\sum_{j=1}^{n} w_{ij}(y_i - \bar{y})(y_j - \bar{y})}{\sum_{i=1}^{n}\sum_{j=1}^{n} w_{ij}} \tag{1}$$

where i and j refer to different spatial units (i.e., cell centroids) of which there are n (i.e., 2,741 cells), and y is the data value in each (e.g., number of oil spills per cell). The product of differences between two observations (i.e., covariance) determines the extent to which they vary together. The term w_{ij} is the weight matrix, a measure of the potential interaction between two spatial units. This measure varies from -1 (negative spatial autocorrelation) to +1 (positive spatial autocorrelation), and 0 indicating lack of spatial autocorrelation. Positive values closer to +1 indicate high spatial autocorrelation.

Geary's c is very similar to Moran's I, and it is calculated as

$$C = \frac{n-1}{\sum_{i=1}^{n}(y_i - \bar{y})^2} \frac{\sum_{i=1}^{n}\sum_{j=1}^{n} w_{ij}(y_i - y_j)^2}{2\sum_{i=1}^{n}\sum_{j=1}^{n} w_{ij}} \tag{2}$$

The main difference between these two indices is that Geary's c is not relative to the mean which makes it more sensitive to absolute differences between neighboring locations. Geary's c with a value of 1 indicates no autocorrelation, values of less than 1 (but more than zero) indicate positive autocorrelation, and values of more than 1 indicate negative autocorrelation [16]. Positive values closer to zero indicate high spatial autocorrelation.

Moran's I and Geary's c were computed using RookCase [17]. The minimum distance (Euclidean distance) required to ensure that each location (i.e., cell centroid) has at least one neighbor (i.e., 5350.09 meters) was chosen as the threshold distance to construct the contiguity matrix used to estimate global Moran's I and Geary's c. These distances are based on distances between the cell centroids. This criterion was selected as the most adequate because it returned the highest spatial autocorrelation values when they were compared using different threshold distances.

Moran's I and Geary' c measurements for each variable were tested for statistical significance using the random permutation procedure or Monte Carlo test. This test statistic is based on the calculation of the statistic many times to generate a reference distribution. The advantage of a numerical approach is that it is data-driven and makes no assumptions (such as normality) about the data [8]. The disadvantage is that its p-values are dependent on the number of permutations [18]. A Monte Carlo test was performed of 99 simulations.

3.2 Local Measures of Spatial Autocorrelation

Local measures of spatial autocorrelation are appropriate to identify the location and spatial scale of aggregations of unusual values, such as clusters of high values (hot spots) and low values (cold spots) [12]. Furthermore, because local measures generate an autocorrelation index for each data site, this can be mapped providing additional information about the pattern under study [12][13].

The local form of Moran's I, local Moran's I_i, also known as LISA (Local Indicator of Spatial Association) [19], was estimated for each of the variables (i.e., surveillance flight counts, oil spill counts, and vessel movement counts for each vessel type) in GeoDa ^{TM} [20].

Local Moran's I_i is defined as a product of the data site z-score (z_i), and the average z-scores of the surrounding sites [19]

$$I_i = z_i \sum_{j \neq i} w_{ij} z_j \qquad (3)$$

where the z-score at the location i is defined by

$$z_i = \left(y_i - \overline{y} \right) / \left(\sum_i \left(y_i - \overline{y} \right)^2 / n \right) \qquad (4)$$

where w_{ij} is row-standardized (i.e., scaled so that each row sums to 1), and the summation is for all j not equal to i [16]. The interpretation of I_i is analogous to its global from Moran's I.

The same distance based weight matrix (w_{ij}) used to calculate global Moran's I was used to estimate local Moran's I_i, where the definition of neighbor was based on the minimum distance between two cell centroids (i.e., 5350.09 meters). This yielded a measure of spatial autocorrelation for each individual location that can be mapped to show the distribution of spatial clusters or outliers [18].

In GeoDa [TM] a "LISA cluster map" can be constructed, which depicts the four types of spatial association; similar values (positive local spatial autocorrelation) are categorized as *high-high* (i.e., high values surrounded by neighbors of similar high values) or *low-low* (i.e., low values surrounded by neighbors of similar low values), whereas dissimilar values (negative local spatial autocorrelation) are coded as *high-low* and *low-high*: low values surrounded by high neighboring values for the former, and *vice versa* for the later [18][19].

A significance test of the local Moran's I_i was performed based on the conditional permutation (or randomization) procedure [18], which assumes that the value y_i at the data site i is held fixed and the remaining data values are randomly permuted over the remaining $(n-1)$ data sites [12]. Limitations of this test are that it is sensitive to the number of permutations selected and it can lead to slightly different results between permutations [18]. In this study, a randomization test was performed using 999 permutations and at a significance level of $p \leq 0.01$.

The G_i^* is another measure of local spatial autocorrelation, and it is described as the sum of the weighted data values (within a specified distance of an observation i $(w_{ij}(d))$, and including x_i, relative to the sum of all data values for the entire study region [19]. More specifically

$$G_i^* = \sum_j w_{ij}(d) z_j / \sum_j z_j \qquad (3)$$

Positive G_i^* values indicate clustering of high data values, whereas negative G_i^* values indicate clustering of low values. Note that G_i^* and I_i statistics measure different concepts of spatial autocorrelation; for instance, for G_i^* statistic positive values indicate clustering of high values while for I_i positive values indicate clustering of either high similar values or low similar values [19].

In this report, only local Moran's I_i is presented in the Results section. The G_i^* statistic was initially calculated for each variable but the cluster maps depicted very similar patterns as the ones presented by the LISA cluster maps, LISA cluster maps allow the distinction between clusters of high values (high-high) and clusters of low values (low-low).

4 Results

4.1 Spatial Autocorrelation

Estimated global Moran's I for the different variables show significant values ($p \leq 0.01$) of positive spatial autocorrelation; that is the magnitude of I is positive (see Table 3). Oil spill observations have the lowest degree of positive spatial dependency (positive value of I but close to 0), indicating that there are few cells near other cells with similar values. In contrast, surveillance flight cell counts and vessel movement cell counts show significantly higher values of positive spatial autocorrelation, indicating cluster patterns for these variables. NASP flight cell counts present the highest spatial autocorrelation indices, followed by fishing vessel movement cell counts, oil tanker movement cell counts, carrier movement cell counts, tug vessel movement cell counts, and finally cruise ship movement cell counts.

Global measure Geary's c showed the same results as Moran's I; all variables presented positive spatial association, i.e., all values are less than one and higher than 0. Oil spill counts show again as the variable with the lowest degree of spatial autocorrelation with a c value closer to 1. The rest of the variables presented higher values of c, being the NASP flight cell counts presented the highest degree of positive spatial autocorrelation with values closer to 0.

Table 3. Moran's I and Geary's c values estimated for each variable under study

	Moran's I (*p*-values)	Geary's c (*p*-values)
NASP flight counts	0.805 (0.01)	0.172 (0.01)
Fishing vessel movement counts	0.783 (0.01)	0.249 (0.01)
Oil tankers movement counts	0.589 (0.01)	0.369 (0.01)
Carrier vessel movement counts	0.553 (0.01)	0.395 (0.01)
Tug vessel movement counts	0.553 (0.01)	0.371 (0.01)
Cruise ship movement counts	0.518 (0.01)	0.455 (0.01)
Oil spills counts	0.158 (0.01)	0.871 (0.01)

4.2 Spatial Pattern Description – LISA Cluster Maps

LISA cluster map were generated for surveillance flight counts, oil spill counts, and vessel movement counts for each vessel type (see Fig. 4). Each significant location (*p* ≤ 0.01) is color coded depending on the degree of spatial autocorrelation: clusters of black cells represent cells containing high values with neighboring cells with similar high values (positive spatial autocorrelation: high-high); clusters of dark gray cells indicate cells with low values are surrounded by cells of similar low values (positive spatial autocorrelation: low-low); and light gray cells indicate the location of outliers or negative spatial autocorrelation (high-low, low-high).

By mapping shipping traffic densities for each vessel type we can distinguish relatively different cluster patterns for each of the vessel types (Map 4c, 4d, 4e, 4f and 4g), as it was expected when vessel type densities where visualized in Fig. 2. However, they share a similar distribution at the entrance of the Strait of Juan de Fuca, south-west of Vancouver Island. High count values of surveillance flights are also concentrated in the same section of the study region, spreading more homogeneously offshore. Oil spill hot spots are detected between the entrance and the interior of Barkley Sound.

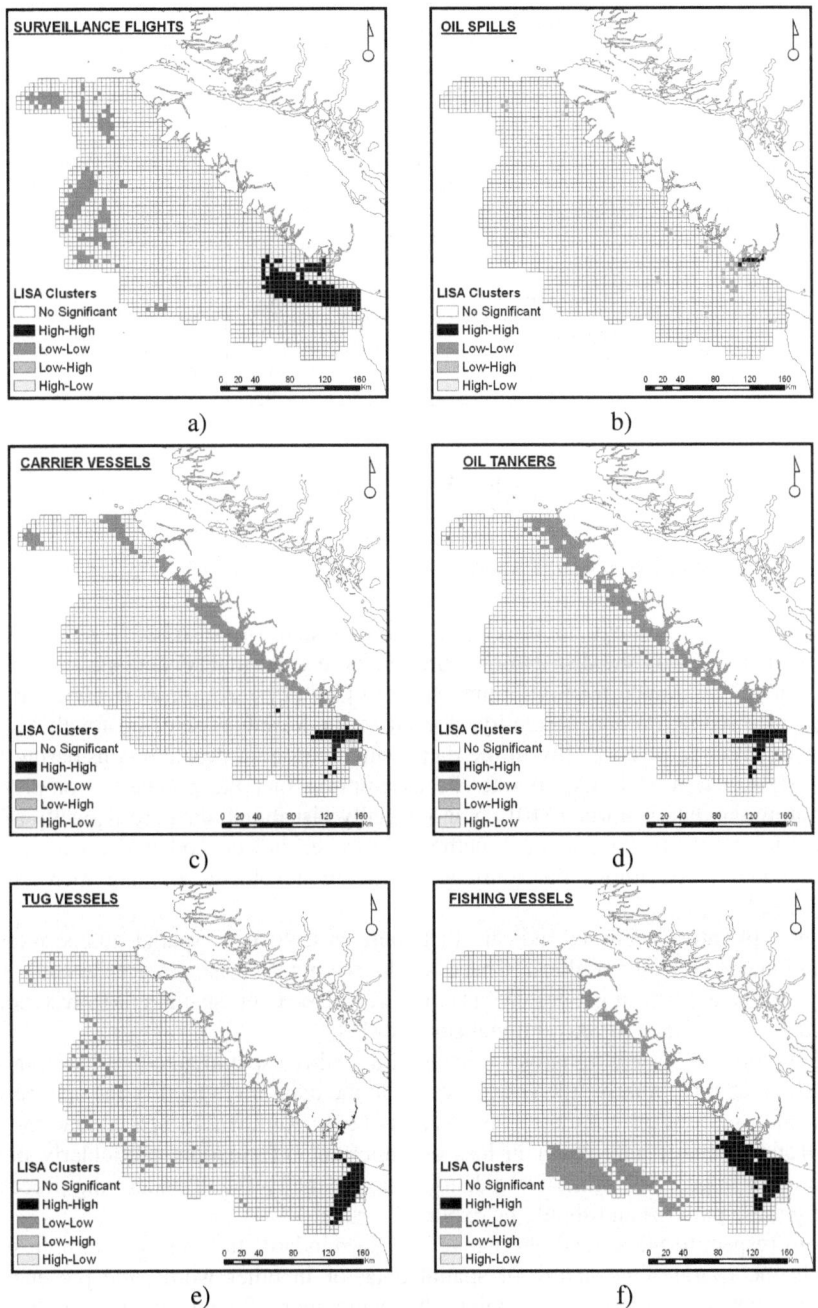

Fig. 3. LISA cluster maps of (a) surveillance flight counts, (b) oil spill counts, (c) carrier vessel movement counts, (d) oil tanker movement counts, (e) tugboat movement counts, (f) fishing vessel movement counts, and (g) cruise ship movement counts. Significance test using 999 permutations and applying a significance filter of $p \leq 0.01$.

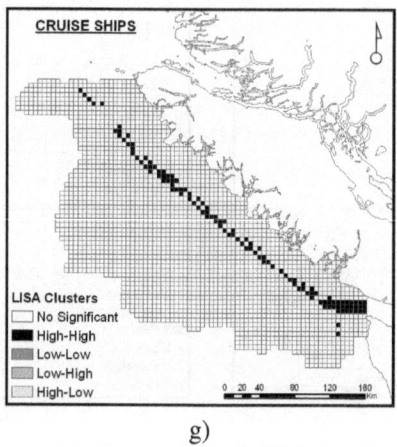

g)

Fig. 3. (*continued*)

5 Discussion and Conclusion

A recommended strategy to develop hypotheses and shape subsequent statistical analysis is to identify the distributional properties (e.g. normality) and spatial dependencies or spatial autocorrelation in the data [10] [16] [21]. An initial overview of the statistical distribution and normality of each of the variables selected for this study indicated absence of normality in the data, which can be a problem considering that many statistical models incorporate the assumption that the population or process generating the data is normal [10]. Non-normally distributed data are typically transformed for conventional statistical methods; however, newer models that can accommodate data with various distributions are being developed and are increasingly utilized [21].

Our exploratory analyses indicate that there is a positive spatial autocorrelation within datasets for all variables. These results are not surprising since "spatial autocorrelation is a given in geography" [16, pp. 28]. However, spatial autocorrelation is considered a nuisance to conventional statistics [9].

There are two basic alternatives to deal with positive spatial autocorrelation in statistical models; (1) remove the correlation from the data or (2) modify the appropriate technique to allow for correlated observations [10]. Removal of spatial autocorrelation from a dataset may result in loss of important information, particularly in the exploratory stage. Any pattern in a dataset may be indicative of underlying processes driving these patterns and should be explored further.

With respect to the second alternative, unlike standard statistics, spatial regression models incorporates the nature of spatial data, or in other words, the presence of spatial dependency of observed points, by being more flexible with modeling the residuals [13]. Unlike most spatial regression models, which are defined as global or semi-local methods, geographically weighted regression (GWR) operate on a local basis; that is, it produces all the elements of a regression model, such as estimated parameters, R^2 values and t-values, for each sampled point of the study area [13][23].

A study carried out by Zhang *et al.* compared six other spatial regression techniques (GWR, Ordinary least-squares (OLS), linear mixed model (LMM), generalized additive model (GAM), multi-layer perceptron (MLP) neural network, and radial basis function (RBF)), and concluded that GWR "produces more accurate predictions for the response variable, as well as more desirable spatial distribution for the model residuals than the ones derived from other five modeling techniques" [23, pp. 175-176]. Hence, GWR might be the most suitable model to predict spatial patterns of illegal oil discharges based on marine commercial vessel type and movement patterns, while controlling for the spatial distribution of surveillance flights.

We also used local methods of spatial autocorrelation as an exploratory analysis to characterize the spatial patterns of the different variables under study. LISA cluster maps allowed the visualization of significant aggregations of high values of oil spill counts, surveillance effort and shipping traffic. However, a weakness of local measures of spatial autocorrelation is that they cannot identify clustering of medium values since mid-range of these measures (i.e., values around zero) can result from either this situation or an absence of clustering of similar values [12]. For example, Barkley Sound and surrounding areas are dominated by tug traffic and commercial fishing; in addition to recreational activities (recreational activities were not included in this study). However, LISA maps do not depict these activities as clearly in this area because the traffic is not as high compared to the entrance to the Strait of Juan de Fuca. Tug and fishing vessel traffic in Barkley Sound fall within the mid-range of density values.

Other problems rising from univariate analysis is the modifiable areal unit problem or MAUP. Spatial autocorrelation (global and local) is affected by the scale of aggregation, also known as the 'smoothing effect', where information of spatial heterogeneity is lost as we increase the degree of aggregation [24], or cell size in our case. The MAUP problem was investigated by using different cell sizes to aggregate our variables. We selected a five by five kilometer size cell, because it returned the highest value of Moran's I for oil spill counts. However, a cell size that is optimal for one variable may not be optimal for some other variable [24]. Future research is necessary to investigate this problem in other variables (surveillance flights and marine traffic).

However, LISA cluster maps did reveal new patterns that have led to inferences about processes underlying observed pattern of detected oil spills. Within our dataset, oil spill patterns are most likely influenced by two processes; the distribution and intensity (effort) of the surveillance flights, and the distribution and density of marine traffic. The former affects the probability of detection and the latter affects the risk of an oil spill. These two processes are spatially associated as surveillance flights typically cover areas of higher shipping activities. However, defining an empirical relationship between these variables is not as straight forward. Ignoring the effect of deterrence, one would expect that in areas where there are more marine traffic and more surveillance flights the chance to detect an oil spill would be also higher. However, deterrence can not be ignored as vessel operators will modify their behaviour and likely choose to discharge waste oil in areas where NASP coverage is lower, probably in areas where vessel densities are relatively lower (given that NASP effort and vessel densities are spatially associated). We believe the LISA clusters maps show the effect of deterrence to some extent. For example, oil spill clusters were found in areas of Barkley Sound, whereas clusters of high counts of surveillance flights and marine vessels were mainly found in the entrance to the Strait of Juan de Fuca (see Fig. 4).

Based on our results exploring autocorrelation properties in our data, we plan to explore the degree of correlation between detected oil spills and marine traffic densities for different vessel types (i.e. fishers, tugs, cargo vessels, tankers, cruise ships and ferries). In these future analyses, we will incorporate results from O'Hara *et al.* [25] that defines the relationship between oil spill detection probabilities and surveillance effort. We also plan on exploring the data using a multivariate approach that will include variables such as, season, distances between oil spills and shore and distances to nearest port or marina, and other characteristics that define the vessel type (i.e., flag state, inbound vs. outbound).

Acknowledgments. We thank Louis Armstrong, John Heiler, and Sue Baumeler from Transport Canada for providing original data and other relevant information. Canadian GEOIDE Network Project "Coastal Security and Risk Management Using GIS and Spatial Analysis" and Environment Canada "Birds oiled at Sea" for funding. Finally, the Department of Geography (UVIC) for technical support.

References

1. International Marine Organization: International Convention from the Prevention of Pollution from Ships, 1973, as modified by the Protocol of 1978 relating thereto (MARPOL 73/78),
   http://www.imo.org/Conventions/
   contents.asp?doc_id=678&topic_id=258
2. National Research Council: Oil in the sea III: Inputs, Fates, and Effects. National Academies Press, Washington (2003)
3. Armstrong, L., Derouin, K.: National Aerial Surveillance Program 2001-2004. Final report. Transport Canada Marine Safety, Ottawa, Canada (2004)
4. McGregor, M., Gautier, M.-F.: Enhancing Situational Awareness in the Arctic Through Aerial Reconnaissance. In: Environment Canada, Marine & Ice Services, Ottawa, Ontario, Canada (2008)
5. Serra-Sogas, N., O'Hara, P., Canessa, R., Keller, P., Pelot, R.: Visualization of spatial patterns and temporal trends for aerial surveillance of illegal oil discharges in western Canadian marine waters. Marine Pollution Bulletin 56, 825–833 (2008)
6. Anselin, L.: Spatial Econometrics: Methods and Models. Martinus Nijhoff, Dordrecht (1988)
7. Goodchild, M.F.: Spatial Autocorrelation. In: Concepts and Techniques in Modern Geography, vol. 47. Geo Books, Norwich (1985)
8. Bailey, T., Gatrell, A.: Interactive Spatial Data Analysis. Longman Scientific & Technical. Burnt Mill, Essex (1995)
9. Griffith, D.A.: What is spatial autocorrelation? Reflections on the past 25 years of spatial statistics. L'Espace geographique 3 (1992)
10. Haggett, P., Cliff, A.D., Frey, A.: Locational Analysis in Human Geography 2. Edward Arnold, Great Britain (1977)

11. Tobler, W.: A computer movie simulating urban growth in the Detroit region. Economic Geography 46, 234–240 (1970)
12. Boots, B.: Local measures of spatial association. Ecoscience 9(2), 168–176 (2002)
13. Fotheringham, A.S., Brunsdon, C., Charlton, M.: Geographically Weighted Regression: the analysis of spatially varying relationships. John Wiley & Sons, Ltd, England (2002)
14. O'Hara, P., Morgan, K.: Do low rates of oiled carcass recovery in beached bird surveys indicate low rates of ship-source oil spills? Marine Ornithology 34, 133–140 (2006)
15. British Columbia's Coastal Environment: Alive and Inseparable. BC Government Publications (2006)
16. O'Sullivan, D., Urwin, D.J.: Geographic Information Analysis. John Wiley & Sonds, New Jersey (2003)
17. Sawada, M.: Rookcase: an Excel 97/ Visual Basic (VB) add-in for exploring global and local spatial autocorrelation. Bulletin of the Ecological Society of America 80, 231–234 (1999/2000)
18. Anselin, L.: Exploring Spatial Data with GeoDa: A Workbook. Spatial Analysis Laboratory, University of Illinois, IL (2005)
19. Anselin, L.: Local indicators of spatial autocorrelation—LISA. Geographical Analysis 27, 93–115 (1995)
20. Anselin, L., Syabri, I., Kho, Y.: GeoDa: An introduction to Spatial Data Analysis. In: Spatial Analysis Laboratory, Department of Agricultural and Consumer Economics, University of Illonois, IL (2005)
21. Haining, R.: Spatial data analysis in the social and environment science. Cambridge University Press, Cambridge (1990)
22. Paéz, A., Uchida, T., Miyamoto, K.: A general framework for estimation and inference of geographically weighted regression models: 1. Location-specific kernel bandwidths and a test for locational heterogeneity. Environmental and Planning A 34, 733–754 (2002)
23. Zhang, L., Gove, J.H., Heath, L.S.: Spatial residual analysis of six modeling techniques. Ecological Modelling 186, 154–177 (2005)
24. Jelinski, D.E., Wu, J.: The modifiable areal unit problem and implications for landscape ecology. Landscape Ecology 11(3), 129–140 (1996)
25. O'Hara, P.D., Serra, N., Canessa, R., Keller, P., Pelot, P.: Estimating oil spill rates and deterrence based on aerial surveillance data in Western Canadian marine waters. Marine Pollution Bulletin (submitted, 2008)

Clustering and Hot Spot Detection in Socio-economic Spatio-temporal Data

Devis Tuia[1], Christian Kaiser[2], Antonio Da Cunha[2], and Mikhail Kanevski[1]

[1] Institute of Geomatics and Analysis of Risk, University of Lausanne, Switzerland
Tel.: +4121 692 35 38; Fax: +4121 692 35 35
devis.tuia@unil.ch
[2] Institute of Geography, University of Lausanne, Switzerland

Abstract. Distribution of socio-economic features in urban space is an important source of information for land and transportation planning. The metropolization phenomenon has changed the distribution of types of professions in space and has given birth to different spatial patterns that the urban planner must know in order to plan a sustainable city. Such distributions can be discovered by statistical and learning algorithms through different methods. In this paper, an unsupervised classification method and a cluster detection method are discussed and applied to analyze the socio-economic structure of Switzerland. The unsupervised classification method, based on Ward's classification and self-organized maps, is used to classify the municipalities of the country and allows to reduce a highly-dimensional input information to interpret the socio-economic landscape. The cluster detection method, the spatial scan statistics, is used in a more specific manner in order to detect hot spots of certain types of service activities. The method is applied to the distribution services in the agglomeration of Lausanne. Results show the emergence of new centralities and can be analyzed in both transportation and social terms.

1 Introduction

The metropolization process [1–3] gives birth to new urban patterns organizing the socio-economic urban landscape. Phenomena like peri-urbanization or sub-urbanization have changed the urban landscape, modifying the distribution of socio-economic features within the city and between the city and its countryside. As a consequence, the urban dynamics become more and more complex [4] and difficult to explain with classical analyst tools. Therefore, the understanding of the new city remains crucial for the urban planner, for instance to plan the best transportation system and avoid social unfairness or environmental pollution: the need for new tools to describe urban systems is real.

The first step to understand such a system in terms of socio-economic features is the analysis of their distribution in space: socio-economic features are unequally distributed between the spatial units (for instance municipalities) and have the tendency to group in coherent ensembles. Therefore, it is possible to

M.L. Gavrilova and C.J.K. Tan (Eds.): Trans. on Comput. Sci. VI, LNCS 5730, pp. 234–250, 2009.

group similar units depending on their socio-economic profile in variable space. The features being complex and associated to high numbers of dimensions, the classification of urban spatial units into a small number of classes can be very effective to understand the structure of the urban space and to discover functional relationships, for instance between the centers and their periphery. Classification of such data is a typical unsupervised problem (also called clustering), because the number of classes is not known in advance nor examples to train the model are available. Several clustering models exist, going from hard partitionment methods (k-means [5], SOM [6]) cutting the features spaces into distinct regions, to hierarchical methods aggregating the observations depending on their similarity (hierarchical ascendant classification, HAC [7]).

Once the structure of the urban space has been modeled, a natural second step is to detect whether the distribution of a certain feature is constant in space or if there are outbreaks of areas where the density of such a feature is higher than normal. Cluster detection methods have been developed to answer this kind of questions. Several cluster detection methods exist, including the Local index of spatial autocorrelation (LISA [8]), the Turnbull's Cluster Evaluation Permutation Procedure (CEPP [9]), the Geographical analysis machine (GAM [10, 11]) and the Spatial scan statistics (SSS [12]). Comparison of the methods can be found in [13–15].

In this paper, we propose two methods for the analysis of clustering of urban spatial units depending on socio-economic features: first, we propose a fusion of SOM and HAC for the clustering of urban municipalities depending on their socio-economic profile. Such a fusion can be found in [16] for the classification of oceanic currents behavior. Second, we apply the SSS for the detection of clusters in the distribution of services, in order to question well known facts about definition of centralities.

The paper is organized as follows: Section 2 discusses the hybrid SOM/HAC model and the SSS. Section 3 presents the datasets studied in the applications shown in Section 4.

2   Models

2.1   Hybrid Self-organizing Map / Hierarchical Classification (HSOM)

Self-organizing maps (SOM), also known as Kohonen maps [6], are a type of artificial neural networks (ANN) using an unsupervised learning technique in order to map a high-dimensional space into a lower dimensional space (typically 2D for easy visualization). SOM keeps the intrinsic topology of the input data and allows the emergence of structures present in the data. A SOM is useful for the visualization of multivariate data, but can also be used for classification. In the case of classification, there can be found two different approaches: the first are SOM where each neuron corresponds to a cluster which has been shown to be almost identical to a k-means clustering [18], the second are SOM where the map space is used as a tool for characterizing high-dimensional data [17].

In the latter, the SOM is composed by several thousand neurons describing the feature structure; this type of SOM is called Emergent SOM (ESOM) [18] as it explicitly allows structure to emerge. Since the difference SOM and ESOM is basically the size of the network, we will use the terminology SOM in the following description.

In a self-organizing map, each data sample is mapped to a neuron. Neurons are typically organized in a square or hexagonal grid (Figure 1, left and center). In the square grid configuration, there are four immediate neighbors for each neuron; in a hexagonal grid, each neuron has six immediate neighbors. The number of neighbors decreases at the borders of the grid; resulting in border effects. In order to avoid this undesired effect, one can use finite but border-less map topologies [18]. One possibility is to connect the map edges to form a toroid map space (Figure 1, right), which allows a simple representation in 2D while offering a topology without border effects.

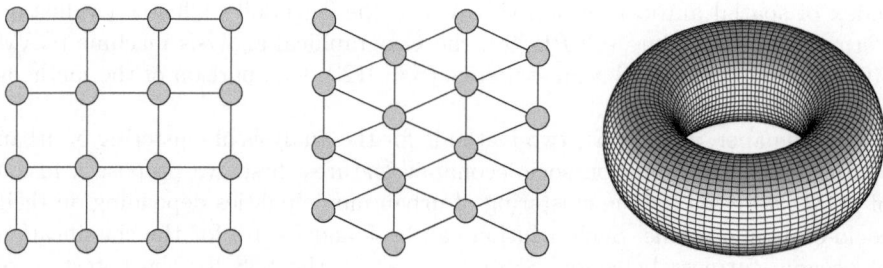

Fig. 1. A square SOM grid (left), a hexagonal one (middle) and a toroid grid (right)

The SOM creation can be divided into several steps.

In the first step, the neurons of the SOM are initialized, associating to each of them a randomly generated data vector of the same dimension as the inputs.

In the second step, the ordering phase, each data sample is compared to the neurons and attributed (mapped) to the most similar one. The data vectors of the selected neuron (the winner or best matching unit) and of its neighbors are updated in order to match better the input data sample. The winner is more importantly updated than the neighbors. This step is repeated iteratively, in order to adjust and match the neurons of the SOM to the input data. The importance of the update (the learning rate) is decreased during the iterations. This process is a competitive learning process, which corresponds to unsupervised learning and can be considered as self-organization of the neuronal map.

The third and final step, the convergence phase, is basically the same as the second step, except that it considers a smaller neighborhood size, a smaller learning rate and a higher number of iterations. It is supposed that after the second step, the neurons are quite well ordered. The third step is just a refinement of the neuron's vectors to better represent the input data set.

One of the problems of the SOM algorithm is the unknown optimal number of neurons. If the SOM is used for clustering, a too small number of neurons will decrease the quality of the cluster boundaries. Large SOM are able to represent better the input data structure. But there is a need to group together similar neurons in the SOM. This can be done manually using the U-Matrix or the P-Matrix, or using another algorithm. The U-Matrix is the canonical display of a SOM [19]. The distance relationship between the neurons in the high-dimensional input space are displayed as a height value, thus creating a 3D visualization of the high-dimensional space [18]. "Mountain ranges" on a U-Matrix point to cluster boundaries while "valleys" indicate cluster centers. While the U-Matrix is a distance-based visualization, the P-Matrix [17] is a density-based visualization showing the local density measures using the Pareto Density Estimation [20].

In this paper, the hierarchical ascendant classification (HAC) is used in order to cluster the neurons of the SOM. By doing so, the SOM acts as a non-linear transform of the original feature space. The combination of the two procedures is the Hybrid SOM (HSOM). The algorithm can be illustrated as follows (see also Figure 2):

1. Prepare the data for use in SOM. The empirical probability distribution of each feature is analyzed and corrected where needed (e.g. log transform). Extreme values should be removed and highly correlated features discarded.
2. Pre-processing of the data with the SOM. This step corresponds to a non-linear transformation and a generalization of the data. The degree of

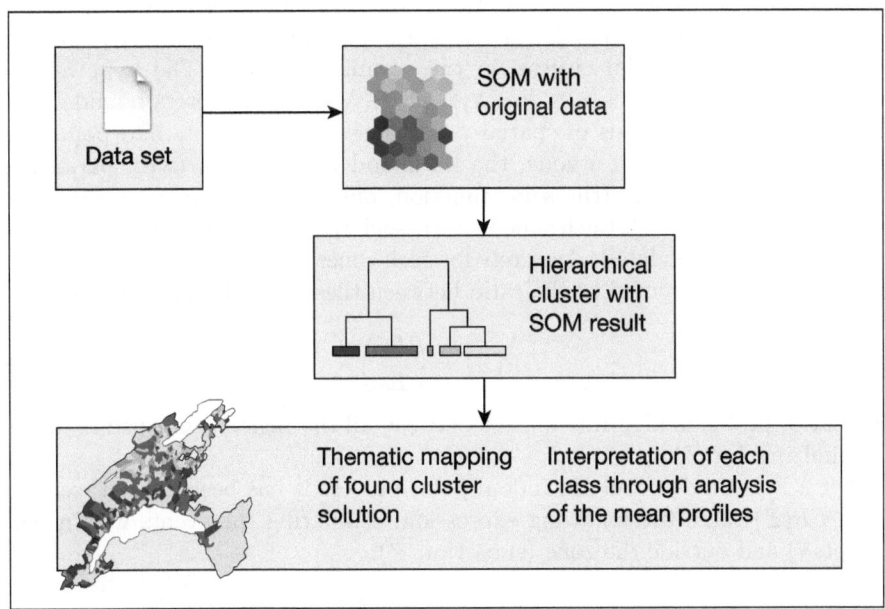

Fig. 2. The classification and mapping process using HSOM

generalization is determined by the grid size; a small grid will generalize more than a bigger one. However, in order to enable emergence of the data structure, the SOM should still have a large number of neurons. In order to avoid border effects and topology error, a border-less topology, e.g. a toroid grid, should be used.

3. HAC of the self-organized map divides the neurons into groups. The number of groups can be determined with standard methods like the analysis of the dendrogram. The result can be compared with the U-Matrix and the P-Matrix.

4. The original data are assigned to a group through the neurons of the SOM. Each data sample is assigned to one neuron which in turn has been assigned to one of the classes. Finally a thematic map of the spatial distribution of the groups is drawn. The analysis of the class profiles enables to label each group.

2.2   Spatial Scan Statistics (SSS)

Contrarily to the SOM, cluster detection methods consider the distribution of a unique process. In this sense, SSS analyzes spatial point processes and searches for over- (or under-) densities in the distribution of the real events by comparison to a process defined for random locations. Several SSS models have been developed so far, the most popular being the Poisson model that applies when the number of events is very small compared to the population considered.

In the Poisson model, a circular moving window scans the area under study defining sub-areas called zones z_i. Each zone is characterized by a number of events c_i and a population p_i, given by the sum of events and of population belonging to the spatial entities in the scanning window. The hypothesis of spatial randomness H_0 is $x \sim Poi(\lambda_0)$, where λ_0 are parameters bounded with respect to the hypothesis of spatial randomness. When events and population have been attributed to a zone, the likelihood functions L_0 (with parameters bounded to λ_0) and L_1 (the same function, but with parameters unrestricted) are computed (see [12]). Each zone being associated to a different population, the parameters are calculated separately for each zone. The Likelihood Ratio $LR(Z)$ for the zone is computed as the ratio between these two likelihood functions:

$$LR(Z) = \left(\frac{L_1}{L_0}\right)_Z \tag{1}$$

The most probable high rate cluster between all the regions analyzed is the one maximizing $LR(Z)$.

For a Poisson distributed random point process it has been proven (see [12]) that $LR(Z)$ takes the following expression, comparing the events within (subscripts i) and outside the zone (subscripts i'):

$$LR_i = \left(\frac{\left(\frac{c_i}{\eta_i}\right)^{c_i} \left(\frac{c_{i'}}{\eta_{i'}}\right)^{c_{i'}}}{\left(\frac{c_{tot}}{\eta_{tot}}\right)}\right) I \tag{2}$$

Where η is the expected number of cases under the H_0 hypothesis and I is an indicator function discarding results when the ratio observed/expected is higher outside of the zone than inside. The analysis is performed for every zone within the region, the scanning window taking every spatial entity as a center and considering different radii (Figure 3). To avoid untreatable computational cost, overlapping windows are not taken into account and a stop criterion can be applied (for instance, maximum radius length).

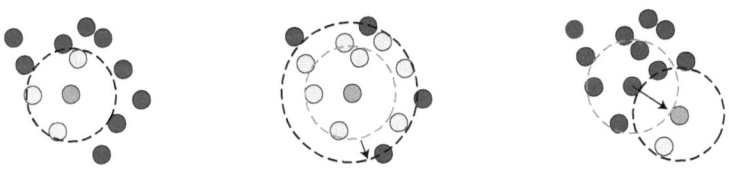

Fig. 3. Scanning window principle in the SSS

Once the most likely clusters have been highlighted, their significance has to be tested. The significance test is done in order to avoid situations where the most likely cluster is in fact included in the interval of confidence of a distribution respecting H_0. Such testing is done using Monte Carlo simulation: a certain number of data simulations respecting the null hypothesis H_0 of spatial randomness (absence of spatial clusters) is generated and the maximum $LR(Z)$ of each replication is compared to the most likely clusters of the real data. If the real clusters are included in the top 5% of the replications, the H_0 hypothesis is rejected and the cluster found are considered as significant.

As stated in the introduction, several statistical methods exist to identify clusters. For urban cluster detection, SSS seems to have a series of advantages:

- Size of clusters is not specified in advance and is discovered by the algorithms maximizing the $LR(Z)$.
- The weighting of events by population avoids the size effects.
- The test statistics is based on a Likelihood ratio and the null hypothesis is clearly stated.
- Since the LR is computed for every zone by taking into account specific events and population within and outside the zone, the method is not sensitive to spatial non stationarity/trends [25].

3   Data and Methods

3.1   HSOM

The HSOM clustering has been applied to the 2896 municipalities of Switzerland. The objective is to map the different socio-economic structures of the country, and to group all municipalities in a small (and thus interpretable) number of coherent classes. The data contain 56 socio-economic variables, composed by 32 economic variables regarding employment per economic sector and position and 24 demographic variables about the age structure for both genders. Data are provided by the Swiss population census 2000. For all variables, the percentage for each municipality has been computed and the values have been reduced to standard scores. When the distribution of the features was skewed, a log transform has been applied. Extreme values have been discarded.

The determination of the SOM grid size is important. In order to enable emergence of the data structures, a sufficiently big grid of 100x60 cells has been chosen. According to Ultsch and Herrmann [21], the ratio of rows and columns should be different from 1, so an arbitrary ratio of 0.6 has been selected. In order to avoid border effects, a toroid topology has been chosen (see Figure 1, right). 6000 neurons are used for about 3000 municipalities. The learning rate decreases from initially 0.5 to 0.1 at the end (linear cooling rate). The Gaussian function has been chosen for the neighborhood function, with a start value of 24 neurons for the radius and a final value of 1.

The SOM obtained is a series of connected neurons representing the input space embedded into a 2-dimensional grid. Therefore, it is possible to visualize the SOM for every variable of the input space (Figure 4 illustrates 4 of the 56 input variables). Note that the SOM is border-less, this means the lower edge is connected to the upper one, and the left edge to the right.

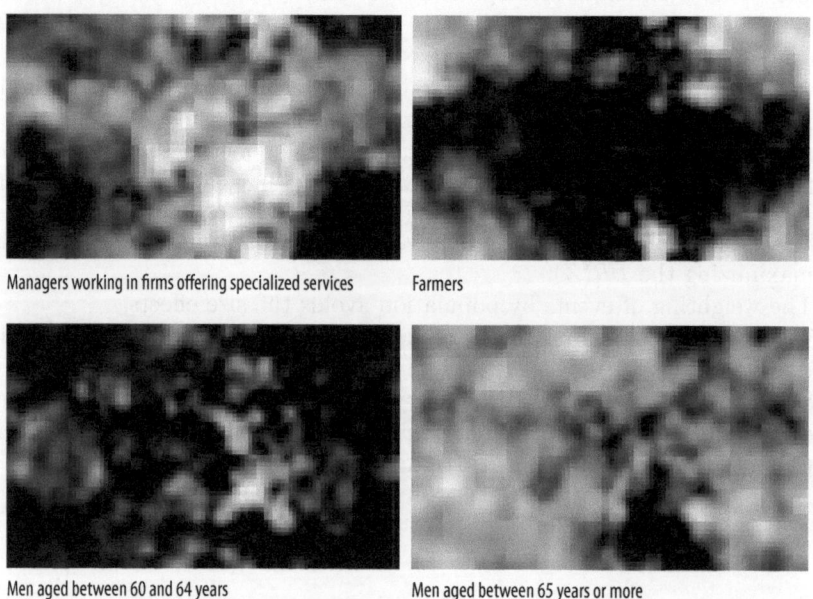

Managers working in firms offering specialized services   Farmers

Men aged between 60 and 64 years   Men aged between 65 years or more

Fig. 4. SOM representation of 4 of the 56 input variables. Dark zones correspond to higher percentage of occurrence.

Finally, the neurons of the SOM are classified using HAC (Figure 5). The dendrogram (Figure 8, in the upper left corner) suggests the creation of 5 classes that, applied to the SOM, give a partition of the embedded space as shown in figure 5a. The classification result can be compared to the U-Matrix and P-Matrix which can be used for "manual" classification. These two matrices give also additional details on the different classes issued from HAC. For example the class represented by lightest gray is located in a zone with high values in both U-Matrix and P-Matrix. This may indicate that this class contains features with a higher variability than some of the other classes. It is then straightforward to map the classification result in the geographical space (see Section 4.1).

As the SOM geometry is based on border-less toroid, upper and lower ends of the grid are in fact connected. Thus, it is possible to represent the SOM by exploiting the coherent regions highlighted by the HAC. In this case, neurons belonging to the same cluster are represented side by side and the grid becomes irregular. Figures 5 and 6 represent the same information, but in the latter, the grid is reorganized by using the HAC clusters. The same transformation has been applied to the U-Matrix and the P-Matrix, which become a U-Map and a P-Map respectively [18]. The classification map could be visualized in 3D using the U-Map as elevation data (see [17] for an example).

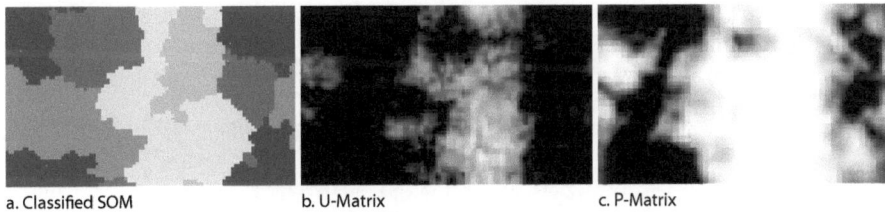

a. Classified SOM   b. U-Matrix   c. P-Matrix

Fig. 5. The classified SOM (left), the U-Matrix (center) and P-Matrix (right)

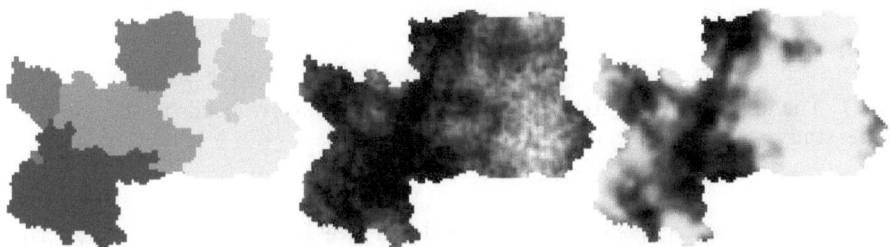

Fig. 6. The classified SOM (left), the U-Map (center) and P-Map (right)

3.2 SSS

The spatial scan statistics has been applied to highlight hot spots in the distribution of services of the agglomeration of Lausanne: this agglomeration counts 300'000 inhabitants for 68 municipalities. The attraction power of Lausanne, as well as the performing transport network, guarantees fast and reliable accessibility to the center of the agglomeration, i.e. the center of Lausanne. The data used come from the Population and Firms Census: the first counts several features related to population at a resolution of one hectare. The latter counts the number of firms and employees for different kinds of activities with the same resolution. Such a high resolution allows to detect clusters at the intra urban scale and to be independent to the political apportionment of the municipalities. As discussed later, such a liberty will be very useful for a big municipality like Lausanne, where the distribution of socio-economic features is very different for diverse neighborhoods. In order to speed up the analysis, original data have been aggregated to a 200 meters grid. This grid represents the spatial entities (Figure 7). Unfortunately, the Census are not available for years previous to 1990: therefore, only the Firms Census 1991/2001 have been used. The analysis has been limited to the emergence of clusters during this period.

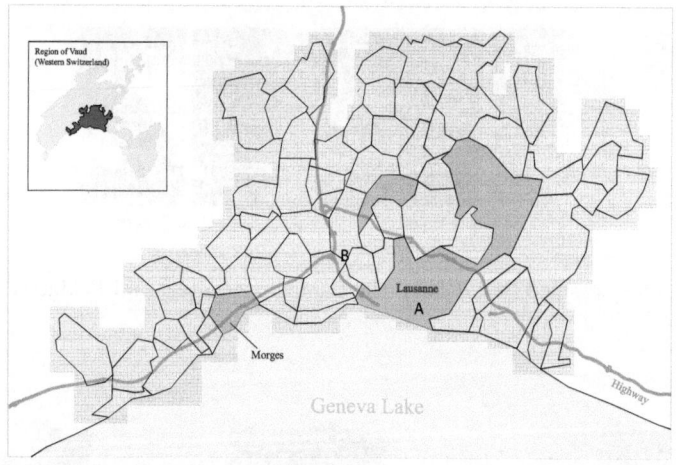

Fig. 7. The urban agglomeration of Lausanne. Gray area corresponds to the grid used in the study. (Source of GIS data: Swisstopo).

Four types of activities have been considered (see Table 1 for detailed description and corresponding NOGA codes[1]): the aim is to detect territorial specialization of the services, to see if there are regions that have specialized creating thus different employment centralities. Such an observation may bring relevant

---

[1] NOGA codes are codes given in the Swiss Census to each type of activity.

information for the debate about the centrality of the city, whose common opinion states the concentration of all services in the city center. Our hypothesis is that the increase of accessibility to transport networks has displaced a series of services toward the suburban areas and that the agglomeration is constructed dynamically around these different poles of employment.

The model used is the spatial Poisson Scan Statistic, the analysis have been performed using the free software SatScan (http://www.satscan.org). As population variable, the use of residential population [27] can be questioned, because the service are used by both the residents and the workers : to take into account this effect, a mixed population accounting for both residential and working population has been used. This way, the weighting variable takes into account the persons that work in the area without living there and the detection of clusters is not affected by the size effect related to them. The neighborhood is defined by circular windows bounded at a radius of 3km. The consideration of large radius results in the inclusion of low rates areas in large hot spots detected, giving a false image of shape and size of the clusters. Therefore, the maximum radius has been kept short, resulting in small circular clusters composing, at a larger scale, the larger irregular clusters presented in Figures 9 and 10.

Table 1. Types of services considered in the study of urban services [26]

Service type	Service category	NOGA codes
	Transport and storing	60-63
Distribution services	Telecommunications	64
	Wholesale trade and retailing	50-52
	Banks	65-67
Production services	Real estate	70
	Informatics, research and development, architects, engineering	72,73,74
	Health and social action	85
Social services	Education	80
	Public administration	75
	Hotels and restauration	55
Personal services	Cultural activities, sports	92
	Other	93

4   Discussion

4.1   HSOM

The HSOM classification map is shown in Figure 8. As stated in the previous section, the neurons of the SOM have been classified into five classes, partitioning the socio-economic landscape of Switzerland. Typical spatial structures of the country can be seen on the map.

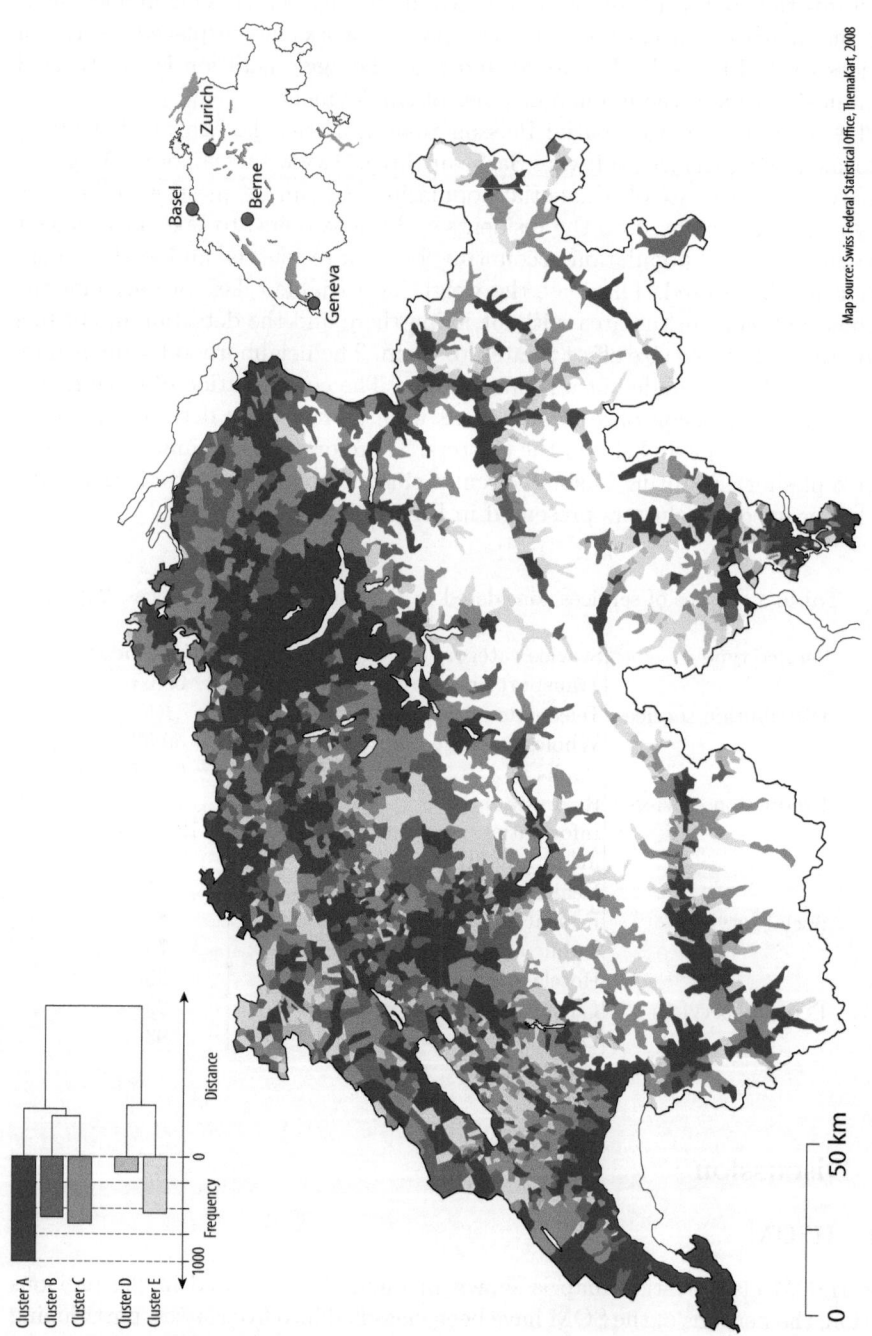

Fig. 8. Thematic map for the resulting classification

Cluster A (dark gray in Figure 8) mainly represents municipalities which are active in production services and which attract a young population. The main urban agglomerations of Switzerland belong to this category, with all the main cities and some touristic regions. Surprisingly, workers in specialized services are not particularly present in this cluster. An explanation to that may be the strength of these services all over the country whereas production services are mainly present in urban areas.

Cluster B is marked by a strong presence of retired people (more than 65 years), and Cluster C by managers and a population aged roughly between 50 to 64 years. Municipalities belonging to cluster C are mainly located in peri-urban areas. Clusters A to C form together the main economic regions of Switzerland. Clusters D and E comprise more agricultural municipalities; cluster E presents however a quite big variety.

It is noteworthy that the cluster boundaries from HAC correspond roughly the visual clusters from the SOM. However, there is no perfect match which reminds us that socio-economic data are complex. However, the method proposes highlights in a very clear way the main urban areas, and the main economic regions are correctly detected.

The results presented allow to have a look on the socio-economic landscape of the country. Such a cartography allows to simplify interpretation about local and regional specificities (the dimensionality of the space has been reduced from 56 original variables to a unique map consisting of 5 groups). In this process, the SOM allows to embed nonlinearly the original dataset into a lower dimension (and easier to interpret) feature space, taking into account the nonlinear relationships to be learned by the algorithm. The coupling of the SOM with the HAC allows to decide easily and rapidly the number of classes and to build mean profiles of the classified neurons for the labeling of the classes.

4.2 SSS

Figure 9 illustrates the hot spots highlighted by SSS for the four types of services in the agglomeration of Lausanne. Contrarily to the commonly accepted theory of the services concentrated in the city center, the patterns emerging from this study show a different configuration of the distribution of the services where 2 main poles are clearly distinguishable. Most of the services, with the exception of the distribution services, are found in the city center of Lausanne (A in Figure 7). This region concentrates the main services of the categories 'Production', 'Social' and 'Personal' by the attraction power of the main city of the agglomeration.

A particular behavior is observed for the distribution services: the main emergent cluster is found in the regions of Crissier (B in Figure 7) and Cugy (highlighted in Figure 9). Although the latter cluster was expected (there is a transportation services pole in this small industrial area), the presence of the cluster in the region of Crissier deserves an explanation. This behavior is related to the creation, in the '90s, of an industrial/services pole in that region, that has attracted the distribution services, in particular peripheral malls related to

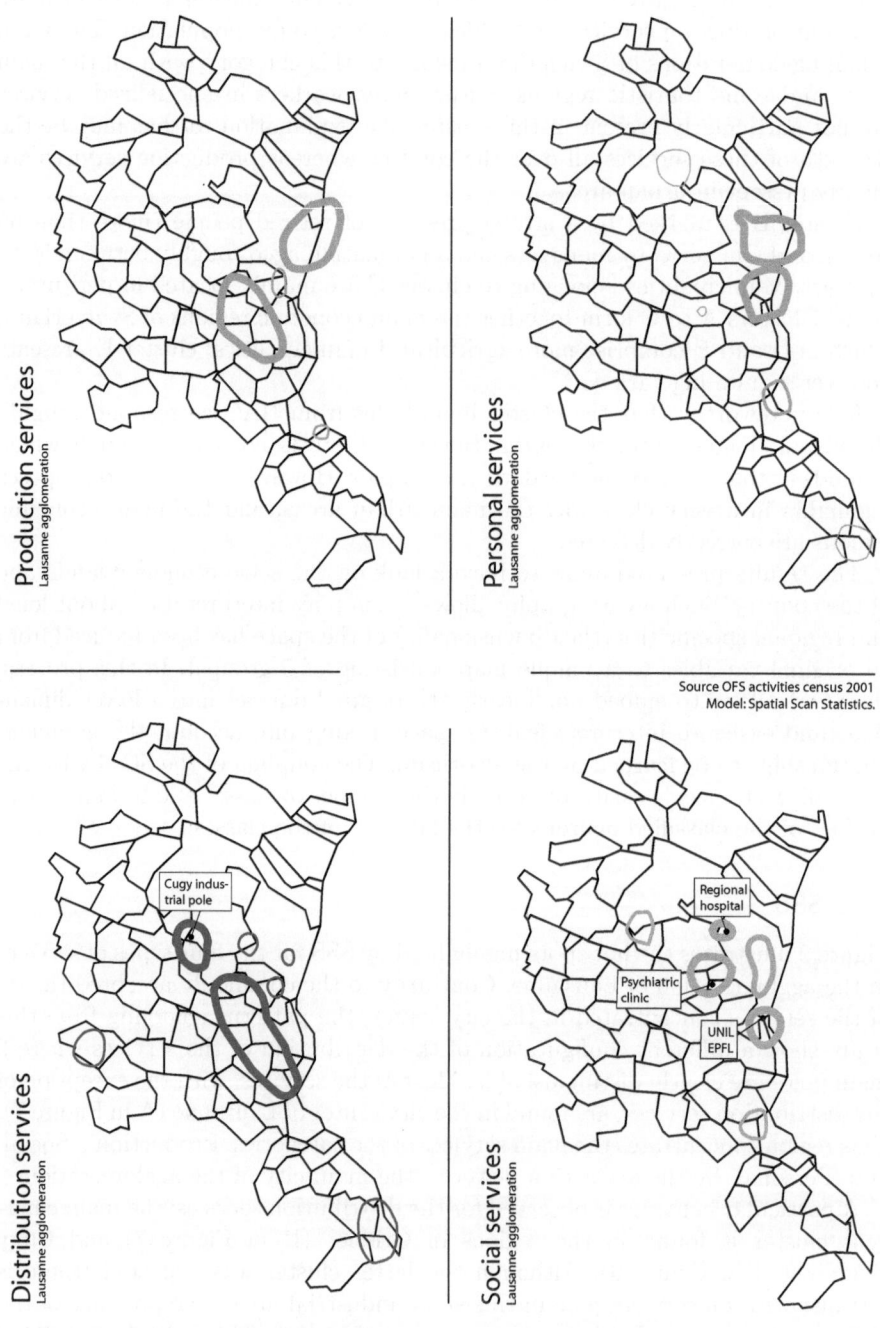

Fig. 9. Clusters of services found by SSS in the agglomeration of Lausanne

Fig. 10. Aggregation of the clusters of services found by SSS in the agglomeration of Lausanne

retail services. Moreover, the proximity to the highway network is a facilitating factor to the development of this kind of services in the region. A cluster of production services is also observed in the region 'B': this is unexpected, because financial services are mainly located in the center. The presence of this cluster can be explained by the development in the Crissier region of a technological pole, where informatics services are present. Since the number of services in these categories was not present in the region before the '90s, this increase has been interpreted by the model as an emergent cluster. For production services, the city center's cluster is mainly related to financial and real estate services, while research, engineering and informatics are the reason of the cluster located in the region B; all these observations show the emergence of a second core of services in the West of the region. This new centrality presents a region not characterized by strong centrality in a traditional sense. The development of peripheral malls and technological centers may be the reason for this behavior.

Regarding social services, the two strong clusters in the North of Lausanne are explained by the presence of the regional hospital and of the psychiatric clinic of Cery, two of the biggest medical centers in the region (highlighted in Figure 9). The university of Lausanne (UNIL) and the Lausanne Institute of technology (EPFL) are the main reason for the Western cluster of social services, as well as for the cluster of personal services, related to the University's sport center.

Figure 10 summarizes the information discussed so far: by superposing the clusters of the four types of services, an image of the concentration of services in the region can be made out. Three main aggregations of clusters can be seen on this image:

1: the Crissier region is the unexpected result of this study. Contrarily to traditional knowledge about distribution of services, this peripheral region has known a strong development of distribution and production services related to the creation of an industrial pole and a series of mall advantaged by their proximity to the main transport axes.

2: the city center of Lausanne, that attracts most of the services of the four types. This cluster confirms the attraction power of the agglomeration center, that drains the population of the region by its high connectivity and by the strong offer, that act as a positive feedback on the number of services.

3: the lakeside of Lausanne. This cluster is mainly related to tourism activities. The East side of the cluster corresponds to a region with strong concentration of hotels and bars, while the West part is related to a regional peculiarity of the region discussed above: the University and the Institute of Technology, that attract services related to resaturation and sports.

5   Conclusion

In this paper, two methods have been presented for the analysis of clustering of the urban space. First, HSOM, an hybrid method based on hierarchical classification of the nodes of a self-organizing map, has been presented for the embedding and clustering of high-dimensional features spaces. Such a method has shown its potential in an application of unsupervised classification of socio-economic profiles of the Swiss municipalities. HSOM has allowed to group the municipalities of the region in five classes defined using the similarities (linear and nonlinear) between the spatial units.

Second, the spatial scan statistics has been proposed for the detection of high rates clusters in space. Analysis of the distribution of services at the agglomeration scale have shown th emergence of a new services pole in the West of the region, characterized by its accessibility to the transport network and the presence of supermarkets and malls. This new pole challenges the city center, where most of the production and personal services are concentrated and construct a bipolar image of the region, in accordance to the local development observed during the last decade.

New challenges for such methodologies would be, for the unsupervised methods, to study deeply the relationship between the size of the SOM and the detection of clusters and the effect of different embeddings or measures of distance. For the cluster detection problem, the implementation of searches for irregular cluster shapes could be of great interest, because (for instance for a region such the one studied) physical constraints prevent the emergence of naturally circular clusters. In this study the size of the clusters has been limited to overcome this problem, but some studies on irregular cluster detection have been published so far (see [28, 29]). Moreover, the effect of mixed population must be studied deeper and will give insight to the relationship between the services and their users.

Acknowledgments

This work has been supported by the Swiss National Science Foundation (projects "Urbanization Regime and Environmental Impact: Analysis and Modelling of Urban Patterns, Clustering and Metamorphoses", n.100012-113506 and "Geokernels: Kernel-Based methods for Geo- and Environmental Sciences (Phase II)", n.200020-121835).

References

1. Schuler, M., Bassand, M.: La Suisse, une métropole mondiale? IREC. Lausanne (1985)
2. Da Cunha, A.: La métropole absente?, IREC, Lausanne (1992)
3. Bassand, M.: Métropolisation et inégalités sociales. Presses Polytechniques Universitaires Romandes, Lausanne (1997)
4. Batty, M.: Cities and complexity. MIT Press, Cambridge (2005)
5. Jain, A.K., Dubes, R.C.: Algorithms for clustering data. Prentice-Hall, Englewood Cliffs (1988)
6. Kohonen, T.: Self-organizing maps, 3rd extended edn. Springer, Berlin (2001)
7. Ward, J.H.: Hierarchical grouping to optimize an objective function. Journal of the American Statistical Association 58, 236–244 (1963)
8. Anselin, L.: Local indicators of spatial autocorrelation - LISA. Geographical Analysis 27, 93–115 (1995)
9. Turnbull, B.W., Iwano, E.J., Burnett, W.S., Howe, H.L., Clark, L.C.: Monitoring for clusters of disease: application to leukemia incidence in Upstate New York. American Journal of Epidemiology 132, 136–143 (1990)
10. Openshaw, S., Charlton, M., Wymer, C., Craft, A.: A Mark 1 Geographical analysis machine for the automated analysis of point data sets. International Journal of Geographical Information Systems 1, 335–358 (1987)
11. Fotheringham, A.S., Zhan, F.B.: A comparison of three exploratory methods for cluster detection in spatial point patterns. Geographical Analysis 28, 200–218 (1996)
12. Kulldorff, M.: A spatial scan statistic. Communications in Statistics 26, 1481–1496 (1997)
13. Lawson, A., Biggeri, A., Böhning, D.: Disease mapping and risk assessment for public health. Wiley, New York (1999)
14. Kulldorff, M., Tango, T., Park, P.J.: Power comparison for disease clustering tests. Computational statistics and Data Analysis 42, 665–684 (2003)
15. Song, C., Kulldorff, M.: Power evaluation of disease clustering tests. International Journal of Health Geographics 2, 1–8 (2003)
16. Leloup, J.A., Lachkar, Z., Boulanger, J.-P., Thiria, S.: Detecting decadal changes in ENSO using neural networks. Climate dynamics 28, 147–162 (2007)
17. Ultsch, A.: Maps for the visualization of high-dimensional data spaces. In: Proceedings of WSOM 2003, Kitakyushu, Japan, September 11-14 (2003)
18. Ultsch, A., Moerchen, F.: ESOM-Maps: tools for clustering, visualization, and classification with Emergent SOM. Department of Mathematics and Computer Science, University of Marburg (2005)
19. Ultsch, A.: Self-organizing neural networks for visualization and classification. In: Proceedings Conf. Soc. for Information and Classification, Dortmund (April 1992)

20. Ultsch, A.: Pareto Density Estimation: Probability Density Estimation for Knowledge Discovery. In: Baier, D., Wernecke, K.-D. (eds.) Innovations in Classification, Data Science, and Information Systems, pp. 91–102. Springer, Berlin (2005)
21. Ultsch, A., Hermann, L.: Architecture of emergent self-organizing maps to reduce projection errors. In: Proc. ESANN, Bruges, Belgium (2005)
22. Kulldorff, M., Athas, W., Feuer, E., Miller, B., Key, C.: Evaluating clusters alarms: A space-time scan statistic and brain cancer in Los Alamos. American Journal of Public Health 88, 1377–1380 (1998)
23. Kulldorff, M., Song, C., Gregorio, D., Samciuk, H., DeChello, L.: Cancer maps patterns: are they random or not? American Journal of Preventive medicine 30, 37–49 (2006)
24. Ceccato, V., Haining, R.: Crime in border regions: The Scandinavian case of resund, 1998-2001. Annals of the Association of American Geographers 94, 807–826 (2004)
25. Coulston, J.W., Riiters, K.H.: Geographic analysis of forest health indicators using spatial scan statistics. Environmental Management 31, 764–773 (2003)
26. Browning, H.L., Singlemann, J.: The emergence of a service society: demographic and sociological aspects of the sectorial transformation of the lbor force in the USA. Springfield, National Technical Information Service
27. Kuhnert, C., Helbling, D., West, G.B.: Scaling laws in urban supply networks. Physica A 363, 96–103 (2007)
28. Conley, J., Gahegan, M., Macgill, J.: A genetic approach to detecting clusters in point data sets. Geographical Analysis 37, 286–314 (2005)
29. Duczmal, L., Kulldorff, M., Huang, L.: Evaluation of spatial scan statistics for irregularly shaped clusters. Journal of Computational and Graphical Statistics 15, 1–15 (2006)

Detecting Alluvial Fans Using Quantitative Roughness Characterization and Fuzzy Logic Analysis

Andrea Taramelli[1,2] and Laura Melelli[2]

[1] Lamont Doherty Earth Observatory of Columbia University, New York,
Route 9W, Palisades, NY 10964, USA
`ataram@ldeo.columbia.edu`
[2] Dipartimento di Scienze della Terra, Università degli Studi di Perugia,
via Faina, 4, 06123 Perugia, Italy

Abstract. This research, based on a similarity geometric model, uses quantitative roughness characterization and fuzzy logic analysis to map alluvial fans. We choose to work in the Italian central Apennine intermountain basins because much human activities could mask this kind of landforms and because the timing of alluvial deposition is tied to land surface instabilities caused by regional climate changes. The main aim of the research is to understand where they form and where they extent in an effort to develop a new approach using the backscatter roughness parameters and primary attributes (elevation and curvature) derived from the SRTM DEM. Moreover, this study helps to provide a benchmark against which future alluvial fans detection using roughness and fuzzy logic analysis can be evaluated, meaning that sophisticated coupling of geomorphic and remote sensing processes can be attempted, in order to test for feedbacks between geomorphic processes and topography.

Keywords: Alluvial Fan, DEM, Roughness, Fuzzy Logic, Curvature, Elevation.

1 Introduction

Most landforms have a well defined descriptions, parameterizations and models in the geomorphologic literature [1], [2], [3], [4], based on the analysis of the relationship between geological and geomorphologic causative factors [5]. The major problem with accurate definitions of landforms is the complexity of the boundaries, which are the result of the interplay of many factors, some of which are known and mappable while others are known but cannot be effectively expressed digitally. Transposing a feature on a map could show very different results and depends on various variables: the scale of the feature (meso -or macro- landform), the scale of the map corresponding to the degree of accuracy, human error introduced by subjectivity and the errors from insight transposing mapping techniques. Landform delineations using traditional surveyor techniques are sometimes inadequate to provide unequivocal results. The quiescent or inactive boundaries of a landform may change over a very short time scale due to non-conservative lithotypes, even thought change may be expected only

M.L. Gavrilova and C.J.K. Tan (Eds.): Trans. on Comput. Sci. VI, LNCS 5730, pp. 251–266, 2009.

over a geomorphologic time scale. Moreover, specific land use and land cover (for example, areas with much human activity) could mask the features. The representation of landforms on thematic maps is thus the first fundamental step to visualize the features and assess associated processes. In this context the terrain roughness is an important parameter in several geomorphic investigation. Examples include characterization and classification of lava flow, alluvial deposits and desert surfaces [6], [7]. Relative ages of alluvial fans or fan units are correlated with variations of the surface roughness [8], [9].

In this research we studied the alluvial fans, depositional features of water-transported material (alluvium) with a longitudinal section cone geometry. They typically form at the base of slopes or at the junction of a tributary stream with a main one, and are characterized by a distinct break in slope [10]. They appear as a cone segment radiating away from a point source (fan apex) with the coarser sediment in the upper fan gradually diminishing toward the plain. Deposits of the alluvial fan may result from debris flows deposition [11], and/or be water-laid [12]. The morphologic and morphometric characteristics of the fan are strictly related to the lithotypes and the shape of the catchment area [13]. Moreover, tectonic control and climatic conditions influence the final shape of the fan [12].

Geographic Information System (GIS) software has made the task of managing spatial data much easier, more interactive and informative. The advent of spatial data in the form of Digital Elevation Models (DEM) and the widespread availability have led, during the past several years, to improve tools for landform delineation and modelling [14], [15], [16]. To provide a sound empirical evaluation of the delineation of landforms using DEM analysis, a strict definition of the landforms being investigated is needed to minimize the error matrix and consequent delineation errors in mapping transposition. In this paper, a fuzzy logic computer-based algorithm that uses the Shuttle Radar Topography Mission (SRTM) DEM attributes and a quantitative analysis of the SRTM radar backscatter data for roughness estimation is employed to investigate the relationship between alluvial fans and the distribution of boundaries.

2   Landforms: Semantic and Geometric Approaches

The goal of defining landforms in geomorphology is usually achieved using two main approaches that are widespread and well known in the scientific literature [1], [2], [17]: the semantic and the geometric approaches. According to the semantic approach, a landform is the result of a classification that simplifies the real world. The classification of the landforms is dependent on the scientists' background and the research context. When we look at a landscape which is made up of the surface of Earth continuously varying in elevation, together with natural and anthropogenic phenomena superimposed on it, the strategy for classification is first based on different conditions (i.e. structural). That strategy can lead to some errors in mapping transposition. The geometric approach (including topological considerations and semantic definitions) highlights the geometric characteristics of a feature related to the topographic surface properties. Some landforms show, independently from the environmental conditions, the same geometric response expressible in terms of angles (slope, curvature), and distance ratio along the principal measurement directions (height, width, thickness).

The attempt to analyze the landscape, using a geometric approach based on the semantic one, has given rise to an area of research which predicts the shape of the surface by delineating a set of rules within a GIS [18], [19], [20]. The advent of GIS software and of spatial data in the form of DEM has made possible the task of generating descriptive statistics of the shape of the surface and locations in the landscape based on the local form of the land surface.

Among the simpler, geometric, and therefore computable set of forms is the assignment of a location of different morphometric parameters to primary or secondary (compound) attributes [4]. Thus, if we consider a regular grid of elevations stored as a DEM, a set of grid cells (or pixels) in the DEM can be assigned to a specific morphometric set of features which people recognize in the landscape. In this context the algorithms traditionally included in most raster processing systems use neighborhood operations to calculate slope, aspect, shaded relief and points of inflection. These calculations are made for each cell, based on the values in the eight cells spatially adjacent in a grid. For instance, a cell that is equal in elevation to all neighbors, meets the criteria to classify it as a member of a flat area. To overcome possible limitations, an automatic procedure has been developed to use interactive spatial techniques as well as a neighborhood operation that can best be visualized as region-growing procedures. They provide an analyst with the ability to extract information from DEMs on morphologic features and properties, specifically topographic depression and curvature. That knowledge recalls the one underlined within the landscape concept that is characterized by a degree of uncertainty, especially in its spatial extent. The ambiguity of the above attributes is probably best matched to the model of vagueness [18]. Despite a precise definition of the different morphometric classes, any location can be allocated to a specific class, but the class to which a location is assigned by this precise process varies, due to the scale of measurement giving rise to ambiguity as to the correct classification and so vagueness. Several researchers have introduced the idea that the vagueness in geomorphology may be appropriate for analysis by fuzzy sets [21], [22], [23], [24] [25], [26], [27] [28] [29]. Algorithms, used to identify such features, are now standard tools within GIS. They can provide a standardized approach to identify landforms [30], [31], [32], [33] [34]. Two methods for defining the membership values of the fuzzy sets have been developed [35], [36]: the semantic import model, based on the a priori knowledge that assigns a value of fuzzy membership to a landscape feature with a particular metric property, such as height [37] and the similarity relation model that uses surface derivatives, such as slope and curvature, as input to a multivariate fuzzy classification which yields the membership values [38]. We choose to use the second model. In this method the similarity representation of different classes in the parameter domain is based on fuzzy logic applied in GIS [22]. Under fuzzy logic, a class at a given pixel can be assigned to more than one geometric class with varying degrees of class assignment [39], [40], [41]. These class degrees assignment are referred to as fuzzy memberships. This fuzzy representation allows a class at each pixel to bear a partial memberships in each of the prescribed classes. Then each fuzzy membership is regarded as similarity measure between the local class and the typical case of the given class. By coupling this similarity representation with a raster GIS data model, allow alluvial fans in different area to take property values of the prescribed classes [42], [43], [44].

3 The Study Area

The Umbria region is located in the central Apennines (central Italy, Fig. 1) and is characterized by different intermontane tectonic basins (with NW to SE direction), now partially filled with alluvial and lacustrine deposits. The very low degree of the slope that characterizes the Umbria intermontane basins is one of the most significant reasons why they are subjected to much human activity. Development may modify or mask the landforms like the large part of the alluvial fans with low values of the longitudinal profile angle that results in a very gentle slope that is not very different from the adjacent alluvial plain. To develop a fuzzy computer-based algorithm that will delineate the alluvial fan landform, two basins and the related alluvial fans were studied.

The Gubbio basin (Fig. 1) is a half-graben located in the northeastern side of Umbria, 20 km long and 4 km width (maximum) where the alluvial fans are made by eterometric deposits, related to flood events. The mean slope is 10°/12°. Several coalescent alluvial fans are recognizable along the NE boundary, with the largest ones in the central part.

The Umbria Valley basin (Fig. 1), located in the central part of the Umbria region, has a NS direction and it is 40 km long. Alluvial fans are located along the eastern boundary of the basin. The human presence is very wide and the boundaries of the fans are very uncertain also because the interaction between the fan deposits with the lower alluvial plain.

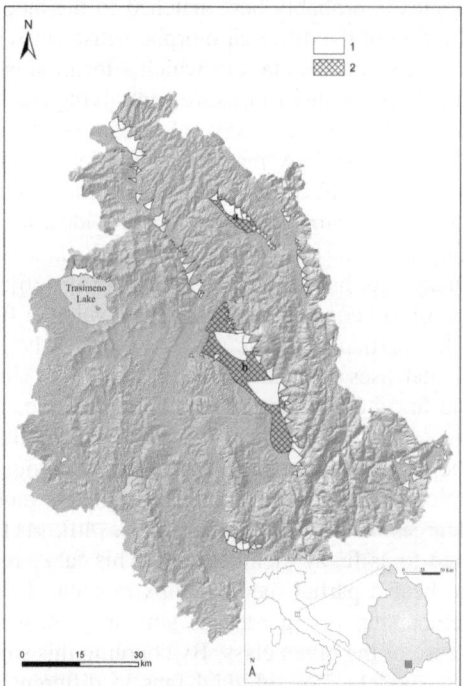

Fig. 1. Location map of the study area: a) the Gubbio Basin, b) the south-eastern branch of the Umbria Valley Basin. 1) Alluvial fans, 2) intermontane basin areas.

4 Methodology

4.1 Estimating SubPixel Surface Roughness Using the C-Band SAR Backscatter from SRTM

The morphology of a geological surface is the result of complex formation and weathering processes. For a quantitative surface characterization, parameters have to be selected that will allow a clear discrimination between different types of morphology. The most obvious parameters are the magnitude and horizontal length scale of surface height variations. If the surface can be modelled as a stationary random Gaussian process, mean and variance of the elevation, and the autocorrelation function (which is related to the horizontal length scale of height variation) provide a complete description of the statistical surface properties. In theoretical models of rough surface scattering, it is often assumed that the surface is stationary with a Gaussian height distribution [45]. In addition, it is assumed that the mean elevation of the surface is subtracted from all height data. In this case, the autocorrelation function is identical to the autocovariance function, and the square root of the height variance is the standard deviation around zero mean. Work in the 1980s and early 1990s by the group at Arizona State University [46], [47] showed that the logarithm of roughness length z_0 depends linearly on radar backscatter power, as measured by normalized radar cross-sections $\sigma°$ expressed on a decibel scale. More recently, radar based estimates of roughness length have been made globally from C-band ERS1/2 radar scatterometer measurements by a group based in the CNRS/Universités Paris VII-XII [48]. Although the spatial resolution is somewhat coarse, the main advantage of ERS1/2 scatterometry lies in its full coverage of the globe every ~4 days with 50 km resolution, so that seasonal and interannual variations in z_0 can be taken into account. The research shows that the scatterometer estimates of z_0 are fully consistent with those of the ASU group's results from the airborne and space shuttle SAR data. After correcting for variable look angle effects, they find a linear relation between the logarithm of z_0 and radar cross section $\sigma°$. More recently, they have developed a method for estimating roughness length z_0 based on bi-directional reflectance distribution functions (BDRF) obtained by the POLDER-1 instrument flown aboard ADEOS 1 platform from October, 1996 through June, 1997 [49]. POLDER-1 was a broad swath instrument that operated in the VNIR, with a zenith viewing angles up to 60°, allowing multiple viewing angles for a given site. Averaging over a 30-day period allowed determination of the BDRF with a nadir spatial resolution of 6 km x 7 km. Small scale roughness elements cast shadows which affect the BDRF. The BDRF data are used to determine a dimensionless quantity called the "protrusion coefficient" PC, and the aerodynamic roughness length z_0 is found through an empirically determined relation to the PC parameter:

$$z_0 = a * \exp(PC/b),. \tag{1}$$

where a is a constant with units of length, and b is a dimensionless constant. Earlier work found that the PC parameter and normalized radar cross-sections $\sigma°$ (in db) are linearly correlated. Furthermore, both parameters are linearly related to the logarithm of the aerodynamic roughness length z_0.

Based on the aforesaid research an higher resolution (~25 m) estimates of z_o can be derived at present for the entire world from the C-band SAR backscatter from the February 2000 Shuttle Radar Topography Mission (SRTM). We analyzed C-band SRTM backscatter data in the form of 1-degree geographic tiles for the Umbria region. Detailed topography, at selected elevation ranges that matches alluvial flood plain and junction areas typical of the intermontane basin, needs to be recognized using useful parameters based on SRTM backscatter data [48], to be able to detect transient surface events that can occur over widely separated geographical locations. The SRTM backscatter analysis leads to the detection of the roughness signature that represents the small-scale variation in the relief of a terrain surface [51], [52], based on the statement that the amount of backscatter is proportional to the roughness of a surface. As a result, the total amount of the increases in the backscatter signal within the areas of the alluvial fans where the other smooth surfaces, like flood plain, reflect most of the microwave energy away from the radar and produce a smaller backscattered portion.

In each 1 degree tile all four SRTM subswaths of backscatter data from trajectories passing through the tile were provided as independent data files, along with corresponding files of radar look angle (nominal incidence angle). The data from the subswaths had to be combined, and the given backscatter power at each pixel needed to be corrected to a standard incidence angle for comparison with the existing map. This was done to train the "signatures" of the roughness.

We processed and analyzed this data, producing roughness maps for the two main basins at full-resolution (90 m) and on a 0.25 degree grid, for comparison with the existing maps of the location of alluvial fans (Fig. 2).

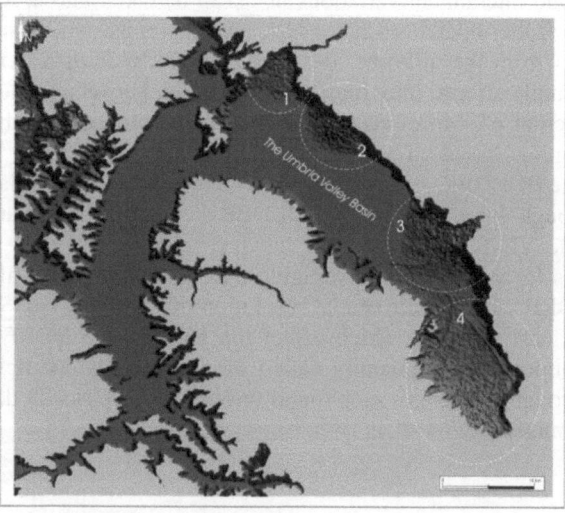

Fig. 2. The southeastern part of the Umbria Valley Basin (blue area) showing regions of great roughness. The roughness signature is compared to the alluvial fan map (Fig. 1) and shows a positive correlation for areas 1 to 3, and a negative correlation for area 4, probably because of the incidence angle. The light area that looks plain in the image represents the bedrock that is not identifiable within the roughness range.

4.2 Populating the Similarity Model: Automatic Geometric Model under Fuzzy Logic

Fuzzy sets and operators are the subjects and verbs of fuzzy logic. These "if-then" rule statements are used to formulate the conditional statements that comprise fuzzy logic. A single fuzzy if-then rule assumes the form:

IF x IS A THEN y IS B

where A and B are linguistic values defined by fuzzy sets on the ranges X and Y, respectively. The if-part of the rule "x is A" is called the antecedent or premise, while the then-part of the rule "y is B" is called the consequent or conclusion. An example of such a rule might be:

IF slope IS inclined THEN area IS suitable

The input to an if-then rule is the current value for the input variable (slope) and the output is an entire fuzzy set (suitable). This set will later be defuzzified, assigning one value to the output. Interpreting an if-then rule involves distinct parts: first evaluating the antecedent (which involves fuzzifying the input and applying any necessary fuzzy operators) and second applying that result to the consequent (known as implication). In the case of two-valued or binary logic, if-then rules don't present much difficulty. If the premise is true, then the conclusion is true. If the antecedent is true to some degree of membership, then the consequent is also true to that same degree.

In this context we first semantic define alluvial fans based on literatures. Then to complete the analysis we had to create a geometric definition of alluvial fans. In this analysis the morphometric class is the geographical scale of measurement. By scale, we mean a combination of spatial extent and spatial detail or resolution of the variation in the extent over which the feature is defined as the basis of the fuzzy membership [53] [54]. Thus geomorphometric measures can indicate the location of a landform by assigning a location in a landscape to an exhaustive set of classes that could be run by an algorithm process. These different classes are illustrated below (Fig. 3):

1) Altitude. The highest value of altitude is equal to the top of the apex of the highest fan; the lowest value of altitude is equal to the lowest value of altitude of the toes of the alluvial fans. Because the range of values of altitudes differ in large areas, the break values to be considered have to be selected for each basin. What may be a specific class at one scale may be another morphometric class at another scale. The alluvial fans must be grouped in a well defined range of threshold altitude for each basin.

2) Convex contour. The shape of the alluvial fan shows typical boundaries along successive breaks of slope on a two-dimensional profile. A convex class referred to slope is recognizable in all the types of the alluvial fan and can be emphasized by contour lines. The same geometric shape is not recognizable in other debris accumulations (i.e. talus heaps) where the convexity attribute is random.

3) Increase in arc circumference. In plan view, the alluvial fans show a cone geometry where the boundary value may approximate natural division. In this context

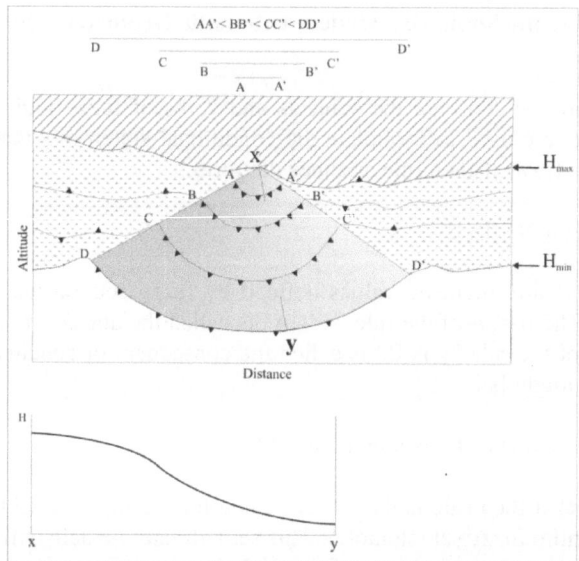

Fig. 3. Scheme of the geometric classes of alluvial fans. The altitude class has a range between H_{max} and H_{min}. The convex contour class is highlighted by the direction of the triangles. The cone geometry is evident from the increase in the arc circumference from the segment AA' to DD'. The longitudinal profile (xy) shows a convex-concave radial shape from the top (H_{max}) to the bottom (H_{min}).

only two parameters, the values for the lower and upper boundaries are needed: the arc circumference increases from the apex (upper fan) to the toe (lower fan). The algorithm uses a linear relationship: while the altitude decreases, the arc circumference increases.

4) Convex-concave radial. Along the longitudinal profile an alluvial fan shows a convex – concave radial shape (from the top to the bottom) because the sediment grain size decreases toward lower altitudes values. Whereas gravels in the upper fans are associated with a convex profile, sands and clays are related to a concave profile in the bottom area. The algorithm first delineates the convex profile within the higher values of the altitudes and then the concave profile within the lower ones.

4.3 Delineation of the Alluvial Fans: Uses of the Similarity Model

Once a set of reliable signatures was created and evaluated, the next step was to perform a classification of the SRTM elevation data with "Parametric/No-parametric" classifier to detect the range of altitude and curvature within the roughness signature using a probability function:

$$Pi = (2\pi)^{-\frac{1}{2n}} |Ci|^{-\frac{1}{2}} e^{-(1/2)[(X-Mi)'C^{-1}(X-Mi)]} \Pr i \qquad (2)$$

Where:

Pi is Maximum likelihood probability of attribution to the class.

n   Number of measurement variables.

Ci   Covariance matrix of the class considered.
Mi   Mean vector of the class considered.
X   Pixel vector.
Pr*i*   Prior probability of the class considered defined from the frequency histograms of the training sets.

$$\mathrm{Pr}\,i = Fr\,/\,Frt \tag{3}$$

Where:

Fr   is the pixel count of the class under examination.
Frt   Is the sum of counts of all the classes.

In fuzzy theory, the algorithm has to define the class which exactly matches the core parameter set to assign a class membership of 1. The membership is assigned by a decreasing real number for classes as they are increasingly dissimilar from that core parameter until they have no similarity at all to the class. At that point the membership is assigned a value 0. The created algorithm has a specific sequence to assign the boundary value of class sets (Table 1).

Table 1. The heuristic rule base for converting initial terrain derivatives into fuzzy landform attributes

No.	Input terrain derivative	Output fuzzy landform attribute	Description of fuzzy landform attribute	Standard Index	Dispersion index
1	Elevation	Near_max	Relatively near maximum elevation	90.0	15.0
2	Elevation	Near_min	Relatively near minimum elevation	10.0	15.0
3	Profile	Concave_D	Relatively convex profile (down)	10.0	5.0
4	Profile	Convex_D	Relatively concave in profile (down)	-10.0	5.0
5	Profile	Planar_D	Relatively planar in profile (down)	0.0	5.0
6	Planar	Convex_A	Relatively convex in plan (across)	10.0	5.0
7	Planar	Concave_A	Relatively concave in profile (across)	-10.0	5.0
8	Planar	Planar_A	Relatively planar in profile (across)	0.0	5.0

The first selected parameter is the range of altitude values. Within this range of values the second assignment chooses only the convex contour shape. As a third boundary the algorithm selects only convex contours with an arc circumference that increases toward lower altitude. Finally, as a fourth boundary, convex-concave radial slope values are chosen. Ideally this approach maximize internal homogeneity and between-unit heterogeneity, and is characterized by unique groups of morphometric parameters. In that way the two studied areas were partitioned into topographic sections, adopting a semi-quantitative approach that combined an unsupervised eight-class cluster-analysis of derivatives. Finally, error matrices (Confusion Matrix) and the usual coefficient (Commission Error, Omission Error, Kappa Coefficient) were evaluated to quantify the different sets of ranges detected:

$$K = \frac{N \sum Xkk - \sum Xk_+ X_{+}k}{N^2 - \sum Xk_+ X_{+}k} \tag{4}$$

5 Results and Discussion

By analyzing SRTM data at different polarizations and look angles for backscatter, potentially derivable parameters were highlighted. While the elevation range is the basic surface pattern to detect roughness, the planar and radial curvature values are the primary attributes to detect alluvial fans surface change. Computation for the fuzzy k-means classification of the stratified random sample cells for the two test areas indicated that for both areas, 8 classes were optimal (Table 2 and Table 3).

Table 2. Cluster results for the eight classes in the Gubbio intermontane basin

Overall Accuracy = (160000/160000) 100.0000%
Kappa Coefficient = 1.0000 Ground Truth (Pixels)

Class	Unclassified	Class 1	Class 2	Class 3	Class 4	Class 5	Class 6	Class 7	Class 8	Total	
Unclassified	0	0	0	0	0	0	0	0	0	0	
Class 1		0	31607	0	0	0	0	0	0	0	31607
Class 2		0	0	20966	0	0	0	0	0	0	20966
Class 3		0	0	0	15769	0	0	0	0	0	15769
Class 4		0	0	0	0	17844	0	0	0	0	17844
Class 5		0	0	0	0	0	18254	0	0	0	18254
Class 6		0	0	0	0	0	0	16587	0	0	16587
Class 7		0	0	0	0	0	0	0	13218	0	13218
Class 8		0	0	0	0	0	0	0	0	25755	25755
Total		0	31607	20966	15769	17844	18254	16587	13218	25755	160000

Ground Truth (Percent)

Class	Unclassified	Class 1	Class 2	Class 3	Class 4	Class 5	Class 6	Class 7	Class 8	Total	
Unclassified	0.00	0.00	0.00	0.00	0.00	0.00	0.00	0.00	0.00	0.00	
Class 1		0.00	100.00	0.00	0.00	0.00	0.00	0.00	0.00	0.00	19.75
Class 2		0.00	0.00	100.00	0.00	0.00	0.00	0.00	0.00	0.00	13.10
Class 3		0.00	0.00	0.00	100.00	0.00	0.00	0.00	0.00	0.00	9.86
Class 4		0.00	0.00	0.00	0.00	100.00	0.00	0.00	0.00	0.00	11.15
Class 5		0.00	0.00	0.00	0.00	0.00	100.00	0.00	0.00	0.00	11.41
Class 6		0.00	0.00	0.00	0.00	0.00	0.00	100.00	0.00	0.00	10.37
Class 7		0.00	0.00	0.00	0.00	0.00	0.00	0.00	100.00	0.00	8.26
Class 8		0.00	0.00	0.00	0.00	0.00	0.00	0.00	0.00	100.00	16.10
Total		0.00	100.00	100.00	100.00	100.00	100.00	100.00	100.00	100.00	100.00

Class	Commission (Percent)	Omission (Percent)	Commission (Pixels)	Omission (Pixels)
Unclassified	0.00	0.00	0/0	0/0
Class 1	0.00	0.00	0/31607	0/31607
Class 2	0.00	0.00	0/20966	0/20966
Class 3	0.00	0.00	0/15769	0/15769
Class 4	0.00	0.00	0/17844	0/17844
Class 5	0.00	0.00	0/18254	0/18254
Class 6	0.00	0.00	0/16587	0/16587
Class 7	0.00	0.00	0/13218	0/13218
Class 8	0.00	0.00	0/25755	0/25755

Class	Prod. Acc. (Percent)	User Acc. (Percent)	Prod. Acc. (Pixels)	User Acc. (Pixels)
Unclassified	0.00	0.00	0/0	0/0
Class 1	100.00	100.00	31607/31607	31607/31607
Class 2	100.00	100.00	20966/20966	20966/20966
Class 3	100.00	100.00	15769/15769	15769/15769
Class 4	100.00	100.00	17844/17844	17844/17844
Class 5	100.00	100.00	18254/18254	18254/18254
Class 6	100.00	100.00	16587/16587	16587/16587
Class 7	100.00	100.00	13218/13218	13218/13218
Class 8	100.00	100.00	25755/25755	25755/25755

Table 3. Cluster results for the eight classes in the Umbria Valley intermontane basin

Overall Accuracy = (57575/57575) 100.0000% Kappa Coefficient = 1.0000 Ground Truth (Pixels)

Class	Unclassified	Class 1	Class 2	Class 3	Class 4	Class 5	Class 6	Class 7	Class 8	Total	
Unclassified	0	0	0	0	0	0	0	0	0	0	
Class 1		0	0	0	0	0	0	0	0	0	0
Class 2		0	0	0	0	0	0	0	0	0	0
Class 3		0	0	0	5627	0	0	0	0	0	5627
Class 4		0	0	0	0	30380	0	0	0	0	30380
Class 5		0	0	0	0	0	15908	0	0	0	15908
Class 6		0	0	0	0	0	0	4512	0	0	4512
Class 7		0	0	0	0	0	0	0	9	0	918
Class 8		0	0	0	0	0	0	0	0	230	230
Total	0	0	0	0	5627	30380	15908	4512	918	230	57575

Ground Truth (Percent)

Class	Unclassified	Class 1	Class 2	Class 3	Class 4	Class 5	Class 6	Class 7	Class 8	Total	
Unclassified	0.00	0.00	0.00	0.00	0.00	0.00	0.00	0.00	0.00	0.00	
Class 1		0.00	0.00	0.00	0.00	0.00	0.00	0.00	0.00	0.00	0.00
Class 2		0.00	0.00	0.00	0.00	0.00	0.00	0.00	0.00	0.00	0.00
Class 3		0.00	0.00	0.00	100.00	0.00	0.00	0.00	0.00	0.00	9.77
Class 4		0.00	0.00	0.00	0.00	100.00	0.00	0.00	0.00	0.00	52.77
Class 5		0.00	0.00	0.00	0.00	0.00	100.00	0.00	0.00	0.00	27.63
Class 6		0.00	0.00	0.00	0.00	0.00	0.00	100.00	0.00	0.00	7.84
Class 7		0.00	0.00	0.00	0.00	0.00	0.00	0.00	100.00	0.00	1.59
Class 8		0.00	0.00	0.00	0.00	0.00	0.00	0.00	0.00	100.00	0.40
Total		0.00	0.00	0.00	100.00	100.00	100.00	100.00	100.00	100.00	100.00

Class	Commission (Percent)	Omission (Percent)	Commission (Pixels)	Omission (Pixels)
Unclassified	0.00	0.00	0/0	0/0
Class 1	0.00	0.00	0/31607	0/31607
Class 2	0.00	0.00	0/20966	0/20966
Class 3	0.00	0.00	0/15769	0/15769
Class 4	0.00	0.00	0/17844	0/17844
Class 5	0.00	0.00	0/18254	0/18254
Class 6	0.00	0.00	0/16587	0/16587
Class 7	0.00	0.00	0/13218	0/13218
Class 8	0.00	0.00	0/25755	0/25755

Class	Prod. Acc. (Percent)	User Acc. (Percent)	Prod. Acc. (Pixels)	User Acc. (Pixels)
Unclassified	0.00	0.00	0/0	0/0
Class 1	0.00	0.00	0/0	0/0
Class 2	0.00	0.00	0/0	0/0
Class 3	100.00	100.00	0/5627	0/5627
Class 4	100.00	100.00	0/30380	0/30380
Class 5	100.00	100.00	0/15908	0/15908
Class 6	100.00	100.00	0/4512	0/4512
Class 7	100.00	100.00	0/918	0/918
Class 8	100.00	100.00	0/230	0/230

Once obtained the classification through the cited method results have been analyzed to qualitative judge (Fig. 4).

With the aforesaid results we observed that, within the detection of alluvial fans on deposits connecting the bedrock slopes and the lower alluvial plain highlight, altitude value are distributed in two main intervals: from a maximum value of 520 m a.s.l. to a minimum value of 440 m for the Gubbio Basin, and from a maximum value of 240 m a.s.l. to a minimum value of 190 m for the Umbria Valley.

These two sets of values could represent either talus or alluvial fan deposits. The first type results from a gravitational morphogenetic process that allows the construction of an homogeneous deposit with values of altitudes similar to the alluvial fans,

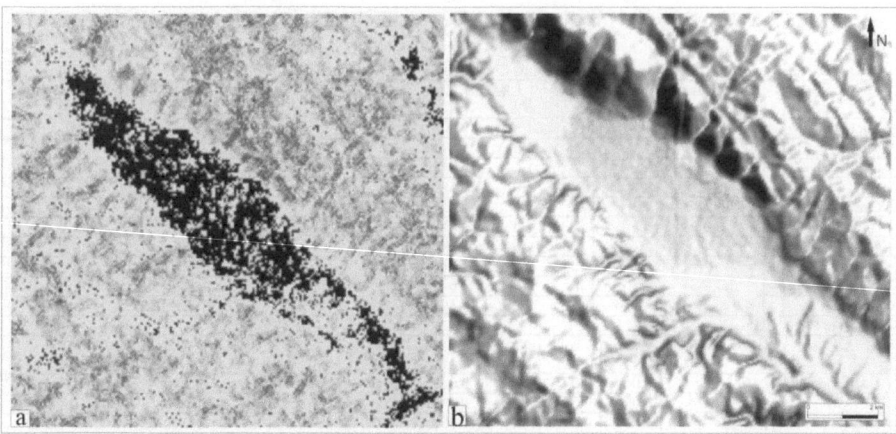

Fig. 4. Results in the Gubbio Basin. (a) Results of fuzzy k-means classification with eight classes for the Gubbio basin – dark blue indicates memberships close to unity for alluvial fans. (b) The final results for Gubbio basin in a 2.5D visualization.

but with very different values of curvature. Whereas talus deposits are characterized by anisotropic values of curvature, fan deposits are characterized by a specific range of curvature. A positive curvature indicates that the surface is upwardly convex at that cell, a negative curvature indicates that the surface is upwardly concave at that cell and a value of zero indicates that the surface is flat. Using the algorithm, three different gradients, -6°, 8° and -0.5° in each direction, were selected as the best results for detecting alluvial fans. These three values correspond to the curvature distributions along the longitudinal profile of the different alluvial fans. Based on the geomorphometric characteristics of the alluvial fans within the intermontane basins of the Umbria region [50], we established that:

- an initial negative value of curvature (-6°) represents the upper fan-head trenching because of the linear channel erosion typical of the alluvial fans in our study area due to the recent regional tectonic uplift and the consequent readjustment of the drainage network;
- a second positive value of curvature (8°) corresponds to upper and medium parts of the fan where the gravel deposits are present and show a convex longitudinal profile.
- a last value (-0.5°) represents the area of the lower fan where lime and clay deposits lay adjacent to flat alluvial sediments.

Even if we minimized the effect of slope in our analysis by including only areas loosely defined, the investigation revealed some correlation between calculated elevation and curvature. Alluvial fans are characterized by a slope angle of at least 5° between the higher slope angle of basement (upward) and the lower slope angle of the adjacent alluvial plain (downward). This is consistent with the results derived from altitude and curvature analysis.

6 Conclusion

This paper presents a new approach to answer the fundamental question of spatial information processing in geomorphology: can the simple knowledge of the spatial extent of a semantic defined alluvial fan with indistinct geographical locations, be improved with fuzzy memberships? This research has shown how alluvial fans can be defined using a novel fuzzy computer - based algorithm to processes the SRTM DEM. The geometric - morphometric analysis does not directly map alluvial fans, but highlights primary attributes classes (roughness, elevation and curvature) of an alluvial fan. Delineation of alluvial fans is then identified within an approximate spatial extent together with fuzzy memberships. This paper has confirmed that fuzzy k-means classification of alluvial fans is then possible using data derived from SRTM DEM. In particular it demonstrates that the procedure based on sampling to obtain training set and a classification scheme could be used with areas with low resolution calls. Problems remain with this method. The method articulated here has been shown to be successful in a limited context: two intermontane basins. This analysis provides new insights and statements that can be made about to what extent type of fans and the morphology are dependent on the physical environment. Future work should examine the integration of more strands of information in recognizing the spatial extents of alluvial fans in other different environments.

References

1. Wood, W.F., Snell, J.B.: A quantitative system for classifying landforms. U.S. Army Quartermaster Research and Engineering Center, Natick, MA, Tech. Rep. EP-124 (1960)
2. Fairbridge, R.W.: The Encyclopedia of Geomorphology. Reinhold Book Corp., New York (1968)
3. Curran, H.A., Justus, P.S., Young, D.M., Garver, J.B.: Atlas of Landforms, 3rd edn. John Wiley and Sons, New York (1984)
4. Speight, J.G.: The role of topography in controlling through-flow generation: a discussion. Earth Surf. Processes Landf. 5, 187–191 (1984)
5. Pike, R.J.: Geomorphometry - progress, practice, and prospect. Z. Geomorph. Suppl.-Bd. 101, 221–238 (1995)
6. Weissel, J.K., Czuchlewski, K.R., Kim, Y.: Synthetic aperture radar (SAR)-based mapping of volcanic flows: Manam Island, Papua New Guinea. Natural Hazard and earth System Sciences 4, 339–346 (2004)
7. Bach, D., Barbour, J., Macchiavello, G., Martinelli, M., Scalas, P., Small, C., Stark, C., Taramelli, A., Torriano, L., Weissel, J.: Integration of the advanced remote sensing technologies to investigate the dust storm areas. In: 8th ICDD Conference, Beijing, February 25-28 (2006)
8. Farr, T.G., Chadwick, O.A.: Geomorphic processes and remote sensing signatures of alluvial fans in the Kun Lun mountains, China. J. Geophys. Res. 101(E10), 23 091–23 100 (1996)
9. Muskin, A., Gillespie, A.: Estimating sub-pixel surface roughness using remotely sensed stereoscopic data. 0-7803-8742-2/04, pp. 1292–1295. IEEE, Los Alamitos (2004)

10. Bull, W.B.: The alluvial fan environment. Progress in Physical Geography 1, 222–270 (1977)
11. Melelli, L., Taramelli, A.: An example of debris-flows hazard modeling using GIS. Natural Hazard and earth System Sciences 4, 347–358 (2004)
12. Nemec, W., Steel, R.J.: What is a fan delta and how we recognize it? In: Nemec, W., Steel, R.J. (eds.) Fan Deltas: sedimentology and tectonic settings, pp. 3–13. Blackie and Son, Glasgow & London (1988)
13. Leeder, M.R.: Sedimentary basins: tectonic recorders of sediment discharge from drainage catchments. Earth Surf. Processes Landf. 22, 229–237 (1997)
14. Wilson, J.P., Burrough, P.A.: Dynamic modeling, geostatistics, and fuzzy classification: new sneakers for a new geography? Annals of the Association of American Geographers 89(4), 736–746 (1999)
15. Burrough, P.A., van Gaans, P.F.M., MacMillan, R.A.: High-resolution landform classification using fuzzy k-means. Fuzzy Sets and Systems 113, 37–52 (2000)
16. Burrough, P.A.: GIS and geostatistics: essential partners for spatial analysis. Environmental and Ecological Statistics 8, 361–377 (2004)
17. Dehn, M., Gärtner, H., Dikau, R.: Principles of Semantic Modeling of Landform Structures. Computers & Geosciences 27, 1005–1010 (2001)
18. Sainsbury, R.M.: What is a vague object? Analysis 49, 99–103 (1989)
19. Pike, R.J.: Geomorphometry – diversity in quantitative surface analysis. Progress in Physical Geography 24, 1–20 (2000)
20. Robinson, V.B.: A Perspective on Geographic Information Systems and Fuzzy Sets, pp. 1–6. IEEE, Los Alamitos (2002); 0-7803-74614102
21. Robinson, V.B.: Some implications of fuzzy set theory applied to geographic databases Computers. Environment and Urban Systems 12, 89–97 (1988)
22. Zhu, A.X.: A similarity model for representing soil spatial information. Geoderma 77, 217–242 (1997)
23. Zhu, A.X.: Measuring uncertainty in class assignment for natural resource maps using a similarity model. Photogrammetric Engineering Remote Sensing 63, 1195–1202 (1997)
24. MacMillan, R.A., Pettapiece, W.W., Nolan, S.C., Goddard, T.W.: A generic procedure for automatically segmenting landforms into landform elements using DEMs, heuristic rules and fuzzy logic. Fuzzy Sets and Systems 113, 81–109 (2000)
25. Argialas, D.P., Tzotsos, A.: Geomorphological feature extraction from a digital elevation model trough fuzzy knowledge-based classification. In: Ehlers, M. (ed.) Proc. SPIE. Remote Sensing for Environmental Monitoring, GIS Applications and Geology 2, vol. 4886, pp. 516–527 (2003)
26. Petry, F., Cobb, M., Wen, L., Yang, H.: Design of system for managing fuzzy relationships for integration of spatial data in querying. Fuzzy Sets and Systems 140, 51–73 (2003)
27. Robinson, V.B.: A perspective on the fundamentals of fuzzy sets and their use in geographic information systems. Transactions in GIS 7, 3–30 (2003)
28. Dubois, D., Prade, H.: What are fuzzy rules and how to use them. Fuzzy Sets and Systems 84(2) (1996)
29. Zadeh, L.A.: Toward a perception-based theory of probabilistic reasoning with imprecise probabilities. Journal of Statistical Planning and Inference 105(1), 105–119 (2002)
30. Dikau, R.: The application of a digital relief model to landform analysis in geomorphology. In: Raper, J. (ed.) Three-dimensional applications in Geographical Information Systems, pp. 51–77. Taylor and Francis, London (1989)

31. Cobb, M., Petry, F., Robinson, V.: Special issue: uncertainty in geographic information systems and spatial data. Fuzzy Sets and Systems 113(1), 1–159 (2003)
32. Cross, V., Firat, A.: Fuzzy objects for geographical information systems. Fuzzy Sets and Systems 113(1), 19–36 (2000)
33. Ladner, R., Petry, F., Cobb, M.: Fuzzy set approaches to spatial data mining of association rules. Transactions in GIS 7(1), 123–138 (2003)
34. Ahamed, T.R., Gopal Rao, K., Murthy, J.S.R.: GIS-based fuzzy membership model for crop-land suitability analysis. Agricultural Systems 63, 75–95 (2000)
35. Fisher, P.F.: Fuzzy modeling. In: Openshaw, S., Abrahart, R., Harris, T. (eds.) Geocomputing, pp. 161–186. Taylor and Francis, London (2000)
36. Fisher, P.F.: Sorties paradox and vague geographies. Fuzzy Sets and Systems 113, 7–18 (2000)
37. Usery, E.L.: A conceptual framework and fuzzy set implementation for geographic features. In: Burrough, P.A., Frank, A. (eds.) Geographic objects with indeterminate boundaries, pp. 87–94. Taylor and Francis, London (1996)
38. Irvin, B.J., Ventura, S.J., Slater, B.K.: Fuzzy and isodata classification of landform elements from digital terrain data in Pleasant Valley, Wisconsin. Geoderma 77, 137–154 (1997)
39. Burrough, P.A., MacMillan, R.A., Van Deursen, W.: Fuzzy classification methods for determining land suitability from soil profile observations. J. Soil Science 43, 193–210 (1992)
40. Burrough, P.A., van Gaans, P., Hootsmans, R.: Continuous classification in soil survey: Spatial correlation, confusion and boundaries. Geoderma 77, 115–135 (1997)
41. McBratney, A.B., De Gruijter, J.J.: A continuum approach to soil classification by modified fuzzy k-mean with extragrades. J. Soil Science 43, 159–175 (1992)
42. McBratney, A.B., Odeh, I.O.A.: Application of fuzzy sets in soil science: Fuzzy logic, fuzzy measurements and fuzzy decisions. Geoderma 77, 85–113 (1997)
43. Zhu, A.X., Band, L.E., Vertessy, R., Dutton, B.: Deriving soil property using a soil land inference model (SoLIM). Soil Sci. Soc. Am. J. 61, 523–533 (1997)
44. Zhu, A.X., Band, L.E., Dutton, B., Nimlos, T.: Automated soil inference under fuzzy logic. Ecol. Modell 90, 123–145 (1996)
45. Ulaby, F.T., Moore, R.K., Fung, A.K.: Microwave Remote Sensing, vol. 2. Addison-Wesley, Reading (1982)
46. Greeley, R., Blumberg, D.G., McHone, J.F., Dobrovolkis, A., Iversen, J.D., Rasmussen, K.R., Wall, S.D., White, B.R.: Applications of spaceborne radar laboratory data to the study of aeolian processes. Jour. Geophys. Res. 102, 10, 971-10, 983 (1997)
47. Greeley, R., Blumberg, D.G., Dobrovolskis, A.R., Gaddis, L.R., Iversen, J.D., Lancaster, N., Rasmussen, K.R., Saunders, R.S., Wall, S.D., White, B.R.: Potential transport of windblown sand: Influence of surface roughness and assessment with radar data. In: Tchakerian, V.P. (ed.) Desert Aeolian Processes, pp. 75–99. Chapman & Hall, London (1995)
48. Prigent, C., Tegen, I., Aires, F., Marticorena, B., Zribi, M.: Estimation of aerodynamic roughness length in arid and semi-arid regions over the globe with the ERS scatterometer. Jour. Geophys. Res. 110, 12 page (2005)
49. Laurent, B., Marticorena, B., Bergametti, G., Chazette, P., Maignan, F., Schmechtig, C.: Simulation of the mineral dust emission frequencies from desert areas of China and Mongolia using an aerodynamic roughness length map derived from the POLDER/ADEOS I surface products. Jour. Geophys. Res. 110, 21 page (2005)

50. Benallegue, M., Taconet, O., Vidal-Madjar, D., Normand, M.: The Use of Radar Backscattering Signals for Measuring Soil Moisture and Surface Roughness. Remote Sensing of Environment 53(1), 61–68, 89 (1995)
51. Dierking, W.: Quantitative Roughness Characterization of Geological Surfaces and Implications for Radar Signature Analysis. IEEE Transaction on Geoscience and Remote Sensing 7(5), 2397–2412 (1999)
52. Band, L.E., Moore, I.D.: Scale: landscape attributes and Geographical Information Systems. Hydrological Process 9, 401–422 (1995)
53. Wang, F.: A fuzzy grammar and possibility theory-based natural language user interface for spatial queries. Fuzzy Sets and Systems, 113, 1, 147-159(13) (2000)
54. Cattuto, C., Gregori, L., Melelli, L., Taramelli, A., Broso, D.: I Conoidi nell'Evoluzione delle Conche Intermontane Umbre. Geografia Fisica e Dinamica del Quaternario 7, 89–95 (2005)

Evaluating the Use of Alternative Distance Metrics in Spatial Regression Analysis of Health Data: A Spatio-temporal Comparison

Stefania Bertazzon and Scott Olson

Department of Geography, University of Calgary
2500 University Dr. NW, Calgary, AB, T2N 1N4, Canada
bertazzs@ucalgary.ca, smolson@ucalgary.ca

Abstract. A method is discussed to enhance the reliability of multivariate spatial regression analysis: alternative values of the Minkowski distance metric are used in the spatial weight matrix. The method is tested on an analysis of the association between heart disease incidence and a pool of socio-economic variables in Calgary over two consecutive census surveys. The method provides a reliable model, which can guide locational decisions to mitigate present and future disease incidence. The model is underpinned by a quantitative definition of neighbourhood connectivity throughout the city. Such connectivity, usually described by Euclidean distance, can be more effectively described by a specifically calibrated distance metric. The analytical results are meaningful, robust to neighbourhood size, and relatively constant over time. Owing to its effectiveness and simplicity, the procedure is generalizable to other health and socio-economic analysis. An automatic implementation is suggested, to assist in the definition of reliable spatial regression models.

Keywords: Spatial regression, distance metric, Minkowski, neighbourhood connectivity, reliability, health, socio-economic, GIS.

1 Introduction

The outbreaks of SARS (Severe Acute Respiratory Syndrome), West Nile virus, and most recently swine flu are but a few examples from recent headlines that point to the compelling need to develop effective analytical tools to model occurrence, transmission, and causes of disease. Many of the most urgent health concerns of today's society are fundamentally spatial in nature: effective accessibility to health care services; prompt and efficient response to epidemic outbreaks; detection and monitoring of environmental health hazards; and consequent urban planning. Spatial analytical methods can be useful management and policy tools to address these concerns, but their use rests on assumptions that are often violated by empirical processes, with the result that much current applied research fails to bring this toolset to its full potential. Presently, management decisions are often supported by quantitative models, specifically regression models; these are potentially desirable tools that can link, for

M.L. Gavrilova and C.J.K. Tan (Eds.): Trans. on Comput. Sci. VI, LNCS 5730, pp. 267–287, 2009.

example, disease incidence to residents' age, thus providing a realistic picture of where health care services will be most needed in the near future. Unfortunately, current models are often uncertain or unreliable. In the best cases, unreliable models provide decision makers with a realistic, but blurry picture of the factors they need to manage, potentially leading to ineffective decisions; in the worst cases the picture is so blurry that it may lead to management decisions that are not just ineffective, but harmful. The uncertainty stems from two properties of geographical phenomena: spatial non-stationarity (things tend to vary unevenly in space), and spatial dependence (near things tend to be more similar than distant things) [1]. Addressing the limitations of regression models is key to improving the reliability of much current quantitative analysis, if, as noted by Griffith and Amrhein [2], most of the multivariate techniques commonly used by geographers can be formulated or reformulated in terms of regression analysis.

The focus of this application is on spatial autoregressive modelling, a technique that specifically addresses the effect of spatial dependence on regression models [3]. In order to maximize the effectiveness of this technique, one crucial element is the correct specification of a spatial weight matrix, capable of providing an accurate representation of the spatial neighbourhood and hence the configuration of the observed spatial dependence. In turn, the spatial weight matrix is specified by a small number of parameters: an appropriate measurement of distance, a correct definition of the range of the observed spatial dependence, and a correct definition of the distance decay effect within the defined neighbourhood [3]. In this paper, the focus is on the distance measurement, discussing the use of a range of distance metrics known as Minkowski distance.

The effect of alternative distance metrics is evaluated on a case study that is relevant from many applied perspectives, including health care provision and urban management. The spatial regression model analyzes the association of demographic and socio-economic factors with the incidence of heart disease in Calgary, a large Canadian city. Population distribution within the city, clustering of age groups, and socioeconomic pattern are the main factors considered; urban connectivity enters the model through the measurement of distance among spatial units. By optimizing the distance measurement and specifying an appropriate spatial weight metric, this study provides a method for enhancing the reliability of the estimated model parameters. As a consequence, the model represents an effective analytical support tool for policy and planning decisions. The same spatial regression analysis is estimated for two temporal intervals, hinging on two consecutive census surveys (2001 and 2006). The comparison of the analytical results constitutes a preliminary but important exploration of spatio-temporal dynamics within the city, its health and demographic characteristics, and its socio-economic structure. The analytical findings provide the foundation for more advanced computational developments.

All the statistical computations are conducted in Splus 7 and Splus Spatial Statistics 1.5, with the exception of the bivariate Pearson correlations that are computed in SPSS 15. Geographical data management and visualization are performed in ArcGIS 9.1.

Section 2 provides some background information and an introduction to the case study. Section 3 outlines the various aspects of the methodology: the specification of exploratory analysis and spatial autocorrelation analysis; the definition, selection, and estimation of spatial regression models; and the definition of the distance metrics of interest. In Section 4 the results of the spatial dependence analyses and regression models are presented for various distance metrics, and in Section 5, the results of the two survey periods are compared and the critical aspects of the methodology are discussed. The final section offers some concluding remarks and future lines of enquiry.

2 Background and Case Study

Heart disease (myocardial infarction) is one of the leading causes of death in the developed world. In addition to the individual characteristics that correlate with the disease, there are a number of factors that are related to a complex variable usually referred to as "lifestyle". Individual characteristics, such as genetic background or simultaneous presence of other conditions, are known as non-modifiable risk factors, in contrast with modifiable risk factors, which include such factors as physical activity, smoking, and diet. These modifiable risk factors tend to correlate with demographic and socio-economic characteristics of individuals [4]. At most geographical scales, demographic and socio-economic characteristics tend to display a pattern, or spatial clustering; disease prevalence, likewise, presents a characteristic geographical distribution, or spatial pattern. For this reason a spatial regression model is an appropriate tool to analyze the spatial pattern of disease occurrence as a function of localized demographic and socio-economic characteristics. This is also the reason why this type of phenomena tends to manifest spatial dependence and non-stationarity. Spatial autoregressive modelling is therefore an effective analytical tool, capable of providing estimates of the coefficients linking disease prevalence to each demographic and socioeconomic factor. The reliable parameters obtained with this method can later be used for analysis and prediction, to ultimately design proactive policy solutions aimed at alleviating and mitigating disease prevalence over the study region.

The spatial regression models discussed in this paper make use of medical records from the APPROACH Project, an ongoing data collection initiative begun in 1995, containing information on all patients undergoing cardiac catheterization in Alberta, and census variables. The medical records represent disease prevalence, and the census variables represent demographic and socio-economic factors. Cardiac catheterization is an invasive procedure for patients experiencing cardiovascular symptoms, which provides important prognostic information for individuals affected by cardiovascular conditions [5]. For the present analysis, a subset of patient records were selected from the provincial database, obtaining a sample of patients undergoing the procedure in Calgary from 1998 to 2002[1], and from 2003 to 2007[2]. Patient address is released at the postal code level; postal code conversion files (PCCFs) from Census Canada were used to calculate the geographic coordinates in latitude and longitude, which were subsequently converted to easting and northing coordinates prior to performing distance computations.

---

[1] A sample of 11,345 cases, on a total population of 875,245 residents.
[2] A sample of 16,355 cases, on a total population of 988,193 residents.

Socio-economic and demographic variables were drawn from the 2001 and 2006 census surveys, respectively. These variables are available at the dissemination area[3] and census tract levels: for this analysis, the census tract aggregation level was used. At this spatial resolution the spatial dependence is more severe; hence there is a stronger need to implement efficient spatial regression models. In addition, these relatively large units are more meaningful in terms or urban planning and health policies; therefore, a model calibrated at this scale is more useful and applicable than one calibrated at the dissemination area level. The cardiac data were spatially aggregated to match the census tracts, resulting in 182 valid census tract records for the period around the 2001 census, and 186 records for the period around the 2006 census. Fig. 1 shows the Calgary census tracts and the distribution of catheterization cases over the two study periods: Fig. 1a for the 2001 census and Fig. 1b for the 2006 census. It may be worth noting that during the study period, the procedure was available only at the Foothills Hospital, located in the northwest sector of Calgary.

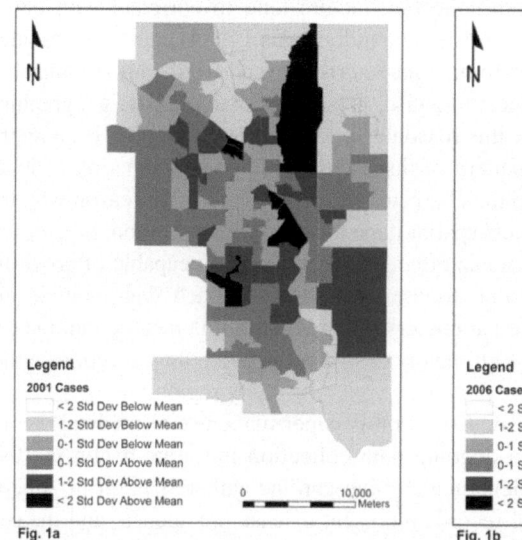
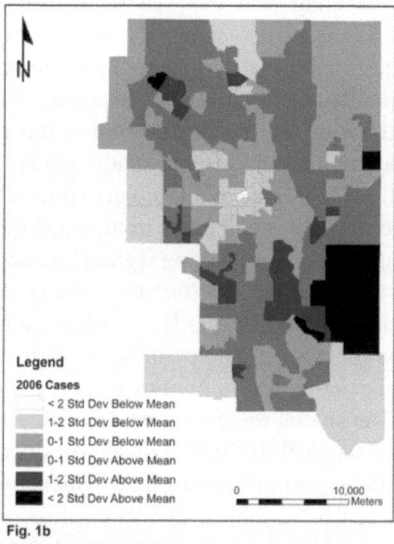

Fig. 1. Spatial distribution of cardiac catheterization cases in 2001 and 2006

Calgary's urban structure is a combination of numerous development episteme. Local patterns of connectivity vary according to local design. For instance, grid pattern road development of the inner city offers different travel options than the circular, cul-de-sac design of its outlying suburban counterparts. Furthermore, large variations in both physical size and shape of neighbourhood form are very apparent in the city. Thus, there is a need to capture how varying urban patterns affect neighbourhood connectivity.

---

[3] A small, relatively stable geographic area composed of one or more neighbouring blocks standardized through uniform population sizes targeted at 400 to 700 persons. These areas are usually delineated by physical features (roads, water, powerlines, etc.) and respect the boundaries of census subdivisions and census tracts (Statistics Canada, 2007).

3 Methodology

The methodology developed in this paper aims at reducing the inflated variance caused by spatial dependence in regression model estimates. The method will be discussed in detail in this section, and can be usefully summarized as follows. Step 1: calculation of distance using a standard metric; Step 2: calculation of spatial autocorrelation using the standard distance measurement; Step 3: estimation of a spatial regression model based on the standard distance measurement; Step 4: assessment of the model variance; Step 5: replication of Steps 1–4 for an array of candidate Minkowski metrics; Step 6: identification of the distance metric that minimizes the model variance.

3.1 Exploratory Analysis and Spatial Autocorrelation

Census variables and medical records consist of incident numbers. For this reason, a normalization of all the variables was conducted, preliminary to any analysis. In most cases the normalization involved the use of the total resident population as the standardizing variable (e.g., number of cardiac catheterizations), while in other cases it involved the use of a pertinent subset of residents (e.g., population over 20 was used to standardize education levels and population over 15 to standardize marital status). On the normalized variables, descriptive spatial and statistical analyses were conducted, to test their normality and other statistical properties. Cross-correlation analysis (Pearson's coefficient) was used to test the strength of the correlation between each explanatory variable and the dependent variable, as well as the cross-correlation among explanatory variables. The latter test was particularly important to ensure the statistical independence of the explanatory variables introduced in the regression model, in order to avoid multcollinearity in the multivariate regression models.

A traditional spatial autocorrelation test, Moran's I [6], is used to test all instances of spatial autocorrelation throughout the analysis presented in this paper, i.e., to assess the spatial dependence in variables as well as in model residuals. The values of Moran's I can vary between -1 and 1, where -1 indicates perfect negative spatial autocorrelation, 0 indicates absence of spatial autocorrelation, and + 1 indicates perfect positive spatial autocorrelation. The computation of this index requires the specification of a model of spatial dependence, defined by a spatial weight matrix. The matrix defined for the calculation of the spatial autocorrelation index will also be used for the estimation of the spatial autoregressive models (Section 3.2). In its simplest form, the matrix is a binary structure, but it is common to use more complex specifications, which include various types of weights to describe distance decay effects. There are several ways of specifying spatial contiguity [7, 8]: a common method is the definition of k orders of spatial neighbours; an alternative method is a threshold distance; a third method is based on shared borders (for areal units only). While some methods are heavily dependent on the topology of the spatial units, the computation of spatial neighbours is a very general method. In all cases, the extent of the spatial dependence must be defined, either via a maximum distance parameter, or via a number (k) of nearest neighbours.

The correct specification of this matrix is key to minimizing the variance of the spatial regression model. Prior to the specification of a regression model, the spatial autocorrelation is estimated for all the model variables. An intermediate objective of the presented methodology, before the final model can be estimated and its variance assessed, is the identification of the spatial weight matrix that produces the highest value of the spatial autocorrelation index for the variable of interest, i.e., the dependent variable of the regression model: this matrix contains the neighbourhood specification that best captures the spatial dependencies in that variable.

3.2 Spatial Regression Models

The number of spatial regression techniques discussed in the academic literature has grown considerably ([9], [10], [11], [12]) in response to the increasing availability of spatial data, easier access to specialized software, and increased awareness of the inadequacy of traditional analytical techniques in dealing with the unique properties of spatial data [12]. Perhaps the most critical of these properties is spatial dependence, which results in a redundancy of information that inflates the variance (uncertainty) associated with the parameter estimates. Large parameter variance also inflates classical inferential tests, resulting in a more frequent rejection of the null hypothesis. As a consequence, inefficient parameter estimates are not only unreliable, but potentially misleading [3].

Various types of spatial analytical methods have been developed to analyze spatial data, including Bayesian approaches [14] and multilevel models [15]. Spatial autoregressive methods include Generalized Least Squares (GLS) and Maximum Likelihood (ML) models; the covariance structure is typically expressed by a conditional autoregressive (CAR), simultaneous autoregressive (SAR), or moving average (MA) specification [10]. In all cases, a contiguity matrix (Section 3.1) determines which units are spatially dependent [7]. The effectiveness of the regression model depends largely upon the choice of the contiguity matrix and the underlying model of spatial dependence. However, defining contiguity remains difficult and subjective, often dependent on the spatial process under consideration [8].

The contiguity—or spatial weight—matrix is used in the computation of spatial autocorrelation indices (e.g., Moran's I) as well as in the spatial regression:

$$Y = X\beta + \rho WY + \varepsilon \qquad (1)$$

where ρ (rho) is the autoregressive parameter and W is the contiguity matrix. Each element of the spatial weight matrix is a spatial weight w_{ij}, which defines the extent of the spatial dependence and the correlation between spatial units as a function of their distance. The spatial autoregressive coefficient, rho, varies between -1 and +1. For rho values approaching 0, the spatial regression model reduces to a standard regression.

The method presented in this paper for the specification of a spatial weight matrix (W) is developed around the nearest neighbour method, where the use of different distance metrics allows for the computation of distance in a way that approximates travel along the road network and actual physical connection, better representing the actual neighbourhood connectivity. The use of alternative distance metrics produces alternative definitions of nearest neighbours (Section 4.1, Fig. 4). The neighbourhood configuration that best represents actual community structure is expected to better capture the spatial dependence, thereby enhancing the effectiveness of the autoregressive component of the model, expressed by the value and significance of the autoregressive coefficient rho (Equation 1). A model that can best capture spatial dependence via an effective autoregressive specification presents lower variance of the estimated parameters, which are therefore more reliable.

In order to identify the regression which can be considered the best model, the criterion applied here is the minimization of the model's variance, evaluated by means of a broad set of indicators. The value of the spatial autocorrelation index in the model's residuals is considered the most important indicator; this index, in fact, indicates whether or not the model was estimated in violation of the hypothesis of independence of the residuals, which determines the properties of all the model's estimates. In addition to the variance minimization criterion, the model's goodness of fit will be weighed heavily, as it indicates the model's effectiveness in explaining the variation of the dependent variable.

3.3 Alternative Distance Functions

Distance can be measured in many ways: travel time and travel cost [16] are very useful in some contexts, but lack fundamental geometric properties (triangle inequality); Mahalanobis distance has a way of accounting for spatial dependencies. Our work focuses on one category of distance metrics, known as Minkowski distance [17]. This array of metrics was chosen because of its flexibility. In fact, once this class of metrics is chosen, a range of parameters can be selected from within that class; therefore, a single yet flexible measurement method can be defined for the optimal estimation of spatial dependence in spatial autoregressive models. The most commonly used distance metric is the Euclidean or straight line distance:

$$d_{ij} = [(x_i - x_j)^2 + (y_i - y_j)^2]^{1/2} \qquad (2)$$

Alternatively, Manhattan distance, also known as City Block Distance [18], is the distance between two points measured along the axes at right angles:

$$d_{ij} = | x_i - x_j | + | y_i - y_j | \qquad (3)$$

The Minkowski distance is described by a general formula, of which Euclidean and Manhattan are special cases:

$$d_{ij} = [(x_i - x_j)^p + (y_i - y_j)^p]^{1/p} \qquad (4)$$

As visually represented in Fig. 2, Minkowski distance can provide intermediate values between Euclidean and Manhattan distance, producing a more realistic overall representation of travel in a city; for example, a road network is typically a mixture of straight-lines, curves, and grid-like patterns [19, 20, 21]. Unlike distances measured empirically along a given road network, the use of a specific distance metric provides a consistent model of distance throughout a city or a region, which provides the benefits of generalization but filters out local detail. These distance measurements have been usefully applied in practice for the effective provision and accessibility of health care services [22]. Preliminary analysis has suggested a value of p=1.54 as the best Minkowski value to represent travel along the Calgary road network [23]. Our purpose is not to mimic the city road network but to select a distance metric that best represents neighbourhood connectivity, which in turn is defined by the interplay of road network and urban design.

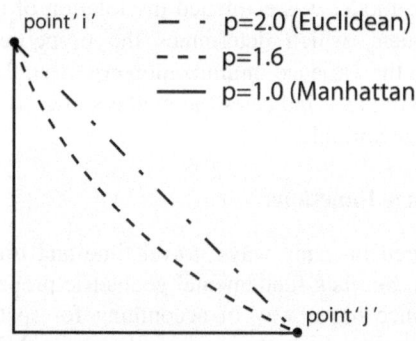

Fig. 2. Travel path for the two extreme and one intermediate Minkowski *p* values

The general Minkowski formula (Equation 4) is used in this application to experiment with a systematic sample of *p* values (i.e., $p = 1.1$, $p = 1.2$, etc.) in the interval [$1 \leq p \leq 2$], to evaluate the performance of alternative spatial neighbourhood definitions based on the various *p* values for a given order of neighbourhood and distance decay functions. The criterion for choosing an optimal *p* value is the minimization of the variance of the estimates of the spatial regression (SAR) model (Section 3.1). The correlation is modeled in the spatial weight matrix not only by a distance metric but also by a distance decay function and a distance range, defined by the number of nearest neighbours. Even though these three parameters are not mutually independent, the method discussed in this paper focuses on the distance metric. Through a series of empirical experiments, the best value of the other parameters was also identified (number of nearest neighbours and distance decay function, respectively). Given these parameters, we determine the distance metric that minimizes the variance of the model estimates. This process leads to the specification of a set of alternative spatial regression models, based on an array of spatial weight matrices and distance metrics. For each survey period, an iterative process guides us to the selection of the metric that, all else being equal, leads to the lowest model variance. It is our intention to extend this line of work to encompass the definition of an algorithm for the selection of the optimal metric.

4 Results

A number of alternative spatial weight matrices were tested for the dependent variable (cardiac catheterization cases) and the independent variables (demographic and socio-economic variables) in both survey periods. Only the results relative to the dependent variable, "Catheterization Cases", will be discussed in detail in the following subsections. Experiments with increasing orders of neighbourhood (i.e., $k = 2$, 3,...., 10) have confirmed that, for all the variables, the spatial autocorrelation index is constantly higher for lower orders of neighbourhood, suggesting that the spatial dependence is more pronounced over short distances. Alternative distance decay functions were also tested: the function that best captures the distance decay effect for most variables is an inverse squared distance function weighted by the area of each spatial unit (census tract)[4]. This analysis of various distance decay functions confirms the indication emerging from the inverse relationship between neighbourhood order and spatial autocorrelation index: spatial dependency is more severe over short distances and decreases sharply as distance increases and more spatial units are considered. These results are consistent for both the temporal periods analyzed.

4.1 Spatial Autocorrelation Index, 2001

Based on the experiments summarized above, systematic analyses were conducted for one and two orders of neighbourhood.[5]

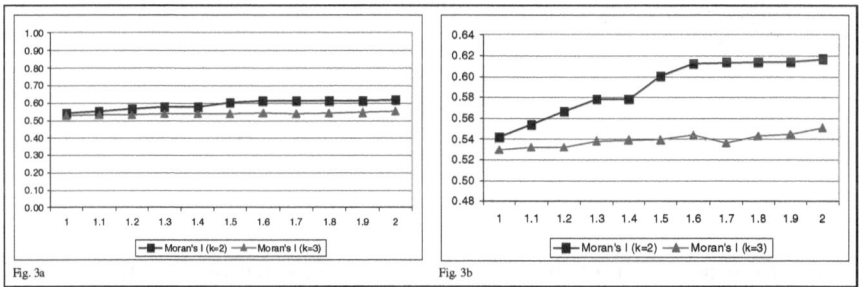

Fig. 3. Spatial autocorrelation in the interest variable for varying p values and distance ranges, 2001

Fig. 3 represents the variation of the spatial autocorrelation index as a function of the p value that defines the Minkowski distance metric. As evidenced in Fig.3a, the difference between one and two nearest neighbours is relatively minor, suggesting that the method is robust with respect to the choice of a neighbourhood order. Fig.3b highlights in greater detail the local features of the spatial autocorrelation function in

---

[4] Inner-city census tracts tend to have smaller areas than peripheral ones; therefore a pure distance weighted specification would tend to under-estimate the neighbourhood connectivity in the suburbs.

[5] $k=2$ and $k=3$, respectively, in Splus.

the interval [1 ≤ p ≤ 2]. For k=2 the function presents an overall increasing trend. A leap upwards is observed at p=1.4; at p=1.6 the line reaches a plateau that remains approximately constant until p=2.0. For k=3 the function presents a relatively stable trend. The line also displays an anomaly, or a peak, at p=1.6, it drops slightly at p=1.7, and then rises again constantly until p=2.0. From this initial analysis, the value p=1.6 emerges as the candidate metric that can best capture the spatial dependence in the variable "Catheterization Cases".

The effect of the distance metric on the selection of nearest neighbours operated by alternative metrics can be better appreciated visually. Fig. 4 shows the comparison between the neighbourhood selection for the extreme p values (p=1 and p=2) as well as for the value p=1.6 that was identified in the spatial autocorrelation analysis (Fig. 3). Fig. 4 presents the analysis conducted for two orders of neighbourhood (k=3), as the visualization results are most effective for this value.

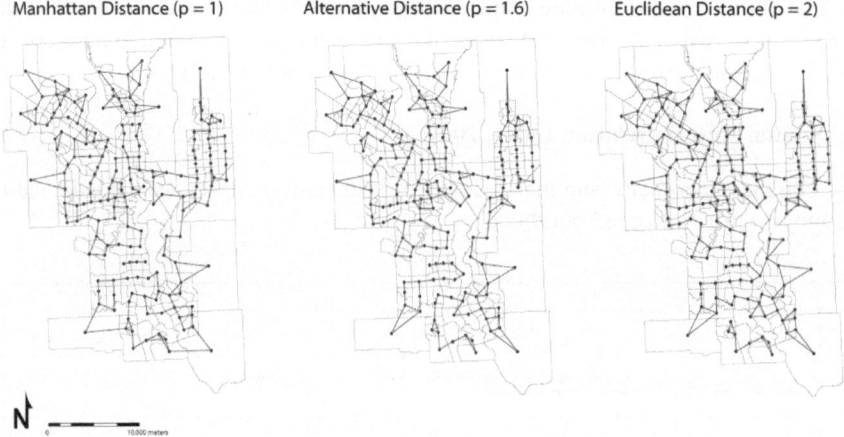

Fig. 4. Second order nearest neighbour connections according to varying distance metrics

A careful examination of the plots in Fig. 4 reveals that the selection of nearest neighbours varies in many parts of the city: individual points selected as nearest neighbours vary for each metric, and these differences become increasingly pronounced as the order of neighbourhood increases.

4.2 Spatial Autocorrelation Index, 2006

The spatial autocorrelation analysis for the second survey period reveals a noticeable change over the 2001 period. Fig. 5 summarizes the variation of the spatial autocorrelation index for the dependent variable as a function of the distance metric.

The main aspect emerging from the analysis summarized in Fig. 5 is the sharp decrease of the spatial autocorrelation value, from an average value[6] of approximately 0.6 in 2001 to a value of approximately 0.3 in 2006. Other analyses conducted on

---

[6] Depending on the number of nearest neighbours.

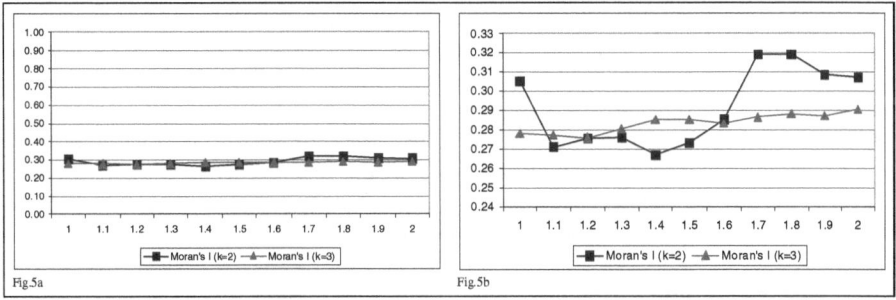

Fig. 5. Spatial autocorrelation in the interest variable for varying *p* values and distance ranges, 2006

these data have suggested a strong correlation between catheterization cases and population over 65 years for the 2001 survey period [25]. Those analyses indeed suggest that the spatial autocorrelation observed in the disease pattern may be driven by the spatial autocorrelation in the distribution of senior citizens. The pattern of cross-correlation observed for the 2006 period (Section 4.3) suggests, instead, a lower correlation between disease and senior citizens, accompanied by higher correlations between disease and younger age groups. These observations can explain the lower spatial autocorrelation observed in the disease pattern: younger age groups tend to be less clustered in Calgary, therefore less spatially autocorrelated. As a consequence also the disease pattern displays a lower spatial autocorrelation for the 2006 survey period.

As for the 2006 census period, Fig. 5a confirms that the number of neighbours has only a marginal effect on the overall variation. As evidenced by Fig. 5b, the variation of the spatial autocorrelation as a function of the distance metric has somewhat changed between the two survey periods. For two nearest neighbours, i.e., *k*=3, the trend is generally similar to the one observed in 2001, with an overall upwards trend in the interval [$1 \leq p \leq 2$]. The most significant increase occurs over the interval between *p*=1.4 and *p*=1.5. After a minor decline at *p*=1.6, a second and higher peak is reached at *p*=1.8; after another minor decline, the final, highest value is reached at *p*=2, suggesting the Euclidean distance as a better metric for this number of neighbours.

A more complex pattern is shown by the function for 1 nearest neighbour, i.e., *k*=2. The spatial autocorrelation index exhibits almost identical values at the two extremes of the interval, i.e., for the Manhattan and Euclidean metrics, respectively. Within the [$1 \leq p \leq 2$] interval, however, the values remain lower between *p*=1.1 and *p*=1.6, while the highest peak is reached for the values *p*=1.7 to *p*=1.8. For these *p* values, the spatial autocorrelation index is a whole decimal degree higher than the values obtained for the Manhattan and Euclidean metrics. Consequently, the single *p* value that maximizes the spatial autocorrelation function for both *k*=2 and *k*=3 is *p*=1.8, for the 2006 period.

The shift from a best *p* value of *p*=1.6 in 2001 and of *p*=1.8 in 2006 is likely related to the shift in the spatial distribution and clustering of the "Catheterization Cases" variable. This may also be due to a shift in the population distribution, and consequently in the disease pattern. The economic boom experienced by the city during the

2003–2007 period caused a massive population increase, mainly constituted by young individuals, which in turn caused an increase in the number of residents, particularly in the suburbs [26]. The spatial autocorrelation index is a joint measure of distance and correlation: these results indicate that it is correlation that drives the change in the spatial autocorrelation index, and therefore the p value of the distance metric evolves, to best capture the new correlation pattern. Over the interval between the two census surveys, Calgary has undergone major social and demographic changes, but lesser physical changes affecting its neighbourhood connectivity.

4.3 Regression Model, 2001

A pool of candidate explanatory variables for the regression model was selected from the census data for each survey period. Table 1 summarizes descriptive spatial statistics and spatial autocorrelation for the dependent variable and the subset of variables used in the 2001 regression models.[7] The standard descriptive statistics evidence the normality of the data, while the spatial autocorrelation index shows that all the variables present significant and generally high spatial dependence.

Table 1. Descriptive statistics and spatial autocorrelation for selected regression variables, 2001

	cases	males	a45.54	a55.64	a65pl	2p.wchld	gr13ls	non.uni	f.m.inc
*** Summary Statistics for data in: Master.CT.Norm ***									
Mean:	1.34	49.77	14.46	7.61	9.64	47.31	30.64	36.68	66.61
Median:	1.28	49.80	13.92	7.24	8.16	48.18	28.47	37.04	63.13
Variance:	0.22	2.76	10.00	6.53	32.66	181.27	122.98	29.90	330.68
Std Dev.:	0.47	1.66	3.16	2.55	5.71	13.46	11.09	5.47	18.18
SE Mean:	0.03	0.12	0.24	0.19	0.42	1.00	0.82	0.41	1.35
Skewness:	0.40	0.01	0.60	0.77	0.81	-0.18	0.64	-0.39	0.61
Kurtosis:	-0.29	2.44	0.40	0.32	0.11	-0.53	-0.34	0.11	-0.56
*** Spatial Correlations ***									
Correlation	0.62	0.47	0.48	0.57	0.73	0.86	0.82	0.37	0.63
Variance	0.01	0.01	0.01	0.01	0.01	0.01	0.01	0.01	0.01
Std. Error	0.10	0.10	0.10	0.10	0.10	0.10	0.10	0.10	0.10
Normal statistic	6.17	4.73	4.80	5.68	7.25	8.60	8.16	3.69	6.25
p-value (2-sided)	0.00	0.00	0.00	0.00	0.00	0.00	0.00	0.00	0.00

Cross-correlation analysis was conducted to assess the correlation between the dependent and each independent variable and the cross-correlation among the independent variables. In Table 2, census variables have been grouped into homogeneous categories and a sample of 2 representative variables for each category is presented.

---

[7] Unless otherwise specified, spatial statistics on any variable are conducted using Euclidean distance (p=2.0) and two orders of nearest neighbours (k=3), following the convention used in Splus: based on the cross-correlations among dependent and independent variables, we select the multiple regression that best expresses the relationship between lifestyle and heart disease incidence).

Table 2. Cross-correlation analysis for dependent and selected independent variables, 2001

	cases	Demographics		Family		Housing		Education		Economics	
	cases	a55.64	a65pl	mar.claw	2p.wchld	owned	s.detach	gr13ls	non.uni	unemp	f.m.inc.k
cases	1.000	.569(**)	.794(**)	-.377(**)	-.495(**)	-.285(**)	-.273(**)	.181(*)	-.235(**)	.171(*)	-.229(**)
a55.64	.569(**)	1.000	.415(**)	0.047	-0.074	0.144	0.054	-0.026	-.292(**)	0.090	.195(**)
a65pl	.794(**)	.415(**)	1.000	-.416(**)	-.555(**)	-.359(**)	-.353(**)	-0.098	-.334(**)	0.045	-0.099
mar.claw	-.377(**)	0.047	-.416(**)	1.000	.819(**)	.909(**)	.871(**)	-.224(**)	0.127	-.367(**)	.665(**)
2p.wchld	-.495(**)	-0.074	-.555(**)	.819(**)	1.000	.818(**)	.803(**)	-0.087	0.044	-.153(*)	.572(**)
owned	-.285(**)	0.144	-.359(**)	.909(**)	.818(**)	1.000	.912(**)	-0.133	0.099	-.347(**)	.647(**)
s.detach	-.273(**)	0.054	-.353(**)	.871(**)	.803(**)	.912(**)	1.000	-0.120	0.093	-.322(**)	.593(**)
gr13ls	.181(*)	-0.026	-0.098	-.224(**)	-0.087	-0.133	-0.120	1.000	.251(**)	.336(**)	-.698(**)
non.uni	-.235(**)	-.292(**)	-.334(**)	0.127	0.044	0.099	0.093	.251(**)	1.000	-.160(*)	-.294(**)
unemp	.171(*)	0.090	0.045	-.367(**)	-.153(*)	-.347(**)	-.322(**)	.336(**)	-.160(*)	1.000	-.366(**)
f.m.inc.k	-.229(**)	.195(**)	-0.099	.665(**)	.572(**)	.647(**)	.593(**)	-.698(**)	-.294(**)	-.366(**)	1.000

The cross-correlations analysis provides an interesting portrait of the socio-economic structure of Calgary, as shown, for example, by the high correlation between "owning a house", "married or in common law", and "single detached home", which suggests a predominant traditional family model, and a widespread wealth. Several high cross-correlation values limited our choice of independent variables[8], but, at the same time, the variables included in the models are also representative of those that could not be directly entered in the regressions.

As evidenced in Table 2, the demographic variables, and particularly those indicating old age, tend to display a very high correlation with the dependent variable. Other multivariate analyses of these data [23, 25] have indicated that the correlation between the dependent and demographic variables is so high that the inclusion of these variables in any regression model results in the exclusion of all the socioeconomic variables. For this reason, in this paper, the demographic variables are excluded from the selection process. Alternative model specifications [25] have also confirmed that age and sex standardization of the disease variable has only a minor impact on the regression results.

After experimenting with various combinations of independent variables, a backwards selection process produced the following regression model:

$$CC = f\,(2\,p.w.chld\,,Uni,F.m.inc,Gr13ls) \qquad (5)$$

where CC is the number of catheterization cases; $2p.w.chld$ is the number of families of two parents with children at home; $Non.uni$ is the number of persons with a post-secondary, non-university degree; $F.m.inc$ is the family median income; and $Gr13ls$ is the number of persons with grade 13 or lower education.

The model parameters and selected diagnostics are summarized in Table 3.

---

[8] Only variables correlated less than +/- 0.7 were included in each model.

Table 3. Multivariate spatial regression model coefficient and selected diagnostics, 2001

	Value	Std. Error	t value	Pr(>ltl)		
(Intercept)	1.7981	0.3648	4.9294	0.0000		
2p.w.chld	-0.0226	0.0032	6.9717	0.0000		
Gr13ls	0.0195	0.0043	2.9257	0.0000		
F.m.inc	-0.0163	0.0056	2.8138	0.0039		
Non.uni	0.0089	0.0032	4.5262	0.0055		
L.Likelihood	Pseudo-R^2	Rho	Sigma^2	Res.Std. Err	Res. Moran
-277.8252	0.3724	0.8504	0.1165	0.3413	-0.0293

The model describes the incidence of heart disease as a function of family structure, education, and income. This set of variables possesses a satisfactory explanatory power (pseudo-R^2 = .37).[9] The negative and highly significant coefficient of families with children (t value) suggests a negative correlation between disease and individuals in young families and appears to be related to fairly young individuals, at early to mid stages of their career, likely with relatively high education and moderately high income. The positive relationship between disease incidence and low education (grade 13 or less) suggests a relationship with old age and fringes of poverty and low social status. The negative coefficient linking disease incidence and areas dominated by post-secondary, non- university education identifies trade workers and professionals: a category with fairly high income levels. Overall, the education variables suggest that higher education levels are associated with greater income and lower disease incidence, suggesting that higher education levels may lead to healthier lifestyle and lower risk of disease. Finally, the positive relationship between disease incidence and income suggests the hypothesis of higher disease incidence in individuals with higher levels of stress and responsibility and appears to be related to more mature professionals, therefore possibly encompassing a latent age factor.

The regression summarized in Equation 5 and Table 3 contains a number of significant explanatory variables, which produce a satisfactory model, from a statistical as well as a conceptual point of view. This model is estimated from a spatial weight matrix based on the standard Euclidean metric for the distance measurement. The model will now be re-estimated, using an array of alternative spatial weight matrices, each based on a different value of the Minkowski metric in the $[1 \leq p \leq 2]$ interval.

The results of this experiment are presented and compared in Fig. 6, which depicts a selection of key regression indicators as a function of the distance metric. The first two values are indicators of the model's performance and goodness-of-fit, i.e., the logarithm of the likelihood and the pseudo-R^2; the rho value (Section 3.1), indicates the relative importance of the autoregressive coefficient; residual standard error and σ^2 (sigma^2) are indicators of the model's variance; finally Moran's I measures the spatial autocorrelation in the model's residuals. All the indicators have been scaled and plotted in one single graph so that they can be compared more easily.

---

[9] Following Anselin [26], the pseudo-R^2 is calculated as the square of the correlation between observations and regression fit.

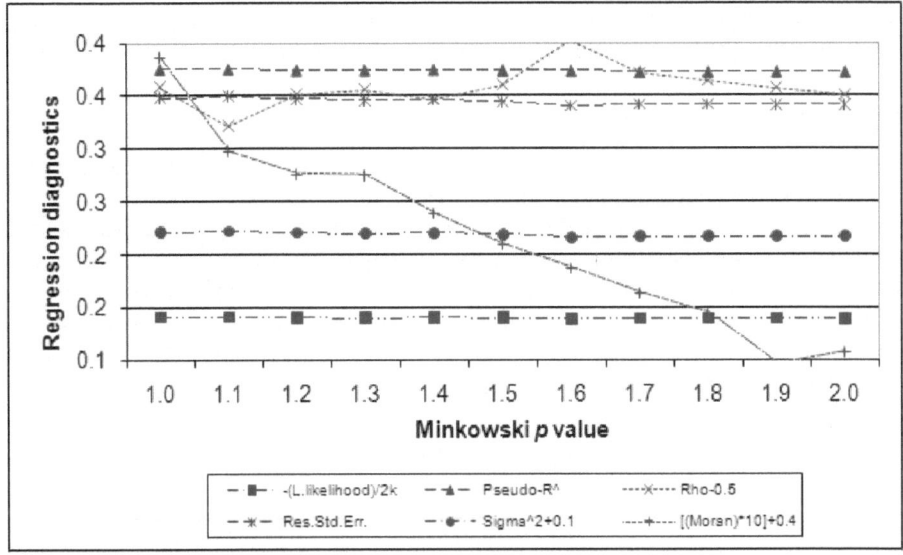

Fig. 6. Spatial regression model selected diagnostics for varying *p* values, 2001

The trends presented in Fig. 5 indicate that the variance indices of the model are affected, to varying degrees, by variations in the distance metric, or *p* value, through the spatial weight matrix. Conversely, the use of alternative distance metrics has only a negligible impact on the goodness-of-fit indicators. The indices that are most affected are the autoregressive coefficient, rho, and the spatial autocorrelation index on the model residuals, Moran's I. These results confirm that an appropriate choice of the *p* value can impact the neighbourhood definition and consequently the model's capacity to effectively capture spatial dependencies, thus ultimately enhancing the reliability of the estimates.

The results cannot be considered conclusive, but from the majority of our tests the value *p*=1.6 continues to emerge as the best candidate for the optimal distance metric. The rho value displays a distinct peak at *p*=1.6, suggesting that with that metric the autoregressive coefficient is most effective at capturing the spatial dependence in the model's residuals. For the same *p* value, a corresponding trough is displayed by residual standard error and sigma square. Visually, the latter features appear to be much less pronounced than the rho value, but range and scaling of these indices should be considered. Somewhat in contrast with the above values, the spatial autocorrelation index in the model's residuals (Moran's I) displays an overall decreasing trend, reaching its lowest value at p = 1.9. The interpretation of this trend presents some difficulties, in that all its values are negative, indicating that negative spatial autocorrelation increases for increasing *p* values. Moreover, the values of residual Moran's I over the $[1 \leq p \leq 2]$ interval are not only insignificant, but very low, as its maximum value is -0.001 for p = 1 and its minimum value is -0.030 for *p* = 1.9. For all these reasons, the trend of the rho coefficient and of the variance indicators will be weighed more heavily than residual Moran's I in the identification of a suitable p value. A p value of 1.6 is therefore the most suitable value emerging from this analysis.

4.4 Regression Model, 2006

Descriptive statistics and spatial autocorrelation index for the variables used in the 2006 regression models are summarized in Table 4.

Overall the changes during the previous period are relatively minor. Most of the variables can still be considered normal, although the variables "Catheterization Cases" and "Male Residents" present very high values of kurtosis. This may be explained by the considerations presented in Section 4.2 on more recent socio-demographic trends. The spatial autocorrelation index presents overall lower values than for the 2001 survey period.

Table 4. Descriptive statistics and spatial autocorrelation for selected regression variables, 2006

	cases	males	a45.54	a55.64	a65pl	single	gr13ls	non.uni	f.m.inc
*** Summary Statistics ***									
Mean:	2.25	49.93	16.35	9.73	10.67	36.08	39.20	27.75	82.01
Median:	2.24	49.77	16.16	9.33	9.65	34.88	36.65	27.78	78.05
Variance:	0.77	4.68	9.81	7.79	30.65	73.30	134.62	25.83	576.84
Std.Dev	0.88	2.16	3.13	2.79	5.54	8.56	11.60	5.08	24.02
SE. Mean	0.06	0.16	0.23	0.20	0.41	0.63	0.85	0.37	1.76
Skewness:	3.07	3.01	0.20	0.76	0.76	0.82	0.61	-0.26	0.96
Kurtosis:	23.27	23.47	0.25	0.61	-0.05	0.38	-0.43	0.41	1.71
*** Spatial Correlation ***									
Correlation	0.29	0.37	0.30	0.35	0.53	0.70	0.80	0.42	0.47
Variance	0.00	0.00	0.00	0.00	0.00	0.00	0.00	0.00	0.00
Std.Error	0.06	0.06	0.06	0.06	0.06	0.06	0.06	0.06	0.06
Normal statistic	4.60	5.84	4.80	5.57	8.32	11.02	12.51	6.56	7.46
p=value(2-sided)	0.00	0.00	0.00	0.00	0.00	0.00	0.00	0.00	0.00

The correlation analysis for the 2006 survey period is summarized in Table 5.

Table 5. Cross-correlation analysis for dependent and selected independent variables, 2006

	cases	Demographics		Family		Housing		Education		Economics	
		a45.54	a55.64	single	2p.wchld	owned	sdetach	gr13ls	non.uni	unemp	f.m.inc.k
cases		.404**	.427**	-.449**	0.046	.363**	.392**	.213**	0.116	0.046	0.106
a45.54	.404**		.544**	-.320**	.333**	.417**	.451**	-0.138	0.034	-0.046	.494**
a55.64	.427**	.544**		-.268**	-0.048	.200**	.160*	-0.025	-0.067	0.056	.216**
single	-.449**	-.320**	-.268**		-.619**	-.870**	-.817**	0.138	-.167*	.218**	-.617**
2p.wchld	0.046	.333**	-0.048	-.619**		.718**	.691**	0.05	.160*	-.146*	.462**
owned	.363**	.417**	.200**	-.870**	.718**		.899**	-.159*	.232**	-.322**	.667**
sdetach	.392**	.451**	.160*	-.817**	.691**	.899**		-0.107	.250**	-.269**	.602**
gr13ls	.213**	-0.138	-0.025	0.138	0.05	-.159*	-0.107		.245**	.257**	-.655**
non.uni	0.116	0.034	-0.067	-.167*	.160*	.232**	.250**	.245**		-.193**	-.176*
unemp	0.046	-0.046	0.056	.218**	-.146*	-.322**	-.269**	.257**	-.193**		-.369**
f.m.inc.k	0.106	.494**	.216**	-.617**	.462**	.667**	.602**	-.655**	-.176*	-.369**	

The 2006 cross-correlation analysis, compared with the 2001 analysis, indicates a different picture of the socio-economic structure of the city of Calgary. The correlation between disease and senior citizens ("age 65 plus") has significantly decreased, and has been surpassed by the correlation between disease incidence and younger age groups, i.e., "age 55–64" and "age 45–54". The variable most highly correlated with disease in 2006 is no longer "age 65 plus", but "single". Most likely this is due to the recent phenomenon of internal immigration (from less wealthy parts of the country) of young persons seeking work in the booming Alberta economy (see also Section 4.2). In addition, availability of new medications and changes in Alberta Health Care policies [28] are likely to have affected the frequency of cardiac catheterization in different age groups.

The model selection procedure results in the model described in Equation 6 and summarized in Table 6.

$$CC = f(Single, Gr13ls) \tag{6}$$

where CC is the number of catheterization cases; $Single$ is the number of persons who were never married; and $Gr13ls$ is the number of persons with grade 13 or lower education.

Table 6. Multivariate spatial regression model coefficient and selected diagnostics, 2006

	Value	Std. Error	t value	Pr(>|t|)	
(Intercept)	2.9993	0.3248	9.2357	0.0000	
Single	-0.0501	0.0075	-6.6779	0.0000	
Gr13ls	0.0267	0.0057	4.6477	0.0000	
L.Likelihood	Pseudo-R^2	Rho	Sigma^2	Res.Std. Err	Res. Moran
-426.9000	0.2751	0.1203	0.5272	0.7261	0.0078

The model consists of only two variables, i.e., "single" (negative coefficient) and "grade 13 and less" (positive coefficient) suggesting that the disease incidence is inversely correlated with family status and directly correlated with lower education. Compared with the 2001 model, this is a simpler, but less informative model. The presence of the variable "grade 13 and less" confirms the importance of education attainment in explaining the disease incidence. The variable "single" may indicate a social evolution, in that the negative correlation between disease and of families with children has been replaced by a negative correlation between disease and singles. These results are likely related to the demographic and social transformations induced by the recent economic boom experienced by the city of Calgary.

The re-computation of this regression with varying spatial weight matrices based on alternative p values is summarized in Fig. 7.

Similar to with the 2001 regressions, the impact of alternative distance metrics is remarkable on the model variance indicators, but negligible on the goodness-of-fit indicators. Again for this census period, the effect of the distance metric is most pronounced on the rho parameter and the index of spatial autocorrelation on the

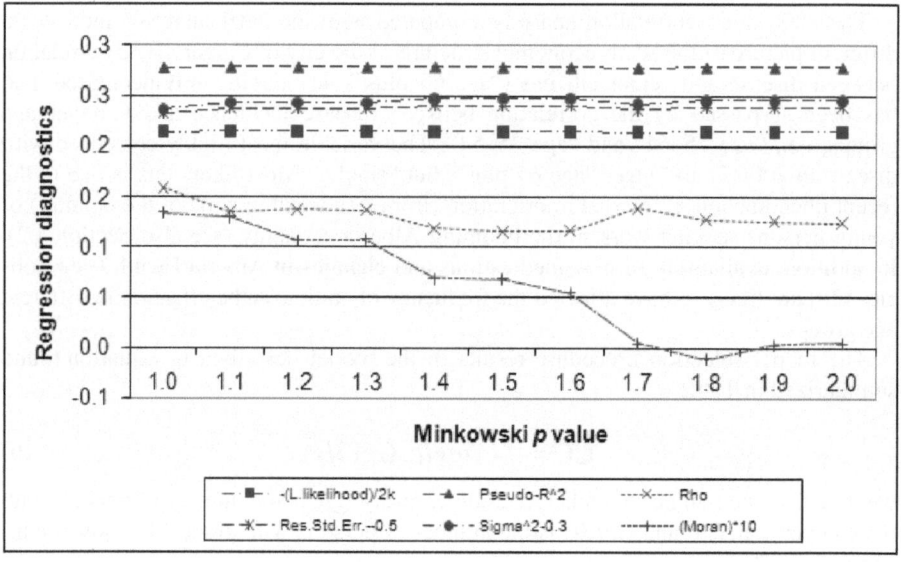

Fig. 7. Spatial regression model selected diagnostics for varying *p* values, 2006

residuals. For the rho value, the highest value corresponds to the *p* value of 1.7. For the spatial autocorrelation index, the lowest value is reached at *p*=1.8. The latter is a negative value, (-0,011), while in absolute value the lowest value is 0.003 corresponding to *p*=1.9.

Once again, the *p* value that maximizes the spatial autocorrelation index for the dependent variable emerges also as the best value for the regression model. These results provide important elements to evaluate the value and applicability of the methodology presented.

5 Discussion

The methodology presented in this paper hinges on the proposition that different ways of measuring distance can affect the definition of the spatial weight matrix, and, ultimately, the reliability of multivariate spatial regression models. The methodology is simple and, owing to its simplicity, presents a strong potential for implementation in an automated procedure. A few critical aspects require an in-depth discussion.

The analysis is limited to the consideration of the Minkowski distance, as this method allows for the identification of parameters that can provide a model of empirical distance, or travelling mode, on a given road network. This is a practical decision, presenting the advantage of limiting the number of possible metrics to consider and, at the same time, allowing for the choice of a specific parameter, i.e., *p* value, within a given interval. Following this method, it is possible to maintain a uniform framework, and simply modify the *p* value to reflect variations in the correlation pattern or the physical network connectivity, as shown by the shift in the best *p* value from *p*=1.6 to *p*=1.8 over the interval between the 2001 and the 2006 census surveys (Section 4.2).

Such simple adjustments of the p value can be rapidly implemented to maintain the highest reliability of the spatial regression model.

The subsequent steps of the process also follow a simple process: initially, a set of 11 spatial weight matrices are defined, based on regular increments of the Minkowski p value in the interval $[1 \leq p \leq 2]$. This series of spatial weight matrices is initially used to compute the spatial autocorrelation index for the variable of interest. This analysis identifies a potential best p value that maximizes the value of the spatial autocorrelation index for that variable. The consequent estimation of a series of spatial regression models has consistently confirmed that the p value preliminarily identified by the spatial autocorrelation analysis is also the best p value for the multivariate spatial regression models, as indicated most strongly by the variation of the spatial autoregressive coefficient (rho). This is a vital point, which suggests that one can simply identify an optimal p value from the spatial autocorrelation analysis, and then proceed to the definition and selection of the spatial regression model using the spatial weight matrix based on that p value.

Indeed, the analysis of the regression model diagnostics (Sections 4.3 and 4.4) confirmed that modification of the spatial weight matrix (i.e., through modifications of the p value) can have a significant impact on the model's variance, most noticeably evidenced by the values of the autoregressive coefficient (i.e., the rho parameter) and the index of spatial autocorrelation in the model's residuals. Conversely, the impact of such variation is constantly negligible on the indicators of goodness-of-fit. These results corroborate the indication that the spatial regression model can be specified and selected independently of the spatial weight matrix specification, thereby suggesting a simplified procedure that can pursue the two main processes, (1) spatial weight matrix definition and (2) model specification and selection, independently.

With reference to the case study presented, the p values identified for the two subsequent census surveys appear to be meaningful values when cross-checked from different perspectives: they are consistent with other findings [23]; they are robust to variation in the distance threshold (i.e., k value, or number of nearest neighbours) considered; and remain relatively consistent over time. Moreover the transition to a higher p value appears to be explained by the major social and demographic transformations experienced by the city over this period.

Overall, the methodology presented in this paper provides an effective tool to enhance the reliability of multivariate regression by altering the distance metric that is used in the spatial weight matrix. The method is simple and can be split into a small number of easy steps. For this reason, a semi-automatic procedure can be envisaged, to select the p value that, all else being equal, minimizes the model's variance. This procedure can accompany the more delicate tasks of model definition and selection, which are not substantially altered by the spatial weight matrix definition, but require judgment and subjective choices, and should therefore continue to be performed manually.

6 Conclusion

The analysis presented in this paper strongly supports the proposition that the choice of a distance metric affects the definition of the spatial weight matrix and can thus lead to the specification of a more reliable spatial autoregressive model. The research presented provides one comprehensive approach to the solution of one of the most common and serious problems affecting the analysis of spatial data: spatial dependence. By estimating efficient regression parameters, the research provides effective support for explicitly spatial policy decisions.

The methodology has been tested on the same variables over two consecutive census surveys: not only have the tests produced consistent results, but the specific differences between the two periods are explained by the major social and demographic changes that have affected the city over that period. The results have also proven meaningful and robust to variations in other parameters of the spatial weight matrix, i.e., distance range. Further analysis of the spatial regression diagnostics have confirmed that variations in the spatial weight matrix can substantially affect the model's variance, but have a negligible effect on its goodness-of-fit.

The analysis confirms the value of the methodology presented. The desirable characteristics of the method, along with the simplicity and modularity of the procedure support the feasibility of its automated implementation to enhance the reliability of applied health and socio-economic spatial regression modelling.

Acknowledgements

We would like to acknowledge the GEOIDE network, and our partners and collaborators for supporting our research project "Multivariate Spatial Regression in the Social Sciences: Alternative Computational Approaches for Estimating Spatial Dependence". We would also like to thank the APPROACH initiative and researchers for providing us with data and support throughout our work. We also appreciate the contributions and suggestions of all the students who helped us with this project, particularly with data preparation and Splus scripting.

References

1. Cliff, D., Ord, J.K.: Spatial Processes. Models and Applications. Pion, London (1981)
2. Griffith, D.A., Amrhein, C.G.: Statistical Analysis for Geographers. Prentice Hall, Englewood Cliffs (1991)
3. Anselin, L.: Spatial Econometrics: Methods and Models. Kluwer Academic Publisher, New York (1988)
4. Kaplan, G.A., Keil, J.E.: Socioeconomic factors and cardiovascular disease: a review of the literature. Circulation 88(4), 1973–1998 (1993)
5. Ghali, W.A., Knudtson, M.L.: Overview of the Alberta Provincial Project for Outcome Assessment in Coronary Heart Disease. Canadian Journal of Cardiology 16(10), 1225–1230 (2000)
6. Getis, A.: A history of the concept of spatial autocorrelation: A geographer's perspective. Geographical Analysis 40(3), 297–309 (2008)
7. Getis, A., Aldstadt, J.: Constructing the Spatial Weights Matrix Using a Local Statistic. Geographical Analysis 36, 90–104 (2004)
8. Bertazzon, S.: A definition of contiguity for spatial regression analysis in GISc: Conceptual and computational aspects of spatial dependence. Rivista Geografica Italiana 2(CX), 247–280 (2003)

9. Anselin, L.: Under the hood. Issues in the specification and interpretation of spatial regression models. Agricultural Economics 27(3), 247–267 (2002)
10. Cressie, N.: Statistics for Spatial Data. Wiley, New York (1993)
11. Fotheringham, A.S., Brundson, C., Charlton, M.: Geographically Weighted Regression: The Analysis of Spatially Varying Relationships. Wiley, Chichester (2002)
12. Ward, M.D.: Spatial regression models. Sage, Los Angeles (2008)
13. Openshaw, S., Alvanides, S.: Applying geocomputation to the analysis of spatial distributions. In: Longley, P.A., Goodchild, M.F., Maguire, D.J., Rhind, D.W., et al. (eds.) Geographical Information Systems: Principles and Technical issues, vol. 1, pp. 267–282 (1999)
14. Besag, J., Green, P.: Spatial statistics and Bayesian computation. Journal of the Royal Statistical Society B 55, 25–37 (1993)
15. Duncan, C., Jones, K.: Using multilevel models to model heterogeneity: Potential and pitfalls. Geographical Analysis 32, 279–305 (2000)
16. Bailey, T., Gatrell, A.: Interactive Spatial Data Analysis. Wiley, New York (1995)
17. Haggett, P., Cliff, A.D., Frey, A.: Locational Analysis in Human Geography. Edward Arnold, London (1977)
18. Krause, E.F.: Taxicab geometry. Addison-Wesley, Menlo Park (1975)
19. Apparicio, P., Abdelmajid, M., Riva, M., Shearmur, R.: Comparing alternative approaches to measuring the geographical accessibility of urban health services: Distance types and aggregation-error issues. International Journal of Health Geographics 7, 7 (2008)
20. Laurini, R., Thompson, D.: Fundamentals of Spatial Information System. The A.P.I.C. Series, vol. 37. Academic Press, London (1992)
21. Phibbs, C.S., Luft, H.S.: Correlation of travel time on roads versus straight line distance. Medical Care Research and Review 52(4), 532–542 (1995)
22. Kohli, S., Sahlen, K., Sivertun, A., et al.: Distance from the primary health center: A GIS method to study geographical access to health care. Journal of Medical Systems 19(6), 425–436 (1995)
23. Shahid, R.: GWR in Health: An Application to Cardiac Catheterization in Calgary. In: Proceedings ESRI Health GIS Conference 2007 (2007)
24. Bertazzon, S.: Cardiovascular disease and socio-economic risk factors: an empirical spatial analysis of Calgary (Canada). Rivista Geografica Italiana 116(3) (forthcoming, 2009)
25. Bertazzon, S., Olson, S., Knudtson, M.: A spatial analysis of the demographic and socio-economic variables associated with cardiovascular disease in Calgary (Canada). Applied Spatial Analysis and Policy (2009), doi:10.1007/s12061-009-9027-7
26. Statistics Canada: Report on the Demographic Situation in Canada (2008)
27. Anselin, L.: SpaceStat tutorial. Regional Research Institute. West Virginia University, Morgantown, West Virginia (1993)
28. Alberta Health and Wellness,
    http://www.health.alberta.ca/health-care-insurance-plan.html
    (accessed, 13/05/2009)

Identifying Hazardous Road Locations: Hot Spots versus Hot Zones

Elke Moons, Tom Brijs, and Geert Wets

Transportation Research Institute, Hasselt University,
Science Park 5/6, 3590 Diepenbeek, Belgium
elke.moons@uhasselt.be

Abstract. Traffic safety has become top priority for policy makers in most European countries. The first step is to identify hazardous locations. This can be carried out in many different ways, via (Bayesian) statistical models or by incorporating the spatial configuration by means of a local indicator of spatial association. In this paper, the structure of the underlying road network is taken into account by applying Moran's I to identify hot spots. One step further than the pure identification of hazardous locations is a deeper investigation of these hot spots in a hot zone analysis. This extended analysis is important both theoretically in enriching the way of conceptualizing and identifying hazardous locations and practically in providing useful information for addressing traffic safety problems. The results are presented on highways in a province in Belgium and in an urban environment. They indicate that incorporating the hot zone methodology in a hot spot analysis reveals a clearer picture of the underlying hazardous road locations and, consequently, this may have an important impact on policy makers.

Keywords: Hot spot analysis, Moran index, hot zone, traffic safety.

1 Introduction

Over the past decades, accident figures have become a 'hot' topic in the media, for policy makers, for academics and for the broad audience. As opposed to most of its neighboring countries, Belgium performs still below par concerning traffic safety. 10.8 individuals per billion vehicle kilometers driven died in a crash in Belgium in 2007 [11]. This figure is more than 30% higher when compared to its immediate neighboring countries, so it seems more than natural that traffic safety takes a top priority in the National Safety Plan. And therefore, the States General of Traffic Safety set the ambitious goal to reduce the number of fatal accidents per year by 2015 up to 500 [30].

An important topic of interest in safety analysis is investigating the location of and the reason for the dangerous sites, often referred to as *hot spot analysis*. In general, hot spot analysis can be split up into four phases. First, the hazardous locations (hot spots) need to be identified. Second, the locations will be rank ordered. If required, the severity of the accident can be accounted for in the

M.L. Gavrilova and C.J.K. Tan (Eds.): Trans. on Comput. Sci. VI, LNCS 5730, pp. 288–300, 2009.

ranking (e.g. [5], [6], [22] and [31]). Third, one tries to find an explanation why some locations are hot spots and others not (i.e. profiling of hot spots). Several techniques can be applied for this purpose. These explanations can be determined based on the analysis of manoeuvre diagrams, conflict observation studies and methods based on information from the traffic accident records [8], [9], [12], [13], [26]. Finally, a selection needs to be made about the sites to treat [23]. This is often a decision of the policy makers and the final selection may depend on many factors: e.g. the financial supplies, whether or not it is preferred to look at a group of locations or at every site separately, on a cost-benefit analysis [2], [14]. In this paper, the focus lies on the first of the four phases, i.e. on the identification of hot spots and consequently of the hot zones.

Hot spot safety work can be described as the task to enhance traffic safety by adapting the geometrical characteristics and the environmental features of problematic locations in the existing road network. The main aim is to reduce the number of accidents, but therefore one needs to know where concentrations of accidents occur. Thus, the geographical dimension plays a major role in describing and tackling the problem of traffic accidents in order to indicate the most critical hazardous locations in a scientifically sound and practical, workable way. Although, one acknowledges the importance of the geographical aspect, very often statistical - non-spatial - regression models are used to model the number of accidents with respect to some explanatory variables, hereby ignoring the existing geographical relationship between the different locations [29], [10], [21]. Hierarchical Bayesian models such as the Poisson-gamma or Poisson-lognormal models are well-established in traffic safety literature [16], [17], [19], [20], [27]. In this paper, the geographical aspect is directly taken into account by applying a local indicator of spatial association to identify hazardous locations alongside the road network. Next to the identification of hot spots, the danger posed by some contiguous road segments may well be more dangerous to the road user than a hot spot alone. These contiguous hazardous locations are called *hot zones*.

The use of the chosen indicator for identifying hot spots and the determination of hot zones will be explained in Sect. 2. Section 3 describes the available accident data on highways in a province of Belgium and in an urban environment, while Sect. 4 reports the results of the application of the hot spot and hot zone methodology. Finally, Sect. 5 ends with conclusions and some ideas for future research.

2   Methods

2.1   Moran's I for Identifying Hot Spots

Recently, in traffic safety literature, next to applying statistical (Poisson) regression models to determine locations with a high number of accidents [29], [10], there is a tendency to use spatial data analysis techniques [8]. This enables to account for the spatial character of a location. In this paper, a spatial autocorrelation index is used. A spatial autocorrelation index aims at evaluating the level of spatial (inter-)dependence between the values of a variable under investigation among spatially located data [15].

In general, a distinction is made between global and local spatial autocorrelation. The global measure investigates globally if locations that belong to the study area are spatially correlated. Next to the global measure, which gives an idea about the study area as a whole, it is often interesting to limit the analysis to a part of it. On the one hand, it might happen that parts of the study area show a spatial autocorrelation, which was not noticed in the global measure. On the other hand, if global spatial autocorrelation is present, the local indexes can be useful to point at the contribution of smaller parts of the area under investigation. These local indexes are considered to be *Local Indicators of Spatial Association (LISA)* [1], if they meet two conditions:

- it needs to measure the extent of spatial autocorrelation around a particular observation, and this for each observation in the data set;
- the sum of the local indexes needs to be proportional to the global measure of spatial association.

The global version of Moran's I was first discussed in [25], though in this paper its local version will be applied. The local version of Moran's I at location i that satisfies the above two requirements can be written down as follows:

$$I_i = \frac{n}{(n-1)S^2}(x_i - \bar{x}) \sum_j w_{ij}(x_j - \bar{x}) \tag{1}$$

with

$$\begin{cases} x_i \text{ the value of the variable under investigation, } X, \text{ at location } i, \\ \bar{x} \text{ the average value of } X, \\ w_{ij} \text{ a weight representing the proximity between location } i \text{ and } j, \\ \quad \text{with } w_{ii} = 0 \text{ for all locations,} \\ n \text{ representing the total number of locations, and} \\ S^2 = \frac{1}{(n-1)} \sum_{i=1}^{n}(x_i - \bar{x})^2, \text{ the variance of the observed values.} \end{cases}$$

A nice property of Moran's I is the fact that it regards the variable under study in a relative way, i.e. with respect to an average value. Because of computational issues, it is often impossible to compute the index for the study area as a whole in one time, and one needs to split up the study area in smaller parts (as will be the case in Sect. 3). By plugging in the average of the entire study area as $\bar{x}$ (instead of just the average of the smaller part), all results can easily be combined. So, $\bar{x}$ might serve as a reference value for the study area under investigation.

Anselin [1] derived the mean and variance of I_i under the randomization assumption for a continuous X-variable. The expected value of I_i is, for example [28]:

$$E[I_i] = \frac{-1}{n-1} \sum_{j=1}^{n} w_{ij} \ .$$

The exact distributional properties of the autocorrelation statistics are elusive, even in the case of a Gaussian random field. Anselin [1] recommends random-ization inference, e.g. by using a permutation approach. However, Besag and Newell [3] and Waller and Gotway [32] note that when the data have heteroge-neous means or variances, a common occurrence with count data such as traffic accidents, the randomization assumption is inappropriate. Instead, they recom-mend the use of Monte Carlo testing. This has been applied in [24]. In this paper, a location is considered to be a hot spot if the local Moran's I value has similar high values between the location under study and its contiguous locations and if it exceeds the 95th percentile (i.e. if $I_i > P_{95}$, where P_{95} is determined from the simulated distribution).

It can be observed that some kind of proximity measure w_{ij} is used to denote the distance between location i and location j in the calculation of Moran's I. In general, geo-referenced x and y coordinates are attached to each location and distances are determined by means of a bird's-eye view. However, accidents take place on a road network, and it may happen that locations are very close to each other in space, though, via the network, they cannot be reached easily (e.g. because one of them is located in a one-way street). A logical step is then to consider the distance traveled alongside the road network. Every location can be pinpointed at the road network map and distances can be determined via the network. This also takes care of junctions and on and off ramps in a proper way and encapsulates the whole network structure in its measure. The inverse of the squared distance is now used as a proximity measure between two locations. This means that the less nearby a location is to the location under study, the less weight is given to that 'distant' location when computing the autocorrelation. Note that at the end of a road, one only accounts for the neighbors that exist. This is one of the adaptations to Moran's I in order to apply it to a traffic safety setting. For more information, the reader is referred to [24].

This local measure of spatial association can also be regarded as being a traffic safety index, since for each hectometer of road (the basic spatial unit for analysis on highways in Belgium) the local Moran index can be regarded as a measure of association between the hectometer under study and the neighboring hectometers. A negative value of the local autocorrelation index at location i indicates opposite values of the variable at location i compared to its neighboring locations. A positive value, on the contrary, points at similar values at location i and its neighborhood. This means that location i and its weighted neighborhood can both have values above the average value or both can have values below the average. In the application area of traffic safety, however, one is only interested in locations that have

a. a high number of accidents in regard to the total average number of accidents ($x_i - \bar{x} > 0$),
b. and where the neighborhood also shows more accidents than was expected on average ($\sum_j w_{ij}(x_j - \bar{x}) > 0$).

One might argue that it is also important to look at locations with a high number of accidents at location i and a very low number in the surrounding area. In this case, very negative values of Moran's I would occur. However, this gives very contradictory effects as will be illustrated by an example. E.g. suppose that the global average equals one accident ($\bar{x} = 1$). If 9 accidents occurred at location i and none in its surrounding, this would lead to a negative value of Moran's I. However, adding one accident to every surrounding point of location i, hence making the surrounding area more hazardous, would lead to a Moran's I of zero. This reasoning entails that a more hazardous location has a less significant Moran's I when compared to a more 'safe' location. This is really counterintuitive, so therefore it is opted to look only at points where the location and its surrounding area reinforce each other in a positive way.

2.2 Hot Zone Methodology

Although hot zones are hazardous to road users, in contrast to hot spot analysis, there is no systematic methodology to identify hot zones on a road network. Most of the studies so far are based on a (number of) selected road(s) [4], [8]. Loo [18] developed a three stage methodology to identify hot zones on a road network. The first step involves the validation of crash locations by combining the crash database, the road network structure and the administrative database. The second step is segmenting the networking into basic spatial units (BSUs) and the calculation of the crash statistics within each BSU. It this step, it is also indicated which BSUs should be considered as contiguous for the identification of hot zones. In the final step, the hot zones are identified according to the procedure as outlined in Figure 1 [18]. The first BSU record is examined to investigate whether it's measure (in this paper: the modified Moran index) is larger than or equal to the threshold measure (determined to be P_{75} of the simulated distribution of Moran's I). If the answer is positive, a new working table will be created with an index number equal to one. Then each contiguous BSU is analyzed, and whenever it's Moran index also exceeds the index is increased by one. When all contiguous BSUs are checked, the index number is investigated: if it is larger than one, a hot zone has been identified, and the variable indicating a hot zone in the main data set is updated. This entire process continues until all BSUs in the network have been checked.

3 Data

This Section describes the data required to apply Moran's I for identifying hot spots and hot zones in two different configurations. A first application comprises accidents on highways in Limburg, a province in Belgium. A second data set consists of accidents on regional roads in an urban environment, more precise in the city of Hasselt (capital of the province of Limburg) and its surroundings. Both data sets are provided by the Belgian Federal Police.

Randomness (variability) in the number of accidents on a certain road segment is typical, because of the nature of accidents and because of unpredictable factors, such as the weather. Therefore, it is of great importance that the study period is

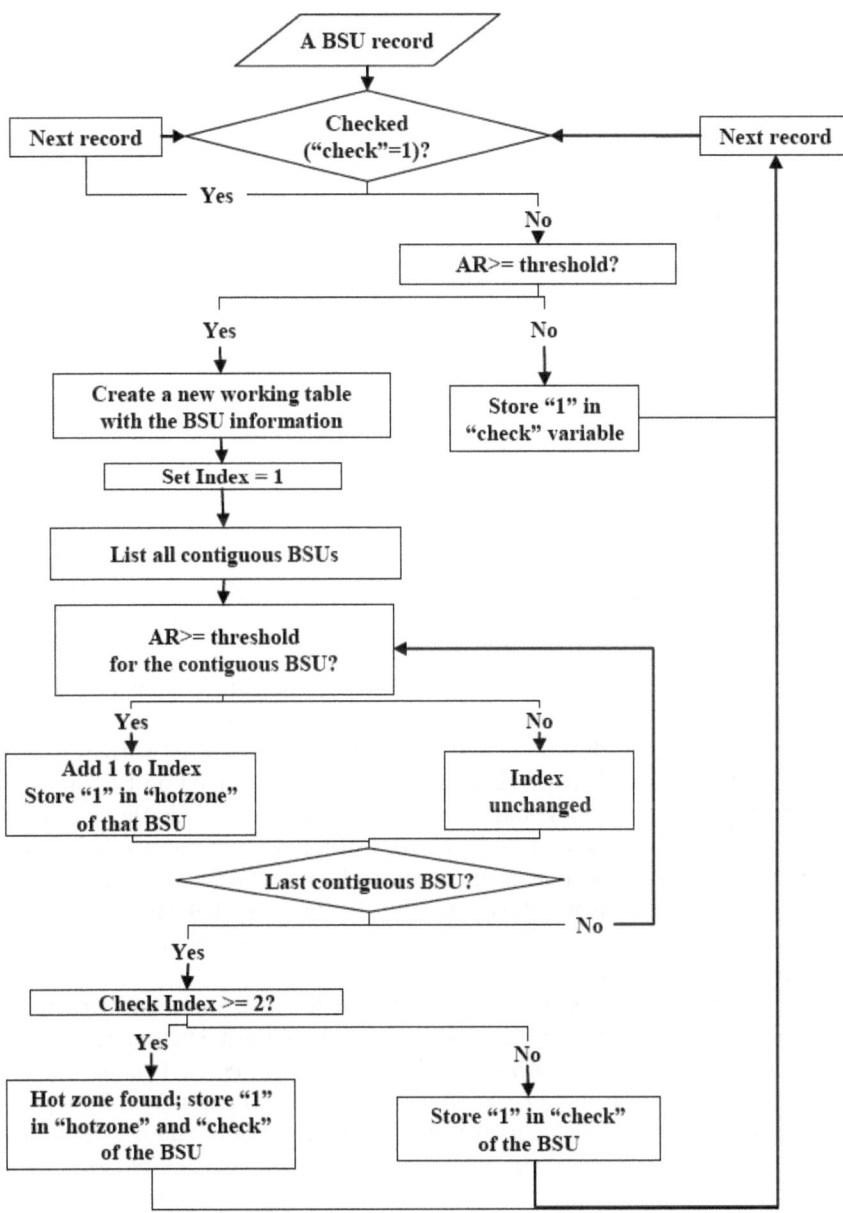

Fig. 1. Identification of hot zones

long enough to ensure representative accident samples. Based on a large number of studies, it is generally agreed upon that a period of three to five years is sufficient to guarantee the reliability of results [7]. For both analyses, data on accidents were collected from 2004 to 2006.

Fig. 2. Limburg within Flanders

The first analysis is carried out on the two highways in the province of Limburg in Belgium. Figure 2 indicates the location of the province of Limburg within Flanders (the upper, Dutch speaking part of Belgium).

The second analysis is carried out on regional roads (they actually comprise of provincial roads, regional roads and highways). Figure 3 indicates the road network of the city of Hasselt and its surroundings, together with the BSUs where accidents occurred. Note that many accidents occurred on the inner and the outer ring way of the city and at the arterial roads towards the city. In the upper left corner, one can observe the clover leaf junction of the two highways in Limburg, E314 and E313. This is expected to be a hazardous location, though one needs to take care in which setting. It may be true that this proves to be dangerous when analyzing highways separately, while on regional roads, it may prove not to be a hot spot after all.

For both settings, the basic spatial unit is defined to be about 100 m. Accidents occurring at highways are assigned to the closest hectometer pole, so they are regarded as BSU, both for highways and for regional roads. As indicated in Sect. 2, the inverse of the squared distance between two locations is used as weight measure, with the distance from one BSU to the next one being determined on the network. The number of neighbors is also distance based. For each BSU, BSUs within a 1 km range from the BSU under investigation are included as neighbors. So, each point, not located near the end of any highway, has approximately 20 neighboring points (more neighbors are possible for the urban environment configuration). Nearby the junction of both highways, it may happen that BSUs from the second highway are within the predefined number of neighbors for a location at the first highway. To account for them in a proper way, distances need to be network-based. In the city environment, this becomes even more important, since there are much more small roads within the neighborhood of each other.

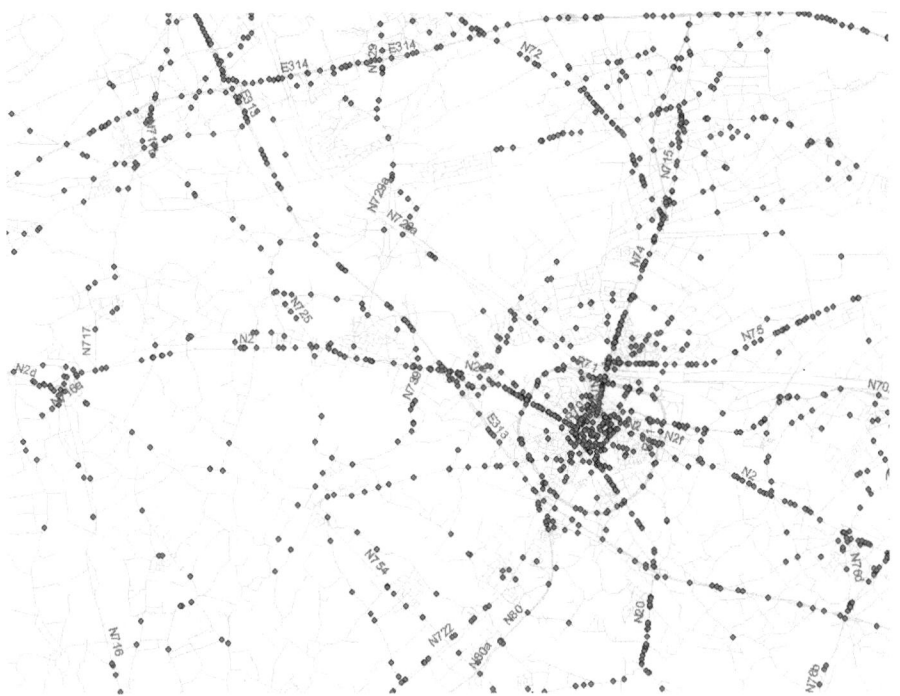

Fig. 3. Road network around Hasselt

Because the idea is to compare the results of Limburg with other provinces in Flanders, the number of accidents for Flanders was set as a reference value ($\bar{x}$) in both analyses.

Limburg has 3,252 hectometer poles alongside its two highways (E313 and E314) and 506 accidents occurred on these highways between 2004 and 2006. This makes Limburg the safest province (regarding highways crashes) in Flanders for the period under study. In the second configuration, 1,678 collisions took place on one of the 3,856 possible hectometer sites. As stated above, the Monte Carlo approach is applied to arrive at the distribution of local autocorrelation statistics [24].

4 Analysis and Results

4.1 Case 1: Accidents on Highways

This Section gives an illustration of the use of the local Moran index for accidents on highways in the province of Limburg in Belgium. As outlined in Sect. 2, to determine the hot spots, it was decided to look only at locations that show a positive reinforcement with their neighbors in the calculation of the local autocorrelation index. The 95% percentile of the simulated distribution, i.e. P_{95}, is

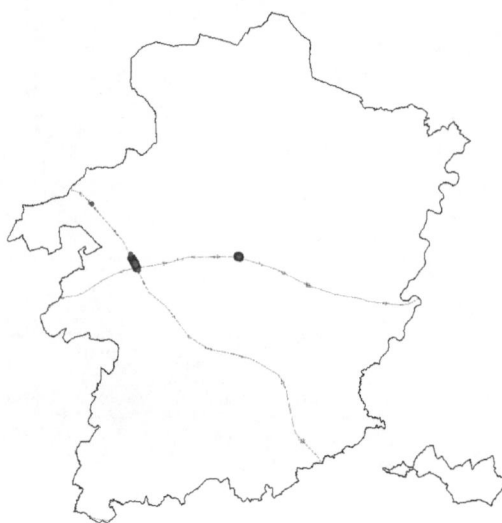

Fig. 4. Hot spots and hot zones in Limburg

determined and utilized as the cut-off value to determine an accident hot spot. If the local Moran index of the location under study exceeded the cut-off value (i.e. if $I_i > P_{95}$), then location i was considered to be hazardous. Of the 3,252 locations, only 6 proved to be hot spots. They are shown as red dots in Fig. 4. The black lines indicate the two highways in Limburg.

In order to identify hot zones, the threshold value was now lowered to P_{75}. This was an arbitrary chosen value, but it is logical that it should be lower than the cut-off value for the hot spot methodology. Note that again only locations are considered where the location under study and the surrounding area positively reinforce each other. Figure 4 also illustrates the hot zones (in blue). It can be observed that the hot zone methodology provides an added value for the hot spot methodology, in the sense that, although most hot spot locations are embedded within a hot zone, this is not necessarily true. Next, the second configuration will show that also the opposite proves to be true: a series of locations can form a hot zone, even if no hot spot is nearby. This clearly indicates the complementarity of a hot spot and a hot zone analysis.

4.2   Case 2: Accidents in an Urban Environment

For the second setting, 48 locations proved to be hot spots, according to a similar 95% percentile definition. They are indicated in red in Fig. 5. The hot zones are again indicated in blue. Three hot spots seem isolated, while the rest is embedded within a hot zone. Next to this, also two extra hot zones are created without the presence of a hot spot.

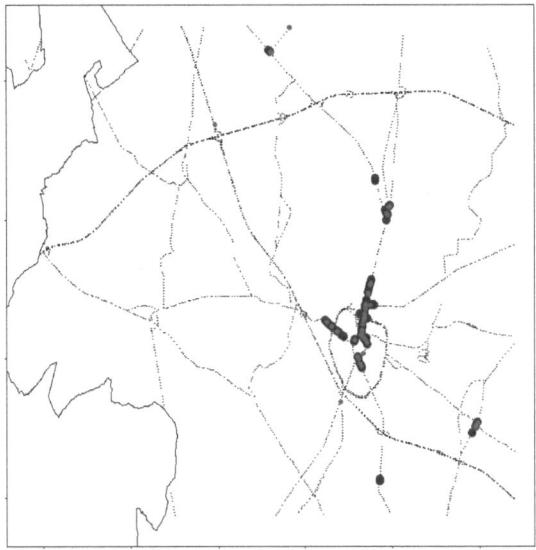

Fig. 5. Hot spots and hot zones in a city environment

In this urban environment, it is clear that an analysis of hot spots, without considering hot zones, would not be sufficient. At least two hot zones would be overlooked, although a large number of accidents occurred on those particular locations. This may lead to a wrong allocation of money that was intended for road safety amelioration.

In this application both the application of Moran's I and the hot zone methodology account for the spatial structure of the data. It would be nice to compare these results with an application of Bayesian models to identify hot spots and hot zones. It is expected that the results will be even more prominent when the hot zone methodology will be applied to e.g. the expected crash rates from a statistical model, since most Bayesian models do not account for the spatial structure in computing the expected crash values.

5   Conclusion

The aim of this paper was to identify hazardous locations on highways and on regional roads by means of a local indicator of spatial association. An adapted network-based version of the well-known Moran index was used to carry out a hot spot analysis. Although a hot spot analysis succeeds in identifying the most dangerous locations, it is often worthwhile to investigate the environment of the hazardous locations some more. It may very well be that not only the hot spot location is particularly accident-prone, but the entire neighborhood can be (moderately) dangerous. Therefore, a hot zone methodology was adopted in this paper to identify hot zones, next to hot spots. The same adapted Moran index was used to identify hot zones, but for identification of hot zones, the threshold value was lowered. If contiguous locations exceed the threshold value, they form a hot zone. The results show that the hot zone methodology supplements the hot spot methodology and it is superior and more flexible in some ways. Apparently, most hot spot locations are embedded within a hot zone. This means that not only a particular basic spatial unit needs to be considered as hazardous, but it is better to thoroughly investigate the entire hot zone before remedial actions are taken on that particular site. Next to this, there may also appear hot zones without the presence of a hot spot. This also requires further investigation of those particular sites. This is a very relevant result for policy makers, since they usually do not have access to an unlimited budget to treat hazardous locations. To allocate their funds in the best possible way, it is important to know the location of the hot spots and the hot zones. The authors therefore recommend to supplement each hot spot analysis with a hot zone analysis to determine the most hazardous locations.

A sensitivity study of the threshold value for identifying (hot spots and) hot zones may shed light on the value and the sensitivity of the hot zone methodology. This is certainly part of further research.

Applying a series of spatial and non-spatial techniques to identify hot spots and hot zones on different road types and to compare the different results is certainly an important avenue for future research.

Acknowledgments

The authors would like to thank the Strategic Analysis Department of the Federal Police for providing the data used in this study.

References

1. Anselin, L.: Local indicators of spatial association – LISA. Geog. An. 27(2), 93–115 (1995)
2. Banihashemi, M.: EB Analysis in the micro optimization of the improvement benefits of highway segments for models with accident modification factors (AMFs). In: El. proc. of the 86th Annual Meeting of the Transportation Research Board, Washington D.C., USA (2007)
3. Besag, J., Newell, J.: The detection of clusters in rare diseases. J. Roy. Stat. Soc. A 154, 327–333 (1991)
4. Black, W.R., Thomas, I.: Accidents on Belgium's highways: a network autocorrelation analysis. J. Transp. Geog. 6(1), 23–31 (1998)
5. Brijs, T., Van den Bossche, F., Wets, G., Karlis, D.: A model for identifying and ranking dangerous accident locations: a case study in Flanders. Stat. Neerl. 60(4), 457–476 (2006)
6. Brijs, T., Karlis, D., Van den Bossche, F., Wets, G.: A Bayesian model for ranking hazardous road sites. J. Roy. Stat. Soc. A 170, 1–17 (2007)
7. Cheng, W., Washington, S.P.: Experimental evaluation of hotspot identification methods. Acc. An. Prev. 37, 870–881 (2005)

8. Flahaut, B., Mouchart, M., San Martin, E., Thomas, I.: The local spatial auto-correlation and the kernel method for identifying black zones - A comparative approach. Acc. An. Prev. 35, 991–1004 (2003)
9. Geurts, K., Thomas, I., Wets, G.: Understanding spatial concentrations of road accidents using frequent itemsets. Acc. An. Prev. 37(4), 787–799 (2005)
10. Hauer, E.: Identification of sites with promise. Transp. Res. Rec. 1542, 54–60 (1996)
11. International Traffic Safety Data and Analysis Group (IRTAD) Risk indicators for the year 2007 (2009),
    http://www.internationaltransportforum.org/irtad/datasets.html
    (accessed May 17, 2009)
12. Jianming, M., Kockelman, K.M.: Bayesian multivariate Poisson regression for models of injury count, by severity. In: El. proc. of the 85th Annual Meeting of the Transportation Research Board, Washington D.C., USA (2006)
13. Jianming, M., Kochelman, K.M., Damien, P.: Bayesian multivariate Poisson-Lognormal regression for crash prediction on rural two-lane highways. In: El. proc. of the 86th Annual Meeting of the Transportation Research Board, Washington D.C., USA (2007)
14. Kar, K., Datta, T.K.: Development of a safety resource allocation model in Michigan. Transp. Res. Rec. 1865, 64–71 (2004)
15. Levine, N.: CrimeStat: A spatial statistics program for the analysis of crime incident locations, vol. 1.1. Ned Levine and Associates/National Institute of Justice, Annandale, VA/Washington, D.C (2000)
16. Li, L., Zhang, Y.: A GIS-based Bayesian approach for identifying hazardous roadway segments for traffic crashes. In: El. proc. of the 86th Annual meeting of the Transportation Research Board, Washington, D.C., USA (2007)
17. Li, L., Zhu, L., Sui, D.Z.: A GIS-based Bayesian approach for analyzing spatial-temporal patterns of intra-city motor vehicle crashes. J. Transport Geogr. 15, 274–285 (2007)
18. Loo, B.P.Y.: The identification of hazardous road locations: A comparison of black-site and hot zone methodologies in Hong Kong. Int. J. Sust. Transp. 3(3), 187–202 (2009)
19. Lord, D.: Modeling motor vehicle crashes using Poisson-gamma models: Examining the effects of low sample mean values and small sample size on the estimation of the fixed dispersion parameter. Acc. An. Prev. 38, 751–766 (2006)
20. Lord, D., Miranda-Moreno, L.F.: Effects of low sample mean values and small sample size on the estimation of the fixed dispersion parameter of Poisson-gamma models for modelling motor vehicle crashes: a Bayesian perspective. In: El. proc. of the 86th Annual Meeting of the Transportation Research Board, Washington, D.C., USA (2007)
21. McGuigan, D.R.D.: The use of relationships between road accidents and traffic flow in 'black spot' identification. Traffic Eng. Contr. 22, 448–453 (1981)
22. Miaou, S.-P., Song, J.J.: Bayesian ranking of sites for engineering safety improvements: decision parameter, treatability concept, statistical criterion, and spatial dependence. Acc. An. Prev. 37, 699–720 (2005)
23. Miranda-Moreno, L.F., Labbe, A., Fu, L.: Bayesian multiple testing procedures for hotspot identification. Acc. An. Prev. 39(6), 1192–1201 (2007)
24. Moons, E., Brijs, T., Wets, G.: Improving Moran's index to identify hot spots in traffic safety. In: Murgante, B., Borruso, G., Lapucci, A. (eds.) Geocomputation and Urban Planning, pp. 117–132. Springer, Heidelberg (2009)
25. Moran, P.A.P.: Notes on continuous stochastic phenomena. Biometrika 37, 17–23 (1950)

26. Pande, A., Abdel-Aty, M.: Market basket analysis: novel way to find patterns in crash data from large jurisdictions. In: El. proc. of the 86th Annual Meeting of the Transportation Research Board, Washington D.C., USA (2007)
27. Park, E.S., Lord, D.: Multivariate Poisson-Lognormal models for jointly modeling crash frequency by severity. In: El. proc. of the 86th Annual Meeting of the Transportation Research Board, Washington, D.C., USA (2007)
28. Schabenberger, O., Gatway, C.A.: Statistical methods for spatial data analysis. Chapman & Hall/CRC Press, Boca Raton (2005)
29. Sørensen, M., Elvik, R.: Black spot management and safety analysis of road networks - Best practice guidelines and implementation steps. TØI report 919/2007, RIPCORD/ISEREST project (2007)
30. Staten-Generaal van de Verkeersveiligheid: Verslag van de Federale Commissie voor de Verkeersveiligheid, p. 38 (2007) (in Dutch)
31. Vistisen, D.: Models and methods for hotspot safety work. Phd Dissertation, Technical University of Denmark (2002)
32. Waller, L.A., Gotway, C.A.: Applied spatial statistics for public health data. John Wiley and Sons, Chichester (2004)

Geographical Analysis of Foreign Immigration and Spatial Patterns in Urban Areas: Density Estimation, Spatial Segregation and Diversity Analysis

Giuseppe Borruso

University of Trieste, Department of Geographical and Historical Sciences
P. le Europa 1, 34127 Trieste, Italy
giuseppe.borruso@econ.units.it

Abstract. The paper is focused on the analysis of immigrant population and particularly on some of the characteristics of their spatial distribution in an urban environment. The attention is drawn on examining whenever there is a tendency to cluster in some parts of a city, with the risk of generating ethnic enclaves or ghettoes, therefore analysing also diversity other than the pure spatial distribution. Methods used in the past to measure segregation and other characteristics of immigrants have long been aspatial, therefore not considering relationships between people within a city. In this paper the attention is dedicated to methods to analyse the immigrant residential distribution spatially, with particular reference to density and diversity-based methods. The analysis is focused on the Municipality of Trieste (Italy) as a case study to test different methods for the analysis of immigration, and particularly to compare different indices, particularly traditional ones, as Location Quotients and the Index of Segregation, to different, spatial ones, based on Kernel Density Estimation functions, as the S index, and indices of diversity, as Shannon (SHDI) and Simpson (SIDI) ones as well as another diversity index proposed (IDiv). The different analysis and indices are performed and implemented in a GIS environment[1].

Keywords: Geographical Analysis, Foreign Immigration, Spatial Segregation, Density Estimation, Diversity Index, Shannon Index, Simpson Index, Trieste (Italy).

1 The Analysis of Immigrants at Urban Level: Qualitative and Quantitative Methods

The analysis on migrations, as correctly observed by Krasna [1] can rely on a mixed combination of methods and tools, both quantitative and qualitative ones. The former ones benefit from the diffusion of spatial analytical instruments and information

---

[1] GIS analysis was performed using Intergraph GeoMedia Professional and GRID 6.1 under the Registered Research Laboratory (RRL) agreement between Intergraph and the Department of Geographical and Historical Sciences of the University of Trieste (Italy).

systems, as well as from the huge availability of digital data and computational power unthinkable since few years ago. That allows filtering data and preparing information for the evaluation to be performed by a scholar on the phenomenon under examination. We can remind Tobler's first law of geography [2], stating that every phenomenon over space is linked to all the other ones, but closer phenomena are more related to each other than farther ones, and therefore understand that the geographical space is capable of being analyzed by means of such quantitative methods, but also that no universal rules can be established, given the different characteristics and peculiarities of places over the Earth's surface. The scholars involved in migration research should therefore rely also on qualitative methods in order to integrate their studies, with the difficult task of interpreting correctly what is happening over space.

With reference to migration studies, and particularly when these are referred to the urban environment, several analyses have been carried out in recent years to examine their spatial distribution, the characteristics of settlements and, more recently, the phenomena of residential segregation and the impact of migrants over the job market and the economy as a whole, in parallel with the migrants' rooting in space as a structural component of society and economy. Researchers have focused their attention on different indicators in order to examine the characters of the spatial distribution of migrant groups, particularly in order to highlight the trends towards concentration rather than dispersion or homogeneity, or, still, the preferences for central rather than peripheral areas. The attention however is in particular focused on the analysis of phenomena related to residential segregation, at risk particularly in areas where a too high concentration of a single immigrant group is present if compared to the local residents such that ghettoes or 'ethnic islands' take place.

Some authors, as recalled by Cristaldi [3], draw their attention on some aspects related to segregation, as particularly the level of residential concentration, assimilation and encapsulation. The indices generally used in the international research concerning geographical mobility and widely applied are the segregation index, based on the dissimilarity index (D) developed by Duncan and Duncan [4], with its different declinations. Its applications generally deal with the possibilities of comparison of the distribution of national groups in the intra-metropolitan area, among cities or on a diachronic scale [5][2], as the main characteristic of the index is of being aspatial, and therefore allowing a direct comparison with other areas but saying little about the internal aspects of dissimilarity or segregation. Although such index has been used for decades and its family actually dominates the literature [10], this is a serious limitation, as it does not provide any indication on the population mix between zones but just within them, thus producing results that depend also on the zoning system chosen [11]. As O' Sullivan and Wong remarks [11], summary indices are useful to portray the level of segregation of a region and for comparing the results obtained for different regions, but they say little about some spatial aspects of segregation, as the possible rise of non-uniformity at local level or the level of segregation across the study area, and do not provide a visualization of segregation

---

[2] Few applications of segregation indices have been done to Italian cases, particularly referred to the cities of Turin, Genoa and Milan [6], Parma, Reggio Emilia [7] and Piacenza [8] and more recently to Rome [5] and Trieste [9]. These two latter cases propose also a disaggregated analysis of segregation at urban level.

inside a region. Another limitation of such an index, shared also by a wealth of other indices, is the use of census data, generally collected and aggregated using a zoning system and that usually consider the different characteristics of population within the zones with little attention to the relations among them. Furthermore, the use of different zoning systems can lead to different values for a same study region, as it will be evident in the case studied in this paper, and therefore the spatial partitioning system strongly affects the evaluation of segregation, with a resulting 'scale effect', causing "smaller statistical enumeration units producing higher measured segregation levels" [12]. This is also related to the fact that moving from a highly disaggregated partition of space into a more aggregated one can lead to a generalization that, although valid for a certain level of analysis, is not valid for another one [13].

2 Measures of Segregation

The measures of segregations applied in the last decades are based on Duncan and Duncan's index of dissimilarity D [4]. The index is expressed in equation 1,

$$D = \frac{1}{2} * \sum \left| \frac{x_i}{X} - \frac{y_i}{Y} \right| . \tag{1}$$

x_i and y_i are the population counts for the two subgroups in the areal unit i, while X and Y are the total counts for the two groups in the study region. The index ranges from 0 to 1, representing respectively the highest dispersion and highest concentration. As noticed before, the results the index D can assume varies with the choice of a zoning system and therefore with the areas and shapes the areal units i will have [11]. Furthermore, the index assumes that people living in a certain areal unit do not mix or interact with other people in neighbouring ones.

Researchers and scholars in geography proposed different methods through years to insert the spatial component within the dissimilarity index or to couple the index with other ones more prone to spatial representation and analysis. Some authors, as Jakubs [14] and Morgan [15] proposed distance-based approaches, Morril [16] and Wong [17] tried to adjust the level of D by introducing a neighbourhood-interaction approach, introducing additional elements into the D equation to consider the interaction among subgroups in neighbouring units. Other authors have coupled the use of the segregation index with other indices to explore more in depth the spatial structure of immigrants, as the Location Quotient (LQ), used to facilitate the analysis of residential segregation in different subunits of an area and therefore allowing mapping the spatial distribution of migrants according to a more disaggregated zoning system of the study region [5].

$$LQ = \left(\frac{x_i}{y_i} \right) \bigg/ \left(\frac{X}{Y} \right) . \tag{2}$$

x_i represents the number of residents of a national group in areal unit i, X the number of residents in the entire study area (i.e., a municipality), y_i the foreign population in areal unit i and Y the foreign overall population in the study region. The location quotient *LQ* equals to unity if the analyzed group holds in the areal unit *i* the same characteristics of the study region; if *LQ* > 1 than it is overrepresented in areal unit *i*, and if *LQ* < 1 than it is underrepresented. Although the drawback still lies onto the zoning system chosen and therefore on the higher or lower levels of aggregation of data, it allows also a visual, geographical analysis of the results obtained.

In the last years a quite wide use have been done of different version of Kernel Density Estimation (KDE) to analyze phenomena expressed as point patterns, both *per se*, providing a visual three-dimensional surface of the spatial distribution of the phenomenon under examination, and to model some other aspatial indicator into a spatial context. Kernel Density Estimation was developed to provide an estimate of a population probability density function from a sample of elements as an alternative to histograms [18], afterwards being extended to the spatial case [19] [20] [21]. The function creates a density surface from a point pattern in space, providing an estimate of the events' distribution within its searching radius, according to the distance to the point where the intensity is being estimated [22].

$$\hat{\lambda}(s) = \sum_{i=1}^{n} \frac{1}{\tau^2} k\left(\frac{s - s_i}{\tau}\right) .$$
(3)

$\hat{\lambda}(s)$ provides an estimate of the intensity of the spatial distribution of events, measured at location s; s_i is the ith events, $k (.)$ represents the kernel function and τ is the bandwidth, varying which it is possible to obtain more or less smoothed surfaces and to analyze the phenomenon at different scales. A wide bandwidth oversmoothes the estimate by including many distant events, while a narrow one tends to overemphasize local variations in the events' distribution [11] [23]. Among the advantages of the function there is the spread of the function all over the study region, obtained by assigning the estimated values over a fine grid overlaid onto the study region, which cells become the places where values are attributed. Another desirable property is represented by the possibility of expressing the results of the Kernel Density Estimation either as density values (i.e., events per square kilometer) or as probability estimates, with the sum of cell values over the study region integrating to unity [24] and therefore allowing a direct comparison with other distributions of events.

The applications of KDE span through Earth's and social sciences, almost in every field where geographical data can be expressed as point events over a study region. In population related studies, Kernel Density Estimation has been widely used particularly in the recent years, from the 'pure' analysis of population distribution [19] [23] [27], to the analysis of immigrant population in urban areas [9] [11] [25] [26]. In this latter case, the estimator can be used together with other, spatial or aspatial, indices to provide an immediate visualization of the phenomenon observed and examine the possible formation of 'hot spots' or clusters of particular ethnic groups, allowing the scholar to go and analyze more in depth those areas of interest for values higher – or lower – than expected. However, Kernel Density Estimation is also used for modeling other elements referred to a spatial extension, in order to allow more sophisticated and complex analyses and host indices in a spatial framework. This is the case of a series of indices being explored in the very last few years, where some aspatial index or qualitative analysis is modeled into the KDE framework.

3 Spatial Indices of Dissimilarity

Three indices are considered here to enhance the analysis on segregation from a more 'spatial' point of view, as well as to examine the qualitative aspects of foreign immigrants settling patterns. The attention in particular is drawn on O' Sullivan and Wong's [11] segregation index S, on the Shannon [36] [37] and Simpson [38] diversity indices and on the diversity index *IDiv* tested by Borruso and Donato [25].

3.1 The Segregation Index

O' Sullivan and Wong present a spatial modification of the Duncan's index D, called S, basically comparing, at a very local level, the space deriving from the *intersection* of the extents occupied by two sub-groups of an overall population and the total extent of the *union* of such areas [28]. Operationally, the calculation of the index for the study region involves the computation of probability density functions by means of KDE for the different population sub-groups of interest. Each reference cell i is therefore assigned a probability value for each subgroup, and for each of the subgroups the probability value in that cell contributes to the integration to unity. The S index for the study region is then calculated as follows:

$$S = 1 - \frac{\sum_i \min(p_{x_i}, p_{y_i})}{\sum_i \max(p_{x_i}, p_{y_i})} . \tag{4}$$

For each cell i minimum and maximum values are computed for the true probability of the two subgroups, p_{xi} and p_{yi}, these are summed for all the i cells, obtaining minimum and maximum values under the two surfaces, and their ratio is subtracted from unity to produce index S. The index obtained is aspatial as well, as it can be obtained for a study region, but the intermediate values, as the differences in maximum and minimum values, can be mapped, giving a disaggregate view of the contribution of each cell to the overall segregation, with lower values indicating areas with some degree of ethnic mixing. As the bandwidth in Kernel Density Estimation determines the level of smoothing, different values produce a decay of the index as bandwidth increases, thus reducing the segregation index overall the study region and still the differences of this behaviour in different regions or for different groups can be analyzed to explore dynamics proper of a territory or group, allowing an analysis not limited to a certain spatial extent but also observable at different distance scales.

3.2 The Shannon and Simpson Diversity Indices

Other than analyzing the characteristics of single ethnic groups and comparing them, an element of interest lays in the analysis of the overall distribution and variability of foreign population in an urban area, therefore considering both quantitative and categorical data concerning the presence of migrants. In such sense the use of indices of diversity can be useful when the focus is also on the different number and type of nationalities in an area, other than their figures. It can therefore be useful to rely on and implement indices taking in account such differences in the population in an urban area.

Diversity indices originates, and are mainly – but not solely – applied to, ecology and biology. In particular such family of indices regards statistics aimed at measuring the biodiversity of an ecosystem. Therefore, the diversity indices can be used to assess the diversity of any population in which each member belongs to a unique species.

Landscape studies [40] as well as social sciences benefit too of this kind of indices [41], where 'species' can be respectively replaced with land cover types rather than ethnic groups or other forms of categorical data, while 'individuals belonging to species' can be thought as individual residents of an ethnic group – as the case to be tackled in the present paper.

The Index of Diversity - also referred to as the Index of Variability - is a commonly used measure, in demographic research, to determine the variation in categorical data. Shannon index [36] [37] in particular represents the information entropy of the distribution, treating species as symbols and their relative population sizes as the probability. The index considers the number of species and their evenness. Its value increases either by adding unique species, or by having a greater evenness of species.

The index can be expressed as SHDI (Shannon Diversity Index):

$$SHDI = -\sum_{i=1}^{N} p_i * \ln p_i \, , \tag{5}$$

where p is the proportion of individuals or objects in a category - the relative abundance of each type or the proportion of individuals of a given type to the total number of individuals in the categories - and N is the number of categories.

As stressed also by Nagendra [35], the index in theory ranges from 0 to infinity, where 0 represents a case where the area where the analysis is carried on presents a perfectly homogeneous population, while higher values are simply representative of higher heterogeneity as the figures increase.

Another index is the Simpson Diversity Index D – here called SIDI to avoid confusion with Duncan and Duncan one - [38]. Actually there is a family of three Simpson Diversity Indices – Simpson's Diversity Indices D, 1-D and 1/ D – as the basic elements for dissimilarities studies. In particular it is used to measure the variation in categorical data. The index can be written as SIDI (Simpson Diversity Index):

$$SIDI = 1 - \sum_{i=1}^{N} p_i^2 \, . \tag{6}$$

Here also p is the proportion of individuals or objects in a category and N is the number of categories. A perfectly homogeneous population would have a diversity index score of 0, while a perfectly heterogeneous population would have a diversity index score of 1 - assuming infinite categories with equal representation in each category. As the number of categories increases, the maximum value of the diversity index score also increases.

Both indices are mostly used in ecology to measure the biodiversity of an ecosystem, and more in general to assess the diversity of a population where members belong to different categories or groups. The two indices are focused on different aspects concerning diversity. The Shannon index is particularly aimed at highlighting the richness component, while Simpson one is more concerned with evenness and the analysis of dominant types [35]. According to some authors [39] the Shannon diversity index is more sensitive to changes occurring in the importance of the rarest elements, while Simpson index seems to respond to changes in the proportional abundance of the most common community [35].

The indices are adapted in this study to the different ethnic groups and their figures, focusing the attention on different spatial level. As in the case of the Location Quotient analysis, census units have been used to aggregate individual data concerning population that can lead to a reduction of the information deriving from the data itself, given the arbitrariness in the area-based analyses.

Data available at address-point level were aggregated within the census units that partition the Municipality of Trieste area, so that each unit displays the figures for the nationalities present within its area, therefore forming a matrix structure with rows identifying the census units and columns the nationalities. Figures for residents belonging to the different nationalities are placed at the intersections of each row and respective column – when data available. That allows the computation of the two indices where the proportion of each nationality contributes to the final value of the index (Shannon or Simpson).

In order to limit the drawbacks concerning the choice of a zoning system and to rely on non-homogeneous area units, the diversity indices are also computed continuously over the study area, performing an analysis not relying on pre-defined type of zoning or spatial delimitation. This approach relies on the concept of "moving window", also at the basis of local scan statistics and Kernel Density Estimation, revealing local variations of spatial phenomena. That allows obtaining indices that can be visualized continuously over the entire study area. A moving window enables to obtain metrics at a more detailed level than using a zoning system. Differently from analysis like Kernel Density Estimation, where the searching function is generally computed over each cell of the grid over the study region, it was decided to consider only the built environment of the city, therefore choosing all the address-points in the municipality of Trieste as sampling locations. The indices were therefore attributed to such point features and successively interpolated[3] in order to obtain a smooth 3D-like surface.

3.3 The Index of Diversity (IDiv)

The Kernel Density Estimation (KDE) has been also used to host *Index of Diversity* to observe if characteristics of homogeneity or differentiation between ethnic groups rise in the urban space. The Index of Diversity (*IDiv*) [25] can be expressed as:

$$IDiv_i = N_i \left/ \left(\left(\frac{y_i}{x_i} \right) * 100 \right) \right. . \tag{7}$$

The index *IDiv_i* is referred to each sub-area *i*, this being either a census block or a grid cell, and N_i is the number of countries represented in sub-area *i*, y_i the immigrant population in zone *i* and x_i the overall resident population in sub-area *i*. The index is then processed by means of KDE to be transformed in a three-dimensional density

---

[3] The interpolation was performed using the Inverse Distance Weighting (IDW) algorithm.

surface to visualize its continuous variation over space. The index is not focused on the possibility of highlighting areas segregated or with high values of a single group's concentration: the Index of Diversity must be seen as an effort to represent synthetically some of the differences and peculiarities of migration in the urban areas and to highlight some elements of potential union and cohabitation rather than segregation and division. The index is built starting from the census blocks, transformed in point elements for the further elaboration by means of Kernel Density Estimation. The index is based on the computation of the number of countries represented in each census block. Then this value is multiplied by the percentage of foreign residents in the census block. The index therefore considers both the diversity of countries represented and the number of immigrants. High values of the index indicates the presence of several immigrants coming from different countries, while lower values represent a small number of immigrants and little diversity of countries. The index is not alternative to other indices but aimed at highlighting some qualitative characteristics of the spatial distribution of migrants in urban areas. The development of such indicators should move towards the implementation of entropy based methods [29] in order to consider, as a characteristic to map, the diversity and variety in the distribution of population, rather than focusing on elements of division and separation.

As already noticed in the previous section, the different indices observed here present limitations deriving from the availability of census data and their level of detail. In both cases the indices require the use of point data to be processed by the Kernel Density Estimation. This is generally done mainly in an intermediate stage of the process, as data are generally aggregated at census blocks level, therefore adopting a certain zoning system. In fact, the higher is the level of disaggregation of data, the more refined can be the geographical analysis and minor the error propagation effect. On the contrary, a higher level of data aggregation leads to a dilution of the message linked to the data itself and more difficult is its transformation in true information for the researcher. The data available are not always so disaggregated - i.e. referred to address points, not always available for population data – but collected using areal spatial units as census blocks or enumeration districts, etc. Such areas generally present non homogeneous shapes and dimensions as well as being subject to possible re-aggregation and modification, what can lead to misinterpretations of the phenomena under examination, as a same measurement of the phenomenon can depend on the unit chosen.

These difficulties are partially minimized when a sufficiently fine zoning is used and these are then converted in geographical coordinate pairs, generally referred to areas' centroids that become point elements. This allow studying for instance the density of a phenomenon using homogeneous grouping methods or transforming the point dataset into a 'density surface' as in the Kernel Density analyses[4]. Still another issue, not yet completely tackled, lies in the multivariate nature of population data that, other than being characterised by their geographical locations, consist also on qualitative and quantitative attributes, these including the ethnic group, gender, age, origin, therefore becoming also sometimes difficult to manage and analyze.

---

[4] However, in previous analysis on spatial distribution of population, very little difference on the overall pattern drawn by a Kernel Density Estimation was noticed between applications on point data based on address-points and those considering census blocks' centroids [26].

4   Foreign Population in Urban Areas: The Spatial Distribution and Characteristics in the City of Trieste (Italy)

4.1   The Study Area and the Data

The indices described above have been tested on the Municipality of Trieste as a study area on a selected set of ethnic groups, considering 2005 data referred to those people registered as residents in the city. The applications considered the more traditional, aspatial indices and the spatial ones, as well as the density based methods applied for the analysis of the distribution of the phenomenon and for incorporating other indices to portray spatially some of the characteristics of migration, as clustering, tendency to segregation and diversity.

Some considerations on the overall characteristics of migrants in the area however need to be done before the quantitative analysis by means of the indices examined.

The immigrant population as a whole, thus not considering residents coming from advanced developed countries seems to present elements of uniformity when observed as a scatter plot at address point level over the municipality of Trieste, however presenting a preference for central areas. A unique pattern of the spatial distribution of population however cannot be highlighted, and the analysis of single ethnic groups can reveal the diversities in the structure of the resident population.

People coming from countries once belonging to former Yugoslavian were quite relevant in the past in shaping the population distribution of the city and tend to concentrate historically in densely populated and ethnically diversified sub-areas of the city, not properly 'central' but located around the city centre. This trend of concentration of foreign groups into properly urban and central areas of a city was observed in different national contexts, given the higher attraction played by the city for new immigrants, in terms of job opportunities and housing market [8]. In the case of Trieste it is necessary to separate the analysis on migration at different levels. On one side the migrants from highly developed countries tend to distribute in residential areas as local ones do; on another side migrants from former Yugoslavia form a 'backbone' of the city itself, with migrations dating back from the 19th Century on, while a third group can be identified in the 'new ethnic groups' that characterize the city with higher numbers, although still low if compared to other historical groups, particularly in the last few years. These latter two groups tend to concentrate in the proper urban districts of the city[5].

In the first stages of the present analysis a selection of ethnic groups has been done, these including those groups of more recent migration in the history of the city itself. These are therefore the Albanian, the Chinese, the Romanian and the Senegalese ones. Their immigration in larger figures dates back in the recent past, and particularly for the extra-Europe groups, as Chinese and Senegalese ones, the tendency to concentration is more evident from previous research [25] [26].

In the second stage of the work carried on, particularly when analyzing the diversity structure of the population, the data considered in the analysis consisted on the

---

[5] This is confirmed by the slightly higher value of the percentage of immigrant residents in the urban districts than those registered in the Municipality as a whole. Urban districts present a 5.28 % and the Municipality displays an average 5.06 % value (2005 data).

overall database of resident population living in the Municipality of Trieste - 208,731 georeferenced individuals of which 10,562 foreign nationals from 115 countries - with detail at address point level of nationality. Such data were both aggregated at census unit level – or higher – and elaborated as a point dataset.

4.2 The Segregation Index

One of the first analyses carried on is therefore focused on the segregation index (SI), derived from D seen before, applied to the above mentioned ethnic groups. In this first version the index was computed using data aggregated by the census units of the Municipality of Trieste, considering a selection of the 929 units in the Municipality (Fig. 1)[6]. Such areas are among the smallest one in a zoning system at Municipality level. The index is not mapped as it derives from a sum of the values obtained for the single census units and allows a first exam on the segregation.

The results are portrayed in Fig. 2a. A general trend of high clustering of the different ethnic groups considered can be noticed, with particular reference to the extra-European groups as Chinese and Senegalese.

The index was also calculated aggregating data at municipal district level rather than using census blocks. As the district zoning system provides bigger areas, this caused a dramatic decrease in values of the segregation index as expected after the consideration drawn before. However, the Chinese group still maintains higher figures if compared to the other ones (Fig. 2b)

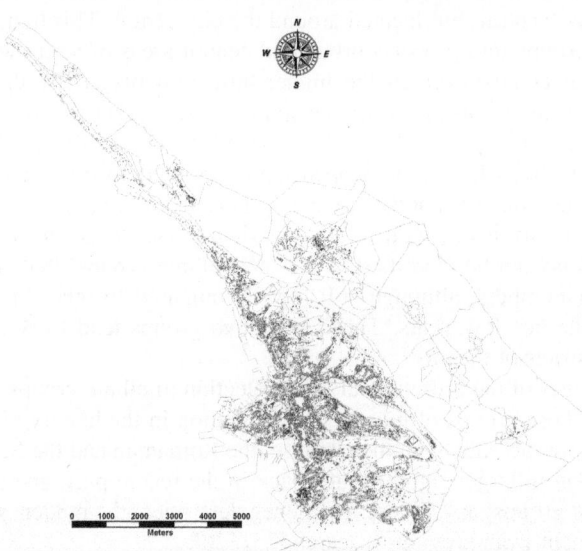

Fig. 1. Spatial distribution of resident population in the Municipality of Trieste. The map displays census blocks (gray boundary areas) and residents' locations at address-point level (black dots).

---

[6] Census data were also available at street-number level and this will be seen particularly in the S index and standard KDE computation for single ethnic groups.

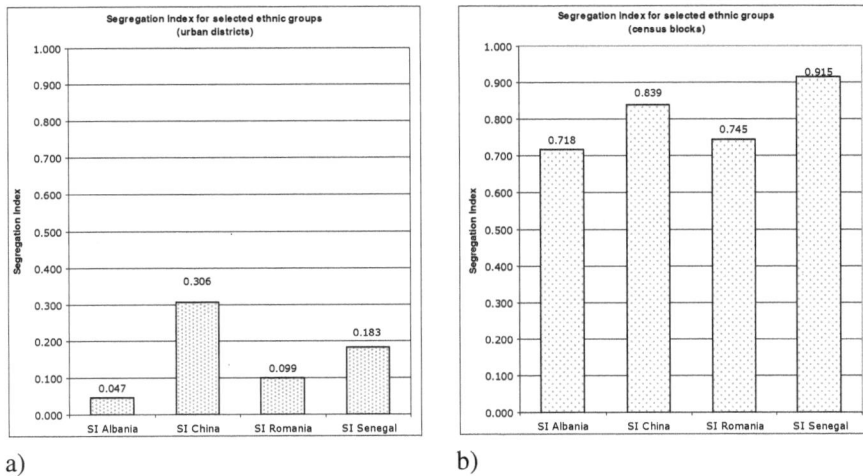

a)   b)

Fig. 2. Segregation Index computed on area units using urban districts (*a*) and census blocks (*b*) zoning systems for the Municipality of Trieste. Note that changing the scale and using wider administrative units (urban districts) the index decreases remarkably with reference to the ethnic groups considered.

4.3   The Location Quotients

Location Quotients were also computed for the selected ethnic groups in order to provide a cartographic visualization of the phenomena of concentration or diffusion. Also in this case the quotient was computed using census blocks. Albanian and Romanian nationals are quite sparse over the urban area (Fig. 3a and c), with several census blocks where the Location Quotient is considerably higher than 2, therefore denoting a high level of concentration.

On the contrary, both Chinese and Senegalese groups are more clustered in some parts of the central city blocks, covering approximately neighbouring and non-completely overlapping areas. An area characterized by high concentration of these two groups is the one close to the railway station and a part of the city centre characterized by a lower density of resident population and by economic activities carried on in daytime (CBD). For these two groups the preferred areas are those close to the railway station and the city centre, as well as those close to the main access routes to these areas [25] [26]. Results obtained applying the Location Quotient to different subgroups can be confirmed by means of the Kernel Density Estimation.

4.4   Density Analysis by Kernel Density Estimation

This method was applied to the different events' distributions characterized by the spatial distribution of the four ethnic groups starting from their address-point locations (Fig. 4). The three-dimensional surfaces obtained allow confirming and better visualizing the information obtained with the Location Quotient.

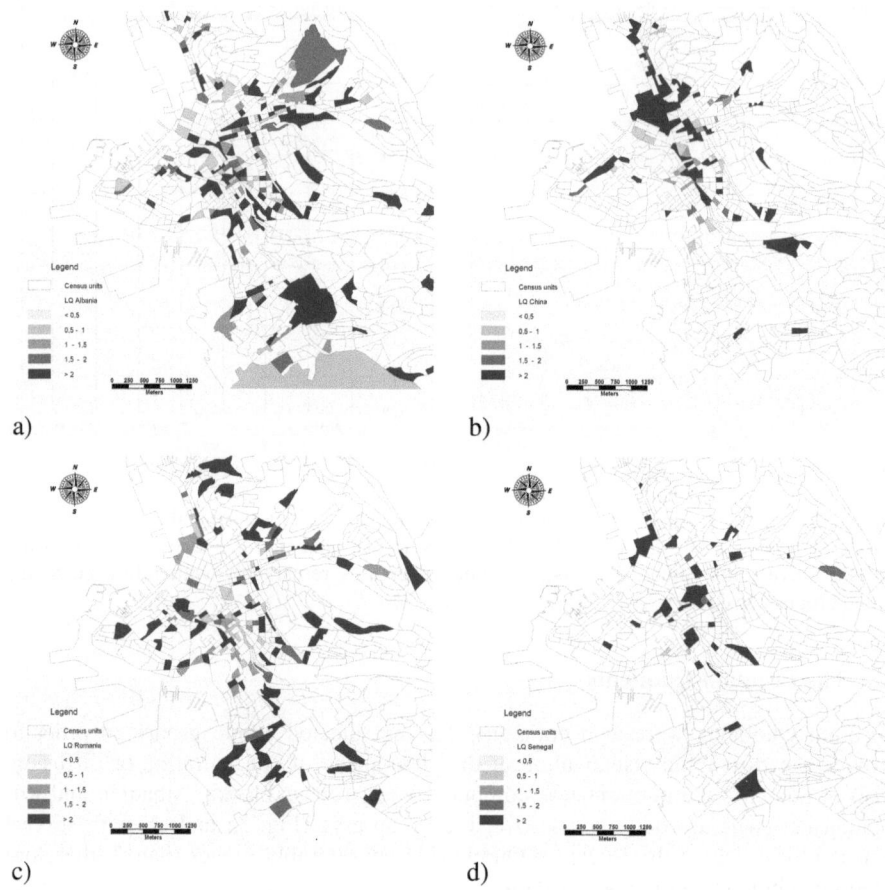

Fig. 3. Maps – zoomed in the central area of the city – of Location Quotients computed on census blocks for different ethnic groups. Index computed for immigrant residents from Albania (*a*), China (*b*), Romania (*c*) and Senegal (*d*). Note that Chinese and Senegalese residents present a more clustered pattern of the index.

With particular reference to the locations of Chinese and Senegalese groups (Fig. 4b and d) it is possible to notice their clustering in different parts of the city and an almost overlapping area in one of the peaks in the two distributions, corresponding to the less populated central area of the city[7].

---

[7] It has been often suggested to test different bandwidths according to the data distribution, particularly on the size of the study area or the number of points, and according to a researcher's aim and scale of observation [22]. Recently some authors [30] tested a k-nearest neighbor approach, with the bandwidth related to the mean nearest neighbor distance for different orders of k.

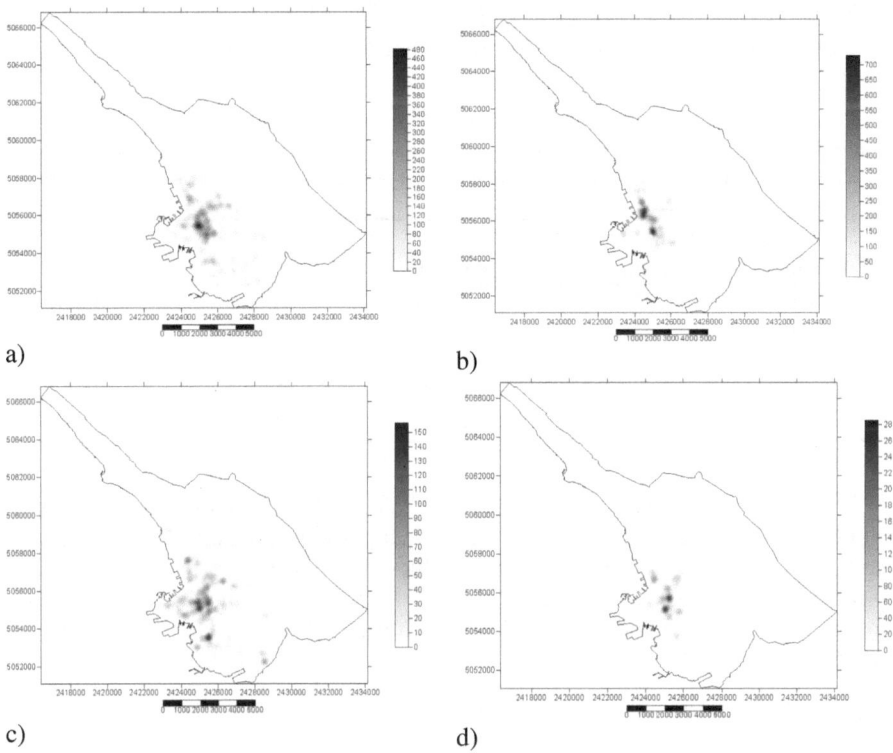

Fig. 4. Maps of Kernel Density Estimation calculated for four selected national groups in the Municipality of Trieste: Albanese (*a*), Chinese (*b*), Romanian (*c*) and Senegalese (*d*). The function used is quartic with a 300-m bandwidth and a 50-m cell size and the results are expressed as probability density distributions. Map scale in meters (Reproduced from [42]).

Quartic Kernel Density Estimations have been performed using a 300-m bandwidth and 50-m cell size[8]. The other two national groups are more dispersed over the Municipal territory, presenting therefore lower values in terms of intensity in the area of their higher presence.

4.5 The *S* index of Segregation

The index of segregation *S* was then computed again following O' Sullivan and Wong [11] procedure, that implying using a quartic Kernel Density Estimation over the study region with different bandwidths and summing the values obtained at grid cell size level. The index obtained is less dependent on the zoning system chosen, as it is based on aggregated values from uniform 50-m grid cells used as sampling locations. A difference with O' Sullivan and Wong method is given by the data chosen for the

---

[8] With reference to the choice of the grid, generally a resolution substantially smaller than the bandwidth by a factor of 5 or more and minimally by a factor of 2 has little effect on the density estimate [11].

analysis, as here the Kernel Density Estimation for the different ethnic groups was performed over data available at address-point level rather than on census blocks[9]. Table 1 shows the results obtained for the index S using different bandwidths.

Table 1. Segregation values measured in the Trieste Municipality for different ethnic groups. Note the different decreasing values of the index for different groups as bandwidth increases.

| Kernel Bandwidth (m) | Segregation, S | | | |
	Albania	China	Romania	Senegal
150	0.757	0.868	0.774	0.814
300	0.641	0.803	0.625	0.815
450	0.583	0.771	0.548	0.766
600	0.546	0.744	0.495	0.729
900	0.502	0.731	0.438	0.731

One element to be noticed is the starting values of the 150-m bandwidth, not dissimilar to those already observed in Fig. 2, ranking very similarly the four groups to each other. It can be also noticed that values decrease as the bandwidth is incremented, not an unexpected results as we already pointed out the higher distance tend to dilute the data into wider areas, similarly to what observed in the 'traditional' segregation index (Fig. 5).

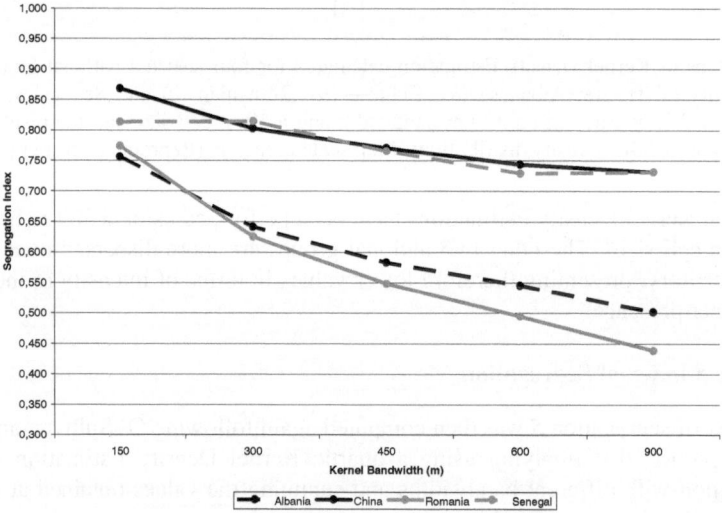

Fig. 5. Variation of the Segregation values measured in the Municipality of Trieste for different ethnic groups as bandwidth increases (Reproduced from [42])

---

[9] For the Municipality of Trieste the differences in the three-dimensional surfaces obtained from the two different spatial elements were minimal. On bandwidth's choice, O'Sullivan and Wong [11] propose calculating the average nearest neighbour distances between census blocks.

However, interestingly decay functions characterize the different groups, with Chinese and Senegalese presenting the higher initial values and with a smooth decrease of the index as the bandwidth increases, while both Albanian and Romanian people present a sharper decrease in the index when bandwidth increase. This says something more about the characteristics of settling of the different nationals, with a higher possible mix for the European nationals considered here if compared to the Asian and African ones.

The computation of index S gives also the opportunity to produce intermediate maps, providing a disaggregated view of the local contribution of each cell of the grid covering the area to the overall segregation. Maps in Fig. 6 present the numerator of equation 4, showing maximum minus minimum probability density values for each location between Italians and the four other national groups analyzed in this paper. Darker areas are those with higher concentrations of immigrants from the different countries and these areas highly contribute to the overall segregation, while light-color regions are those more prone to showing a higher level of ethnic mix.

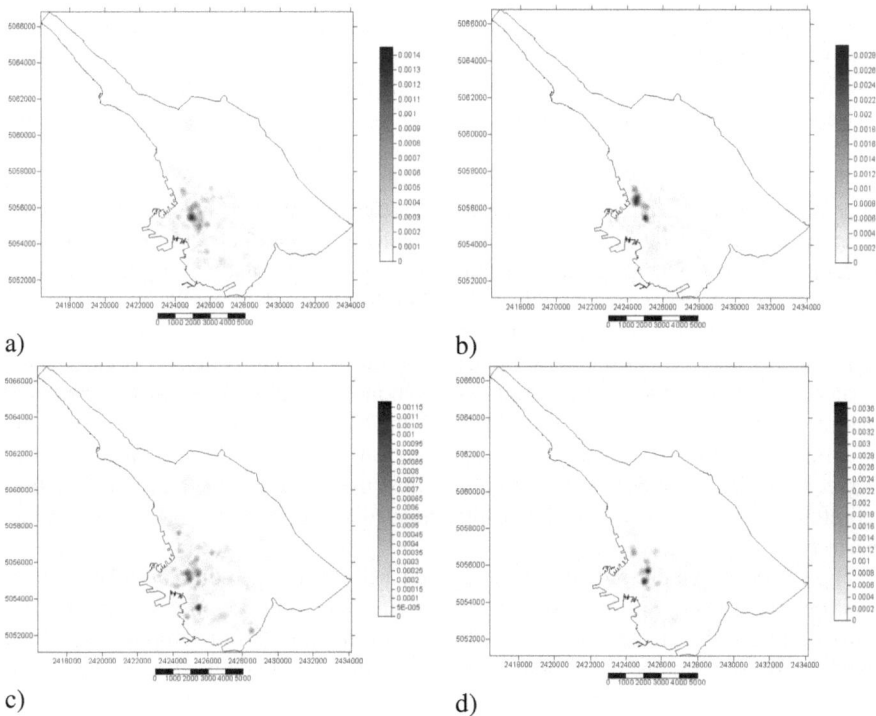

Fig. 6. Maps of maximum population proportions minus minimum population proportions for the four ethnic group considered in the Municipality of Trieste. Maps are derived from data as in Fig. 4 and the density analysis performed over Italian nationals (not portrayed in a map here) and present respectively the difference maps for Albanian (*a*), Chinese (*b*), Romanian (*c*) and Senegalese (*d*) groups. Dark areas are those that most contribute to the overall segregation white lighter ones can be interpreted as those presenting a higher population mix. Map scale in meters (Reproduced from [42]).

4.6 The Shannon and Simpson Indices of Diversity

Shannon and Simpson indices of diversity were then computed in order to analyze more in detail some of the qualitative characteristics of the foreign population in the urban area, and particularly the presence of different nationalities and their figures. The indices previously observed maintain some limitations, due to the limited number of categories that can be analyzed at a time. For instance, both location quotients and Kernel Density Estimation can be performed on one migrant group at a time, while more complex realization of the segregation index need to cross different datasets repeating several times similar elaborations.

Diversity indices on the contrary consider the variability of the categories in the dataset and can be applied on area features. In fact a first level of analysis implied aggregating the individual data at census unit level. In order to have a flavor of the characteristics of the population under exam, it was decided both to observe the behavior of the two indices considering the overall population (Fig. 7a and 7b) and also to consider just the foreign nationals (Fig. 7c and 7d) to evaluate the possible formation of ghetto areas.

Shannon and Simpson diversity indices present high values in central parts of the city where we already noticed the concentration of the selected four ethnic groups considered before and generally of the foreign population [25] [26], that is the area between the city centre and partially overlapping the CBD, the major sea-side routes and the popular districts - North and South-East from the city centre. Lower values can be registered in fringe and suburban residential areas. Diversity values are high in these areas when the analysis is computed on the overall nationals, dramatically increasing when considering diversity between foreign nationals. In such sense it is difficult to detect a neat 'local' variation in census units and the figures present a widespread mix and diversity at all level and in different urban contexts.

A second analysis implied a more intensive computation, as the two indices were performed on the overall database of residents in the Municipality of Trieste and estimated over a continuous space, following a concept common in density analysis as quadrat count and Kernel Density Estimation [24], that implies gridding the study area and assigning an estimate of the computation to each cell. The database consists of records of individuals – only containing information as sex, age and nationality – georeferenced at address-point level, that needed to be aggregated in a continuous-like way.

The estimates of the indices are computed using a fixed radius circular "moving window" for each sampling point, here represented by address-points of the Municipality of Trieste, as stated before, in order to limit the number of sampling locations - therefore displaying the progressive transition from a diversified zone to a more homogeneous one, not depending on any type of spatial zoning system.

In the first step for each location a circular window is drawn, in this case using a 150-m radius[10], population data at address-point level are counted, summarized by

---

[10] It would have been interesting to experiment different radii in order to explore the variation of the indices at different scales, as results provided vary when the size of the window is modified. However, at this stage of the research the interest was particularly on micro-scale analysis and on methodological aspects, chosing not to consider more generalised results deriving from an increase in the window size.

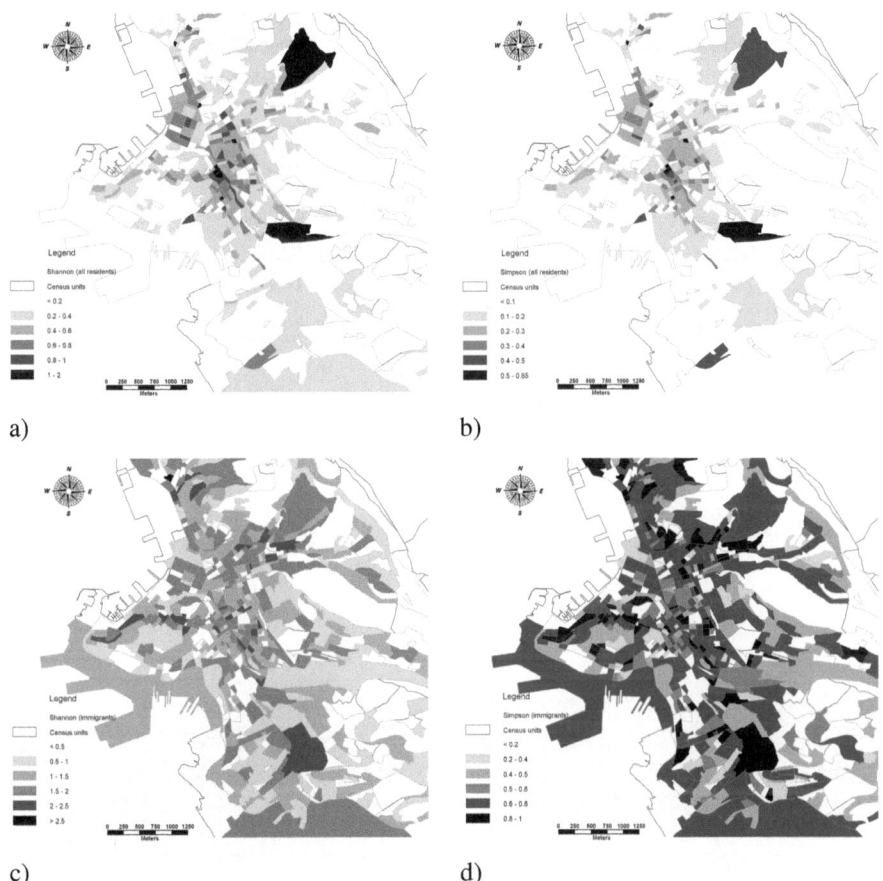

Fig. 7. Maps – zoomed in the central area of the city – of the Shannon Diversity Index (a) and Simpson evenness Index (b) computed over all the residents in the Municipality of Trieste, compared to the same two indices (c, d) computed only on foreign residents. The indices are computed after aggregating population data at census units' level. Dark areas present high number of countries represented and high presence of migrant population, while lighter ones have less countries and a lower number of foreign national as well. Map scale in meters.

nationality and assigned to the centre of the window - in our case the address point used as a sampling location. The sampling locations present therefore a matrix-like structure similar to that observed for the census units, although the nationality counts are referred to those people living within 150-m from the sampling location, therefore allowing a smoothed transition through zones, as the moving window keeps information on adjacent areas.

In a second step the values metrics are interpolated and a raster grid is created. For visualization purpose the grid cell size was set at 50-m

Also in this case the Shannon and Simpson diversity indices were computed on both the overall population and on a subset of the database, consisting of the non-Italian nationalities. Here also the aim was to study the overall characters of the spatial distribution of population also from a categorical point of view and to observe if foreign nationals tend to form non-mixed clusters or enclaves – ghettoes.

The results are visible in Fig. 8, with Fig. 8a and 8c representing the Shannon index computed on the overall population and on foreign nationals, while Fig. 8b and 8d offer results for the Simpson index at the two levels of analysis.

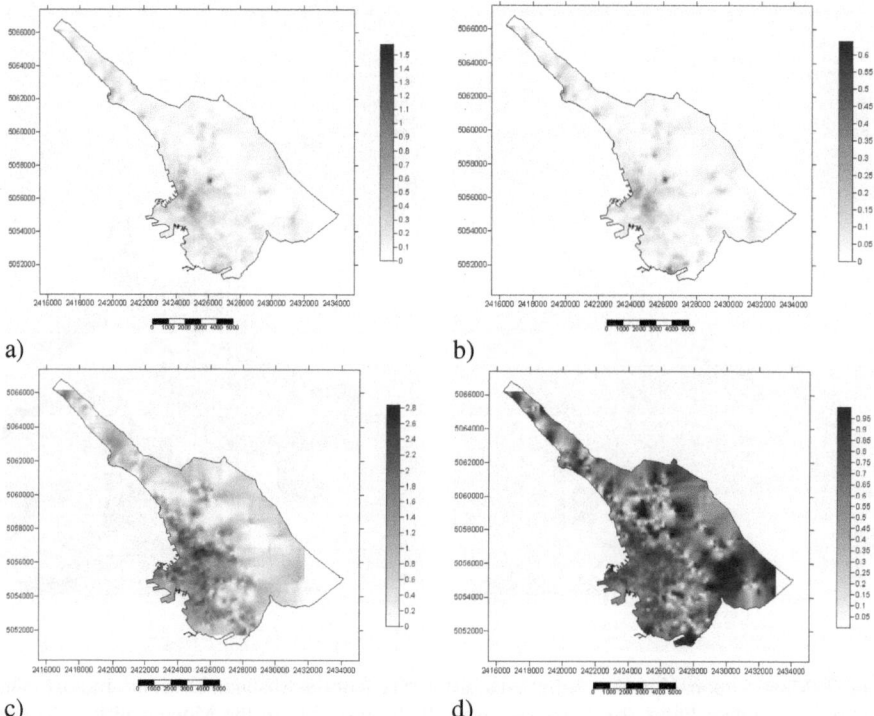

a)   b)

c)   d)

Fig. 8. Maps of the Shannon (a) and Simpson Diversity Indices (b) computed over all the residents in the Municipality of Trieste, compared to the same two indices (c, d) computed only on foreign residents. The indexes are computed starting from address-points as sampling locations, using a 150-m moving window sampling all residents for each population group. Dark areas present high number of countries represented and high presence of migrant population, while lighter ones have less countries and a lower number of foreign national as well. Map scale in meters.

The results confirm and enhance the message already suggested by the indices computed over the census units. It can be noticed that the indices of diversity present quite similar visual results when applied to the complete set of nationalities (Fig. 8a and 8b), displaying an overall trend of high values of diversity on both Shannon and Simpson indices, higher than those obtained by the analysis of census units (Tab. 2). The peaks in the values can be noticed in two areas of the city where foreign population tend to concentrate, one in the city centre, approximately close and partially overlaying the CBD limit and characterized by lower population density but a certain number of foreign nationals, and one North and South-East from the city centre, where a high density of population – both Italian and migrant – is registered. Out of the central districts of the city diversity indices values, as well as population density, decrease. A partial conclusion is that at this scale of analysis we register high values and therefore a limited tendency to single clusters or ghettoes. This is also confirmed by the analysis on the foreign nationals (Fig. 8c and 8d), therefore excluding Italians by the computation. Results are not so visually appealing and easy to detect, although what can be noticed is that the two indices present generally very high values in the entire municipality, therefore both in areas of high or low population densities and in areas where foreign nationals are both numerous and scarce. Although we have noticed in some ethnic group a tendency to cluster, these grouping patterns can be in some part overlapping, therefore diluting the risk of forming enclaves or ghettoes. Absolute diversity values display a general heterogeneity mainly in areas when migrants are numerous and a higher homogeneity in residential districts where Italian nationals live.

Table 2. Shannon and Simpson diversity indices computed over different units – moving window on address point data, census units and Municipality area. Note close results in maximum values in census units and address-point analysis.

Summary statistics (address-point data)	SHDI (All nationalities)	SIDI (All All nationalities)	SHDI (foreign nationals)	SIDI foreign nationals)
min not null	0.0101	0.0026	0.2237	0.1107
max	1.9864	0.7975	3.1296	1.0000
max (census units)	1.1610	0.6160	2.5275	0.9100
municipality values	0.3336	0.0974	2.6805	0.8143

4.7 The Index of Diversity (IDiv)

The last index implemented is the IDiv to test and represent synthetically some of the differences and characteristics of the immigration phenomenon as a whole and, instead of measuring the elements of separation it considers those potentially representing union and cohabitation as the cultural and ethnic mix.

As in the case of the Shannon and Simpson indices, the index implemented here is not limited to the four ethnic groups analyzed, but considers the overall foreign residents in the Municipality of Trieste, counting both the number of countries represented and the number of residents for each country. Fig. 9 portrays the results for the index of diversity computed over the data, according to equation 5, aggregated at census block level.

Two different bandwidths were analyzed, using nearest neighbor index over census blocks' centroids to determine the different orders of medium average distance of blocks. A 177-m bandwidth, corresponding to grade $k = 2$ (Fig. 9a) and a 281-m one, referred to $k = 5$ (Fig. 9b) were experimented to see the ethnic mixing at two different

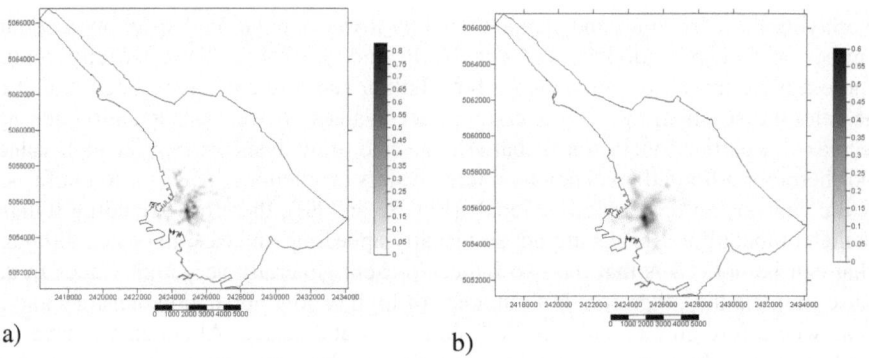

a)   b)

Fig. 9. Maps of the Index of Diversity (*IDiv*) computed over the immigrant groups in the Municipality of Trieste. The index highlights dark areas with high number of countries represented and high presence of migrant population, while lighter ones have less countries and a lower number of foreign national as well. Map scale in meters (Reproduced from [42]).

scales. It is interesting to notice the high value of the index in areas where some of the groups tend to cluster. This is also the more populated area, and the index helps in highlighting also population diversity in terms of ethnic groups represented, therefore limiting some of the possible suspects of true segregation. Furthermore it can be noticed that also other areas where segregation values for some ethnic groups are high, still present some mix that, given the lower number of population living in it, can still be attributed to a noticeable presence of different ethnic groups in the area.

5 Conclusions and Discussion

In this paper a summary of some of the most used indices for measuring segregation or diversity in the distribution of migrant groups at urban level have been proposed, considering in particular the spatial aspects of such indices and the need to examine more in depth the articulated structure and characteristics of the population. Some problems still need to be addressed. On one side limitations can be noticed in the availability of disaggregated data, as individual nationals from different countries are often aggregated according to some zoning system that can affect the results from further analyses. However, if the zoning system produces sufficiently small areas some analytical methods reduce such problem. With reference to the indices used, an issue is still concerned with the choice of the bandwidth or distances of observation, although efforts in this direction are under exam [11] [25]. Furthermore, qualitative, multivariate attribute of population data should be considered. There is also the need to go more in depth with entropy-based diversity indices, as well as to examining the relations between economic activities, residential locations and segregation [25] [31] [32] as emerging migration issues to analyse [33] [34]. There is still the need to consider qualitatively the presence of foreign nationals, as the analysis carried on here, particularly in terms of diversity analysis, did not discriminate between 'population at risk' and foreign nationals having a consolidated relation with the particular urban area or coming from industrialized countries (USA, EU, etc.), whose housing and working spatial behaviours can be similar to Italian nationals, whose characters of heterogeneity differ from those that characterize deprived areas where another kind of migration arise. A partial conclusion concerning the use of quantitative methods, and particularly those based on density or on a choice of distances, it is worth noting that these must be refined and that not 'easy' solutions of the problem at stake can be found. Nevertheless they provide a good starting point for more in depth and local analysis by the researchers that can focus their attention over a micro scale of analysis, going further than the administrative divisions of space and reducing the minimum distance of observation to examine locally the dynamics at urban scale. Diversity measures appear to be promising in helping scholars in addressing issues concerning migrations but also open to other fields of human activity. Particularly it seems interesting to continue the research on the implementation of diversity indices over a continuous space, possibly extending the procedure presented and applied in this paper with density analysis - KDE in particular - in order to combine the power of the different methods.

References

1. Krasna, F.: Alcune considerazioni critiche sull'evoluzione delle teorie e dei metodi di analisi dei processi migratori. In: Nodari, P., Krasna, F. (eds.) L'immigrazione straniera in Italia. Casi, metodi e modelli. Geotema, vol. 23, pp. 129–134. Pàtron, Bologna (2006)
2. Tobler, W.R.: Smooth pycnophylactic interpolation for geographical regions. Journal of the American Statistical Association 74, 121–127 (1979)
3. Cristaldi, F.: Roma città plurale: dal diritto alla casa alla segregazione spaziale degli immigrati. In: Nodari, P., Krasna, F. (eds.) L'immigrazione straniera in Italia. Casi, metodi e modelli. Geotema, vol. 23, pp. 16–25. Pàtron, Bologna (2006)
4. Duncan, O.D., Duncan, B.: A Methodological Analysis of Segregation Indexes. American Sociological Review 20, 210–217 (1955)
5. Cristaldi, F.: Multiethnic Rome: Toward residential segregation? GeoJournal 58, 81–90 (2002)
6. Petsimeris, P.: Urban decline and the New Social and Ethnic Divisions in the Core Cities of the Italian Industrial Triangle. Urban Studies 3, 449–465 (1998)
7. Miani-Uluhogian, F.: Considerazioni geografiche sulla transizione multirazziale. In: Brusa, C. (ed.) Immigrazione e multicultura nell'Italia di oggi, Franco Angeli, Milano, vol. I, pp. 338–362 (1997)
8. Miani, F., Fedeli, K.: Aree urbane e immigrazione: la divisione etnica nella città di Piacenza. In: Brusa, C. (ed.) Immigrazione e multicultura nell'Italia di oggi, Franco Angeli, Milano, vol. II, pp. 400–413 (1999)
9. Borruso, G., Donato, C.: L'immigrazione straniera a Trieste – I principali impatti sulla situazione socio-economica e sul tessuto urbano. Quaderni del Centro studi economico-politici. "Ezio Vanoni" Trieste, 3–4 (2003)
10. Massey, D.S., Denton, N.A.: The dimensions of residential segregation. Social Forces 67, 281–315 (1988)
11. O' Sullivan, D., Wong, D.W.S.: A Surface-Based Approach to Measuring Spatial Segregation. Geographical Analysis 39, 147–168 (2007)
12. Wong, D.W.S.: Spatial Dependency of Segregation Indices. The Canadian Geographer 41, 128–136 (1997)

13. Haggett, P.: Locational Analysis in Human Geography. Edward Arnold, London (1965)
14. Jakubs, J.F.: A Distance-Based Segregation Index. Journal of Socio-Economic Planning Sciences 15, 129–141 (1981)
15. Morgan, B.S.: An Alternative Approach to the Development of a Distance-Based Measure of Racial Segregation. American Journal of Sociology 88, 1237–1249 (1983)
16. Morrill, R.L.: On the Measure of Geographical Segregation. Geography Research Forum 11, 25–36 (1991)
17. Wong, D.W.S.: Spatial Indices of Segregation. Urban Studies 30, 559–572 (1993)
18. Silverman, B.W.: Density Estimation for Statistics and Data Analysis. Chapman Hall, London (1986)
19. Brunsdon, C.: Analysis of Univariate Census Data. In: Openshaw, S. (ed.) CensusUsers Handbook, pp. 213–238. GeoInformation International, Cambridge (1995)
20. Diggle, P.J.: The Statistical Analysis of Point Patterns. Academic Press, London (1983)
21. Diggle, P.J.: A Kernel Method for Smoothing Point Process Data. Applied Statistics— Journal of the Royal Statistical Society Series C 153, 349–362 (1985)
22. Bailey, T.C., Gatrell, A.C.: Interactive Spatial Data Analysis. Longman, Harlow (1995)
23. Levine, N.: CrimeStat III: A Spatial Statistics Program for the Analysis of Crime Incident Locations. Ned Levine & Associates, Houston, TX, and the National Institute of Justice, Washington, DC (2004)
24. O' Sullivan, D., Unwin, D.J.: Geographic Information Analysis. John Wiley & Sons, Chichester (2003)
25. Borruso, G., Donato, C.: Caratteri localizzativi dell'immigrazione straniera a Trieste: i principali aspetti della struttura demografica e abitativa. In: Nodari, P., Rotondi, G. (eds.) Verso uno spazio multiculturale? Riflessioni geografiche sull'esperienza migratoria in Italia, pp. 129–163. Pàtron, Bologna (2007)
26. Borruso, G., Schoier, G.: Metodi di analisi e visualizzazione di fenomeni immigratori. In: Nodari, P., Krasna, F. (eds.) L'immigrazione straniera in Italia. Casi, metodi e modelli. Geotema, vol. 23, pp. 105–114. Pàtron, Bologna (2006)
27. Bracken, I.: Population-related social indicators. In: Fotheringham, S., Rogerson, P. (eds.) Spatial Analysis and GIS, pp. 247–259. Taylor & Francis, London (1994)
28. Wong, D.W.S.: Geostatistics as Measures of Spatial Segregation. Urban Geography 20, 635–647 (1999)
29. Reardon, S.F., O'Sullivan, D.: Measures of Spatial Segregation. Sociological Methodology 34, 121–162 (2004)
30. Williamson, D., McLafferty, S., Goldsmith, V., Mollenkopf, J., McGuire, P.: A better method to smooth crime incident data. ESRI ArcUser Magazine (January – March 1999)
31. Glasmeier, A.K., Farringan, T.L.: Landscapes of Inequality: Spatial Segregation, Economic Isolation, and Contingent Residential Locations. Economic Geography 83, 221–229 (2007)
32. Ellis, M., Wright, R., Parks, V.: Geography and the Immigrant Division of Labor. Economic Geography 83, 255–281 (2007)
33. Nodari, P., Krasna, F. (eds.): L'immigrazione straniera in Italia. Casi, metodi e modelli. Geotema, vol. 23. Pàtron, Bologna (2006)
34. Nodari, P., Rotondi, G. (eds.): Verso uno spazio multiculturale? Riflessioni geografiche sull'esperienza migratoria in Italia. Pàtron, Bologna (2007)
35. Nagendra, H.: Opposite trends in response for the Shannon and Simpson indices of landescape diversity. Applied Geography 22, 175–186 (2002)
36. Weaver, W., Shannon, C.E.: The Mathematical Theory of Communication. University of Illinois, Urbana (1949)

37. Shannon, C.E.: A mathematical theory of communication. Bell System Technical Journal 27, 379–423, 623–656 (1948)
38. Simpson, E.H.: Measurement of diversity. Nature 163, 688 (1949)
39. Peet, R.K.: The measurement of species diversity. Annual Review of Ecology and Systematics 5, 285–307 (1974)
40. Elden, G., Kayadjanian, M., Vidal, C.: Quantifying Landscape Structures: spatial and temporal dimensions (ch. 2). From Land Cover to Landscape Diversity in the European Union (2000), http://ec.europa.eu/agriculture/publi/landscape/ch2.htm
41. Gibbs, J.P., Martin, W.T.: Urbanization, technology and the division of labor. American Sociological Review 27, 667–677 (1962)
42. Borruso, G.: Geographical Analysis of Foreign Immigration and Spatial Patterns in Urban Areas: Density Estimation and Spatial Segregation. In: Gervasi, O., Murgante, B., Laganà, A., Taniar, D., Mun, Y., Gavrilova, M.L. (eds.) ICCSA 2008, Part I. LNCS, vol. 5072, pp. 459–474. Springer, Heidelberg (2008)

Geostatistics in Historical Macroseismic Data Analysis

Maria Danese[1,2], Maurizio Lazzari[1], and Beniamino Murgante[2]

[1] Archaeological and monumental heritage institute, National Research Council, C/da S. Loia Zona industriale, 85050, Tito Scalo (PZ), Italy
m.danese@ibam.cnr.it, m.lazzari@ibam.cnr.it
[2] Laboratory of Urban and Territorial Systems, University of Basilicata, Via dell'Ateneo Lucano 10, 85100, Potenza, Italy
maria.danese@unibas.it, beniamino.murgante@unibas.it

Abstract. This paper follows a geostatistical approach for the evaluation, modelling and visualization of the possible local interactions between natural components and built-up elements in seismic risk analysis. This method, applied to old town centre of Potenza hilltop town, offers a new point of view for civil protection planning using kernel density and autocorrelation indexes maps to analyse macroseismic damage scenarios and to evaluate the local geological, geomorphological and 1857 earthquake's macroseismic data.

Keywords: Geostatistics, Nearest-Neighbour Index, Kernel Density Estimation, Spatial Autocorrelation, Macroseismic data analysis, Southern Italy.

1 Introduction

According to the first principle of geography "Nearest things are more related then distant things" [1], spatial analysis studies the aggregation status of spatial phenomena and their relationships.

There are four main families of spatial techniques: interpolation techniques, density based measures, distance based measures and autocorrelation measures. These techniques consider phenomena as spatial processes, model them as points characterized by a couple of x, y coordinates and study their spatial distribution properties.

A phenomenon is spatially stationary and its events are independent if these properties (expected value of the process) do not vary in the studied region [2].

A phenomenon is no more stationary when it is affected by first and second order effects. The probability to find an event in one point of the space, according to first order effects, depends on the properties of the studied region. This probability, according to the second order effects, also depends on the presence of other events within a distance which has to be defined according to a proximity criterion.

It is very difficult to distinguish and clearly separate first and second order effects in spatial phenomena. In particular, according to traditional literature ([3] [4]), first order effects are studied by density-based measures, while second order effects are studied by distance-based measures.

M.L. Gavrilova and C.J.K. Tan (Eds.): Trans. on Comput. Sci. VI, LNCS 5730, pp. 324–341, 2009.
© Springer-Verlag Berlin Heidelberg 2009

Only recently, the geographical approach to spatial analysis was followed to analyse seismic risk [5], [6]. In particular, the study of first and second order effects in this field of application is useful to better understand how seismic effects vary locally. The authors underline that first order effects are associated to local geological-geotechnical and geomorphological site characteristics, whereas second order effects consider the relationships between single buildings according to their relative position.

The final aim is to have a spatial support, using distance and density measures and autocorrelation indexes, in defining urban areas historically most exposed to seismic hazard.

2   Geostatistical Analysis

2.1   Distance Based Measures: The Nearest-Neighbour Index

Nearest-Neighbour Index is the most common distance-based method and it provides information about the interaction among events at the local scale (second order property). *Nearest-Neighbour Index* considers nearest neighbour event-event distance, randomly selected. The distance between events can be calculated using Pythagoras theorem:

$$d(s_i, s_j) = \sqrt{(x_i - x_j)^2 + (y_i - y_j)^2} .$$
(1)

Nearest-Neighbour Index is defined by the following equation:

$$NNI = \frac{\overline{d}_{min}}{d_{ran}}$$
(2)

The numerator of equation (4) represents the average of N events considering the minimum distance of each event from the nearest one, and it can be represented by:

$$\overline{d}_{min} = \frac{\sum_{i=1}^{n} d_{min}(s_i, s_j)}{n}$$
(3)

where d_{min} (S_i,S_j) is the distance between each point and its nearest neighbour, and n is the number of points in the distribution.

The denominator can be expressed by the following equation:

$$d_{ran} = 0.5\sqrt{\frac{A}{n}} .$$
(4)

where, n is the distribution of number of events and A is the area of the spatial domain. This equation represents the expected nearest neighbour distance, based on a completely random distribution.

This means that when NNI is less than 1, mean observed distance is smaller than expected distance, then one event is closer to each other one than expected. If NNI is greater than 1, mean observed distance is higher than the expected distance and therefore events are more scattered than expected.

2.2 KDE Technique: Concepts and Methods

Given N points s_1, ...s_N, characterized by their x and y coordinates, it is possible to estimate point distribution probability density function, otherwise a naive estimator method by Kernel Density Estimation (KDE) technique, related to spatial cases. Both procedures consider density as a continuous function in the space, even though they calculate it in a different way.

Naive estimator approach can be followed drawing a circle with radius r around each point pattern and dividing point number inside the circle by its area. Accordingly, this result is a function characterized by points of discontinuity. Nevertheless, naive estimator does not allow to assign different weights to events.

KDE is a moving three-dimensional function, weighting events within their sphere of influence according to their distance from the point at which intensity is being estimated [2]. The method is commonly used in a more general statistical context to obtain smooth estimates of univariate (or multivariate) probability densities from an observed sample of observations [7].

The three-dimensional function is the kernel one. It is k(x)≥0 , with

$$\int k(x)dx = 1. \tag{5}$$

It is characterized by unimodality, smoothness, symmetry, finite first and second moments, etc. [8] and it is always non-negative.

The consequence is that kernel density is an always non-negative parameter too, so that it is defined by the following expression (2) in each spatial point:

$$\hat{\lambda}_\tau(s) = \frac{1}{\delta_\tau(s)} \sum_{i=1}^{n} \frac{1}{\tau^2} k\left(\frac{(s-s_i)}{\tau}\right). \tag{6}$$

Such density has been called "absolute" by Levine [9], who identifies other two forms in which density can be expressed: the first one is a relative density, which is obtained by dividing absolute density by cell area; the second one is a probabilistic density, where output raster is obtained by dividing absolute density by point pattern's event number.

For all the reasons discussed above, KDE is a function of the choice of some key parameters, such as grid resolution, kernel function and, above all, bandwidth.

2.2.1 Kernel Function

The choice of kernel function is the first important problem in KD estimation, since how each point will be weighted for density estimation depends on it. Nevertheless, although several authors have discussed this topic, a lot of them think that the kernel function is less important than bandwidth choice [10], [7].

In most cases weight attribution has been carried out on the basis of an Euclidean distance. Only during the last few years a new concept of distance in a network space has been considered [11], [12].

The most important kernel function types are: Gaussian kernel, triangular kernel [13], quartic kernel [14] and Epanechnikov's kernel [15].

Through these functions, events are weighted in an inversely proportional way relative to distance from landmarks and directly proportional relative to the way in which the specific function converges to zero or vanishes. An exception to this logic is represented by the negative exponential function where weight is given proportionally to the distance [9], [16].

2.2.2 Grid Resolution

Grid resolution choice is a similar problem to that of 'bin' choice in histogram statistical representation [7], although it is a less important choice than the one relative to bandwidth, since location effect is negligible [4]. Generally, cell size definition is linked to case study, as it occurs, for example, in network density estimation (NDE) case, which can be determined in order to obtain a grid superimposed onto the road network junctions [11], or in representing the scale of analysed case, or for bandwidth choice. In particular, according to O'Sullivan and Wong [4] a cell size smaller than bandwidth by a factor of 5 or more and minimally by a factor of 2 provides a little effect on density estimation.

2.2.3 Bandwidth

The most important problem in KDE is the choice of the smoothing parameter, the bandwidth, present either in univariate cases or in multivariate spatial ones.

The importance of bandwidth is closely linked to a base concept well expressed by Jones et al. [17]: when insufficient smoothing is done, the resulting density or regression estimate is too rough and contains spurious features that are artefacts of the sampling process; when excessive smoothing is done, important features of the underlying structure are smoothed away.

During the last twenty years, several studies have discussed this topic [18], [19], [20], [21], by which two basic approaches to determinate bandwidth have been used: the first approach defines a fixed bandwidth to study all the distribution, while the second one uses an adaptive bandwidth becoming in the very end a type of fourth dimension of KDE.

As concerns fixed bandwidth, the main problem is defining the right value. One of the most used methods to define this value is the nearest neighbour mean, which represents an attempt to adapt the amount of smoothing to local density of data [7]. Fix and Hodges [21] first introduced nearest neighbour allocation rules for non-parametric discrimination. Afterwards other contributions came by Loftsgaarden and Quesenberry [22], and Clark and Evans [23], who extended Silverman's concept to the use of nearest neighbour of k order, beside the reviews of Cao et al. [24], Wand and Jones [25], Simonoff [26], Chiu [27] and Devroye and Lugosi [28]. A synthesis of main methods for the choice of fixed bandwidth has been developed by Jones et al. [17], who defined two families: the first generation includes methods such as least performance rules of thumb, least squares cross-validation and biased cross-validation; the second one includes superior performance, solve equation plug-in and smoothed bootstrap methods.

Distance analysis among events generally represents an alternative to density-based measures, but in several cases, in particular the Nearest Neighbour Distance, could be an input datum for KDE [29].

The second approach of adaptive bandwidth appears more suitable to adapt the amount of smoothing to local density data, as often occurs, for example, working with human geographical data [30]. Several contributions regarding this topic have been published during the last twenty-five years, such as Abramson [31], Breiman et al. [32], Hall and Marron, [33], Hall et al. [34], Sain and Scott [35] and Wu [36]. The estimate is constructed similarly to kernel estimate with fixed bandwidth, but the scale parameter placed on data points is allowed to vary from one data point to another.

In mathematical terms, density estimation function with adaptive bandwidth becomes [14]:

$$\hat{\lambda}_\tau(s) = \frac{1}{\delta_\tau(s)} \sum_{i=1}^{n} \frac{1}{\tau^2(s_i)} k\left(\frac{(s-s_i)}{\tau(s_i)}\right).$$

(7)

where $\tau(s_i)$ is bandwidth value for the event i.

A lot of authors define two kinds of adaptive kernel density:

- the first one is based on a bandwidth calibrated on the case study;
- the second one is based on point number to be included in bandwidth and, therefore, on k nearest neighbours [7], [8].

2.2.4 Intensity

The simplest explanation of what intensity is, can be achieved considering a sort of third dimension of point pattern connected to the case study nature [37], [3], [13], [38].

It is important to pay attention to the difference between the intensity concept of a single event and the intensity of the estimated distribution with KDE (i.e. the density of the examined process).

While the first one is a measure identifying event strength [39], the second is expressed by the following limit:

$$\hat{\lambda}_\tau(s) = \lim_{ds \to 0}\left\{\frac{E(Y(ds))}{ds}\right\}.$$

(8)

where ds is the area determined by a bandwidth also vanishing, E() is the expected average of the number of events in this region, Y is a random variable. If we are in a two-dimension space this limit identifies the average of the number of events per unit area [14], while this limit will be an expression of the individual event per unit area intensity variation, when considering intensity. Therefore, intensity of the individual event and point patterns tend to coincide only when ds is constant and it is vanishing (which generally does not occur, especially in the case of adaptive bandwidth). Considering now the definition of first and second order properties of spatial distribution, it is quite evident that the intensity of an event can be affected by first and second order effects. For example, KDE produces results influenced by both effects, because it implies the presence of a bandwidth, so it is based on a distance concept, too. Concerning second order effects in KDE, it is important to choose an appropriate bandwidth taking into account intensity of events, which is the 'nature' of the studied problem.

2.2.5 Results Classification

After the application of the two methods discussed above, another important issue concerns result classification, which has not been much discussed in the literature, yet. However, it is a critical topic, because it is possible to highlight the studied phenomenon in a correct way, without overestimating or underestimating density values and area extension determined with KDE only by means of a right definition of meaningful class values, achieved with KDE.

Two methods are useful on this purpose, as suggested by Chainey et al. [40]. The first one is the incremental Standard Deviation (SD) approach; with this method, density SD value becomes the lower bound of the first class in the output raster and next classes are calculated by incrementing it by SD unit. The second method is the incremental mean approach, where average density value is used to make result classification, instead.

2.3 The Spatial Autocorrelation

In spatial analysis points are considered as events, i.e. spatial occurrences of the studied phenomenon and their distribution in the space can follow three different patterns: 1) a random distribution, when the position of each point is independent of the others points; 2) a regular distribution, when points have a uniform spatial distribution; 3) a clustered distribution, when points are concentrated in some building clusters. The third group identifies the presence of *spatial autocorrelation*. This notion is complementary to independence concept: events of a distribution can be independent if any kind of spatial relationship exists among them.

Many indexes exist with the aim to find clusters in a point pattern. There are two kind of indexes: global autocorrelation indexes, useful to understand if a point pattern distribution is random, regular or clustered, and local indicators of autocorrelation to find where clusters are. The better solution to completely characterize the point pattern autocorrelation is to use global and local autocorrelation indexes together.

2.3.1 The Global Autocorrelation: Moran's Index

Moran index [41] is an index of global autocorrelation which takes into account the number of events occurring in a certain zone and their intensity. It is a measure of the first order property and can be defined by the following equation:

$$I = \frac{N \sum_i \sum_j w_{ij}(X_i - \overline{X})(X_j - \overline{X})}{(\sum_i \sum_j w_{ij}) \sum_i (X_i - \overline{X})^2} \tag{9}$$

where:

- N is the number of events;
- X_i and X_j are intensity values in the points i and j (with i≠j), respectively;

- $\overline{X}$ is the average of variables;
- $\sum_i \sum_j w_{ij}(X_i - \overline{X})(X_j - \overline{X})$ is the covariance multiplied by an element of the weight matrix. If X_i and X_j are both upper or lower than the mean, this term will be positive, if the two terms are in opposite positions compared to the mean the product will be negative;
- w_{ij} is an element of the weight matrix which depends on the contiguity of events. This matrix is strictly connected to the adjacency matrix.

There are different methods to determine the contiguity matrix w_{ij}. Two of the most common ones are the Inverse Distance Method and the Fixed Distance Band Method. In the first one, weights vary inversely with the distance among events:

$$w_{ij} = d_{ij}^z$$

where z is a number smaller then 0.

The second method defines a critical distance beyond which two events will never be adjacent. If events i and j are contiguous, w_{ij} will be equal to 1, otherwise it will be 0.

Moran index varies between -1 and 1; the higher it is, the more point pattern is positively autocorrelated. The null hypothesis is theoretically verified when I is equal to 0, even if it usually converges towards the theoretical mean value E(I)

$$E(I) = -\frac{1}{N-1} \tag{10}$$

The significance of Moran index can be evaluated by means of a standardized variable z (I), defined as:

$$z(I) = \frac{I - E(I)}{S_{E(I)}} \tag{11}$$

where $S_{E(I)}$ is the standard deviation from the theoretical mean value E(I).

2.3.2 The Local Autocorrelation: G Function by Getis and Ord and Local Indicator of Spatial Association (LISA)

Both G function and LISA are local indexes of spatial autocorrelation, so they provide disaggregated measures of autocorrelations. Using the same weight matrix explained in Moran's index description, these indexes estimates the number of events with homogenous occurrence and intensity included within a defined distance d.

The Local Indicator of Spatial Association [42] is defined as:

$$I_i = \frac{(X_i - \overline{X})}{S_X^2} \sum_{j=1}^N (w_{ij}(X_j - \overline{X})) \tag{12}$$

where symbols are the same used in Moran's I, except for S_X^2 which is the variance.

The function by Getis & Ord [43] is represented by the following equation:

$$G_i(d) = \frac{\sum\limits_{i=1}^{n} w_i(d)\, x_i - x_i \sum\limits_{i=1}^{n} w_i(d)}{S(i)\sqrt{\left[(N-1)\sum\limits_{i=1}^{n} w_i(d) - \left(\sum\limits_{i=1}^{n} w_i(d)\right)^2\right] \Big/ N-2}} \quad (13)$$

which is very similar to Moran's index, except for $w_{ij}(d)$ which, in this case, represents a weight which varies according to distance.

3 The Case Study

Geostatistical approach with KDE has been applied in order to reconstruct and integrate analysis of macroseismic data. Potenza hilltop town has been chosen as a sample area for this study. Potenza municipality is the chief-town of Basilicata region (southern Italy), located in the axial-active seismic belt (30 to 50 km wide) of southern Apennines, characterized by high seismic hazard, where strong earthquakes have occurred (Fig. 1). In fact, Potenza was affected at least by four earthquakes with intensity higher than or equal to VIII MCS, such as those of 1 January 1826 (VIII MCS), 16 December 1857 (VIII-IX MCS), 23 July 1930 (VI-VII MCS) and 23 November 1980 (VII MCS), of which we have wide historical documentation.

In this work we focus on the analysis on macroseismic effects occurred during the 1857 seismic event [44].

A remarkable search of unpublished data, both cartographic and of attributes, has been implemented in a geodatabase, such as a topographic map (1:4,000), a town-planning-historical map of Potenza downtown area (1875 cadastral survey), a geological map, a geomorphological map, borehole logs, a geotechnical laboratory test, geophysical data, historical macroseismic data at building scale, historical photographs of damaged buildings and plans of rebuilding (19th century). Starting from historical macroseismic data, a damage scenario has been reconstructed (Fig. 1) considering five damage levels (D_{1-5}) according to the European Macroseismic Scale – EMS [45]. We applied KDE on the basis of this scenario, in order to show the geostatistical-territorial distribution of seismic effects but also the possible relationships with substratum depth, geo-mechanical characteristics and morphological features of the site.

The study area is located on a long and narrow asymmetrical ridge SW-NE oriented, delimited along the northern sector by steep escarpments. Geologically, it is characterized by a sequence of Pliocene deposits with an over-consolidated clayey substratum on top of which a sandy-conglomerate deposit lays, which varies in thickness along both west-east and north-south directions.

The following step has been performed in order to choose parameters for KDE according to site conditions and building characteristics.

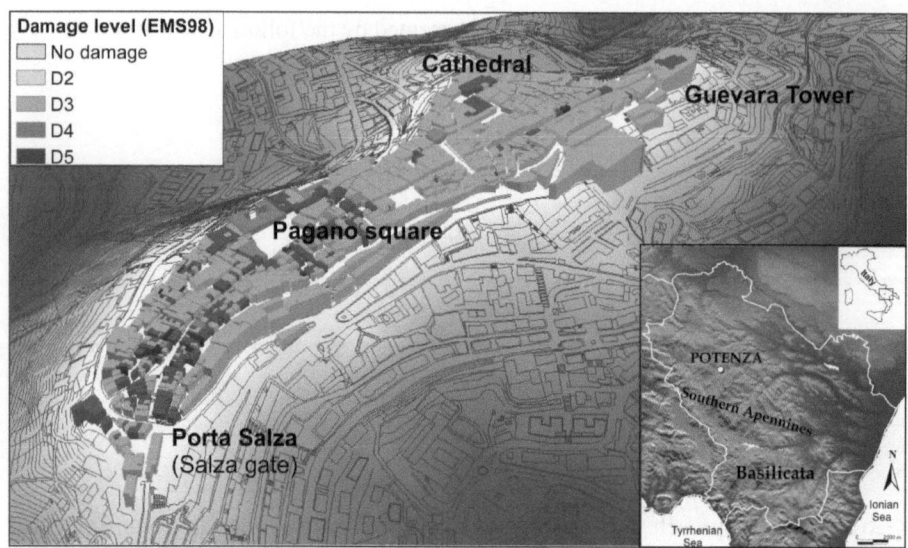

Fig. 1. Geographical location of the study area and 3D representation of the 1857 earthquake macroseismic damage scenario of Potenza hilltop town

3.1 Kernel Density Estimation and Parameters Selection

KDE has been applied as a *Point Pattern Analysis* representing the seismic damage scenario, converting each polygon (damaged buildings) to its centroid. Nevertheless, when choosing bandwidth, one must consider that input data are polygons, characterized by specific area and shape.

Intensity Choice. Intensity is the first parameter to be defined in point pattern, in order to calculate KDE. We considered the EMS98 [45] scale as the starting point to define intensity, even though it offers a limited point of view of the topic. In fact, EMS98 [45] scale considers the damaging effects according to building structure typology, as if it were an autonomous entity; at the same time, it considers that buildings are damaged in a different way, depending on first order properties.

Assigned intensity values must consider both first and second order properties and their physical meaning in this application field.

High damage levels (D4 and D5), in which total or partial structural detachments, collapses and material ejections occur, produce a decreasing in building vulnerability as a function of their reciprocal proximity and, above all, of morphological factors, such as altimetrical drops. According to these remarks, we assigned an intensity value (Table. 1) at damaged buildings (Fig. 2) in the study area of the old town centre of Potenza, as follows:

- equal to EMS levels for all buildings located in the middle-southern sector characterized by sub-horizontal or low gradient morphology;
- increased by one unit only for D4 and D5 buildings located in northern sector characterized by high gradient and steep slopes.

Table 1. Intensity values associated to single points

Damage level	Intensity in middle-southern sector	Intensity in northern sector
D1	1	1
D2	2	2
D3	3	3
D4	4	5
D5	5	6

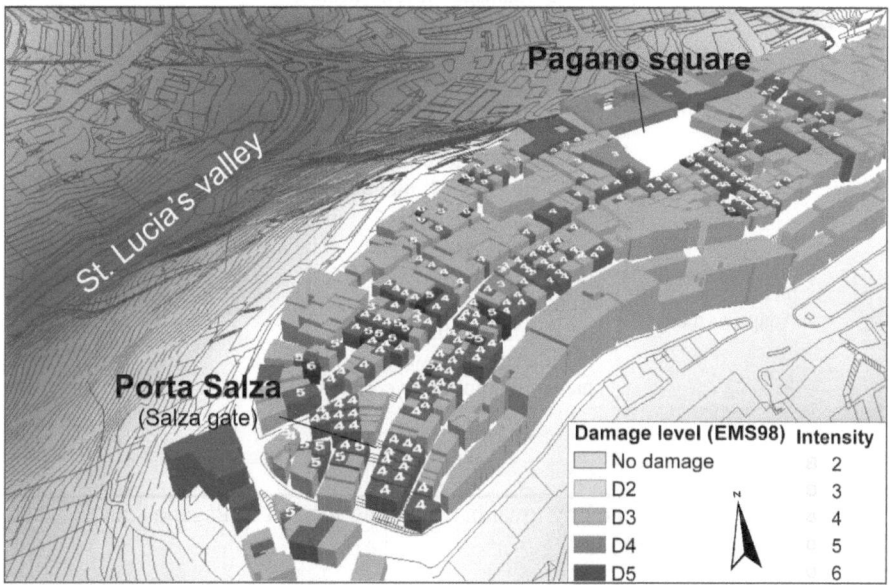

Fig. 2. Overlay between 1857 earthquake macroseismic damage scenario and intensity assigned to each point

Kernel and Cell Size Choice. As for kernel function, we used Epanechnicov's kernel to have a bounded smoothing parameter around buildings; while we adopted a 0,1 m cell size either according to the reference scale or the desired precision of bandwidth.

Bandwidth Choice. Different steps have been performed in order to identify the more suitable bandwidth for the study of point distribution. In the first step (Case 1, Table. 2) fixed bandwidth has been used for the whole distribution calculated by means of the nearest neighbour distance mean method. The τ value, calculated as the mean distance between centroids with the nearest neighbour of order 1, was 6.8 m. So, calculated kernel density map expresses seismic damaging effects not only in terms of first order properties, but of second order ones, too, showing areas where damaged buildings interacted with urban road network and/or with other neighbouring buildings. Examining the final result (Fig. 3a), it is possible to observe that the interaction between buildings has to be differentiated and not equally distributed on the whole point pattern, in order not to have non-null density values also on

Table 2. Synthesis of methods and values adopted for bandwidth (τ) choice

Case	Bandwidth approach	Methods used to estimate τ	τ(m)	KD map
1	Fixed for whole point pattern	Nearest neighbour mean calculated for whole point pattern	6.8	
2	Two different fixed bandwidths	D1-2-3 damage level: average of building minimum semi-dimension.	3.9	Sum of two resultant cs
		D4-5 damage level: nearest neighbour mean calculated for whole point pattern.	6.8	
3	One KDE with Fixed method, one with Adaptive method	D1-2-3 damage level: building minimum semi-dimension	1.4÷9.9*	Sum of resultant rasters
		D4-5 damage level:		
		Building area ≤ mean + sd nearest neighbour mean calculated for whole point pattern.	6.8	
		Building area > mean + sd nearest neighbour mean calculated for whole point pattern multiplied by correction.	4.1	

* An exception is represented by town hall building located in Pagano square, where bandwidth value is 18 m.

undamaged buildings (D0). This unsuitable result is due to proximity to other buildings with damage level of D1, D2 or D3, where partial or total structural collapses do not occur. For these reasons events relative to D1, D2 and D3 damage levels must be independent between them, in contrast with high level damages (D4 and D5).

This last consideration is the base of a second step (case 2, Table. 2) followed in KDE. Two kernel density values with fixed bandwidth have been calculated for both cases discussed above: as regards buildings with D4 and D5 damage level the same τ value used in case 1 (6.8 m) has been adopted, while for D1, D2 and D3 levels, where damage is limited to single buildings, a bandwidth equal to the mean of minimum semi-dimension of a single building can be considered. Afterwards, the two output raster layers have been algebraically summed in order to obtain a single density map. The final raster (Fig. 3b) does not express in a complete way the actual situation, yet, because building dimensions are so variable that the mean of minimum semi-dimension of a single building does not represent this variability; in fact, in some cases it is too wide and in other ones it is too small.

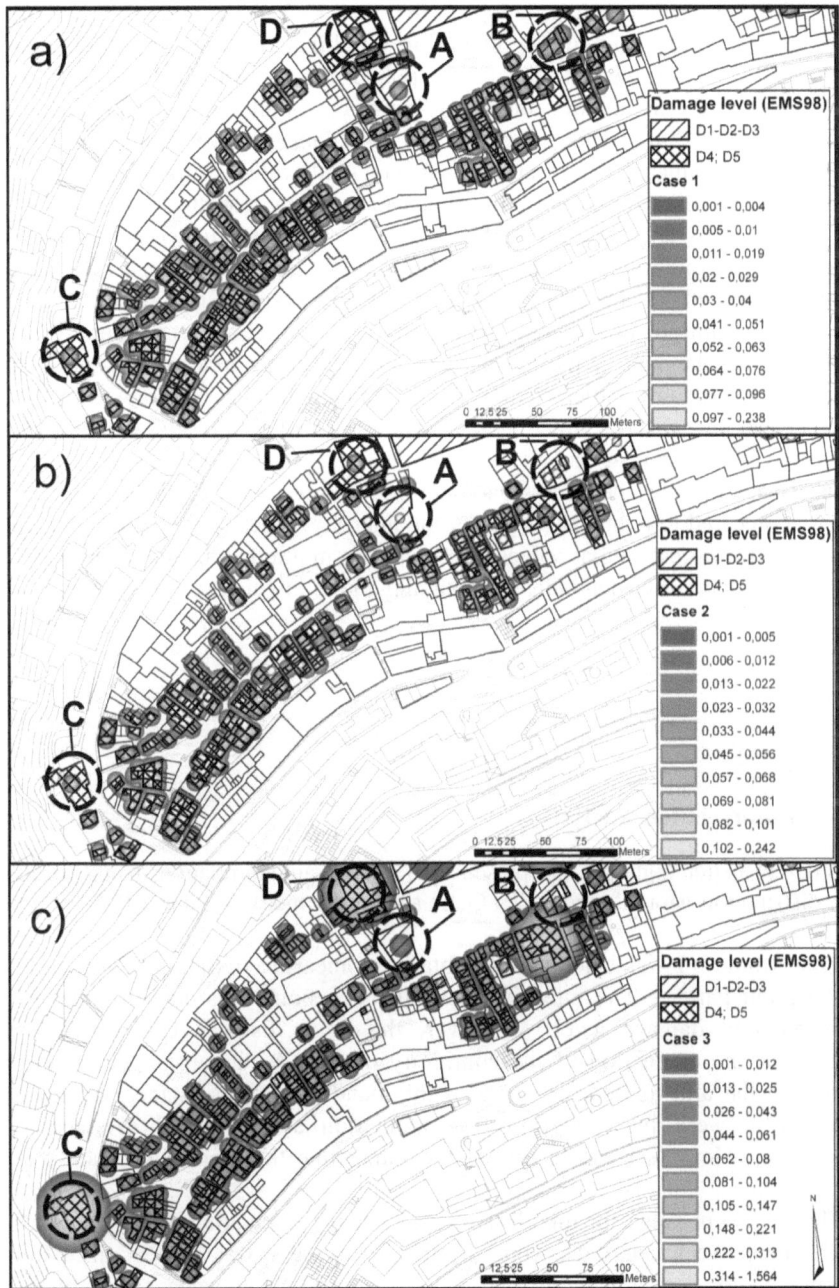

Fig. 3. Comparison among three cases in which different methods of t estimation have been used. Capital letters show some meaningful points to better understand differences among them. The three methods used to estimate t are represented in a), b) and c), respectively.

A third approach has been adopted in order to have a much more sensitive bandwidth, according to building dimension variability: for D1, D2 and D3 damage levels, an adaptive method has been applied to calculate a bandwidth which corresponds to the minimum semi-dimension of each building (values included in a range of 1.4-9.9 m); while, for D4 and D5 damage levels, areas have been preliminarily evaluated before attributing bandwidth value. Some outlier buildings have been identified (buildings with dimensions bigger than those of middle-sized ones) in the study area. Using τ =6.8 m for these outliers, an under-smoothing effect is produced. In this way, the same τ value, used in case 1 (6.8 m), has been adopted for buildings with an area below the sum of the average and the SD of all areas. Besides, we multiplied bandwidth by a corrective factor (Fig. 3c) for buildings with an area above the sum of the average and the SD of all areas. This was obtained dividing mean area of outlier buildings by mean area of other buildings with D4 and D5 damage levels and extracting the square root of the resulting number. The corrective factor obtained was 4.1.

In (2), multiplying the denominator by 4.1^3 involves also a relevant decrease in density compared to points having the same intensity. Since under the same intensity conditions we expect similar values of density, we multiplied the numerator in (2) by the same number of the denominator, obtaining the following expression:

$$\hat{\lambda}_\tau(s) = 69.8 \sum_{i=1}^n \frac{1}{\tau_c^2} k\left(\frac{(s-s_i)}{\tau_c} \right). \tag{14}$$

where $\tau_c = 4.1\,\tau$.

3.2 Autocorrelation Analysis

In order to better highlight the presence of spatial autocorrelation in macroseismic data distribution Moran's index (I), Getis and Ord's index (G) and LISA have been used.

First Moran's I was calculated to obtain the measure of spatial autocorrelation in the point pattern. The intensity value of each point was chosen equal to the corresponding damage level. Moreover, the fixed distance band method was chosen for the evaluation of the contiguity matrix. An iterative calculation was used with the aim to find the value of this band: different distance values, from 8 to 20 m were introduced in the calculation and then the distance value maximising Moran's I and having at the same time the more significant z score was chosen.

The minimum value for this distance was chosen calculating the average nearest neighbour between centroids, while the maximum value was chosen, since from it Moran's index decreases more and more. The following values were found: a) distance value: 10m; b) I: 0.43; c) z score: 15.23.

Considering also that the theoretical mean E(I) is equal to -0,00098, the distribution can be considered clustered.

Afterwards, the calculated distance value was introduced in Getis and Ord's index and in LISA, in order to localize clusters. Indexes values were classified as showed in tables 3 and 4:

Table 3. Getis and Ord's G classification for cluster detection

Class: Damage level	Getis and Ord's range value for positive autocorrelation
D0	-1.44409 ÷ 0.426473
D1-D2-D3	0.426473 ÷ 2.297037
D4-D5	2.297037 ÷ 4.1676

Table 4. LISA classification for cluster detection

Class: Damage level	LISA's range value for positive autocorrelation
D0	-9.9382÷ 0.8745
D1-D2-D3	0.8745 ÷ 11.6872
D4-D5	11.6872 ÷ 22.4999

4 Final Remarks

Starting with the comparison between kernel density map, the damage scenario of 1857's earthquake and DEM of substratum depth of Potenza (Fig. 5), it has been possible to evaluate the relationships existing among damage levels, morphological features and geological-stratigraphycal characters (variation of the substratum depth).

The multilayer point of view allows to distinguish different situations:

- higher values of KD are concentrated in the sector of the old town centre located between Salza Gate and Pagano Square, where the substratum is deeper, the morphological ridge is narrow and D4-D5 damage levels are also more represented; this sector is characterized by seismic amplification factors such as stratigraphy (higher thickness of sandy-conglomerate deposits) and morphology (ridge effect);
- the northern sector, where the cathedral is located, is characterized by local sites with high KD values and high thickness of sandy-conglomerate deposits, but few damaged buildings are there; in this case the geomorphological factor plays an important role because here the ridge is wider, thus reducing possible seismic amplifications;
- the last case is that of Guevara tower sector, morphologically characterized by a long and narrow ridge, where a localized high KD value and high seismic damage level, but the lowest thickness of sandy-conglomerate deposit are measured; here the geomorphological factor is determinant in amplifying seismic intensity.

This interpretation of multilayer analysis seems in agreement with the autocorrelation maps (fig. 4), where the cluster distribution of D4 and D5 point values is shown and it is located along the sector mainly subjected to seismic amplification.

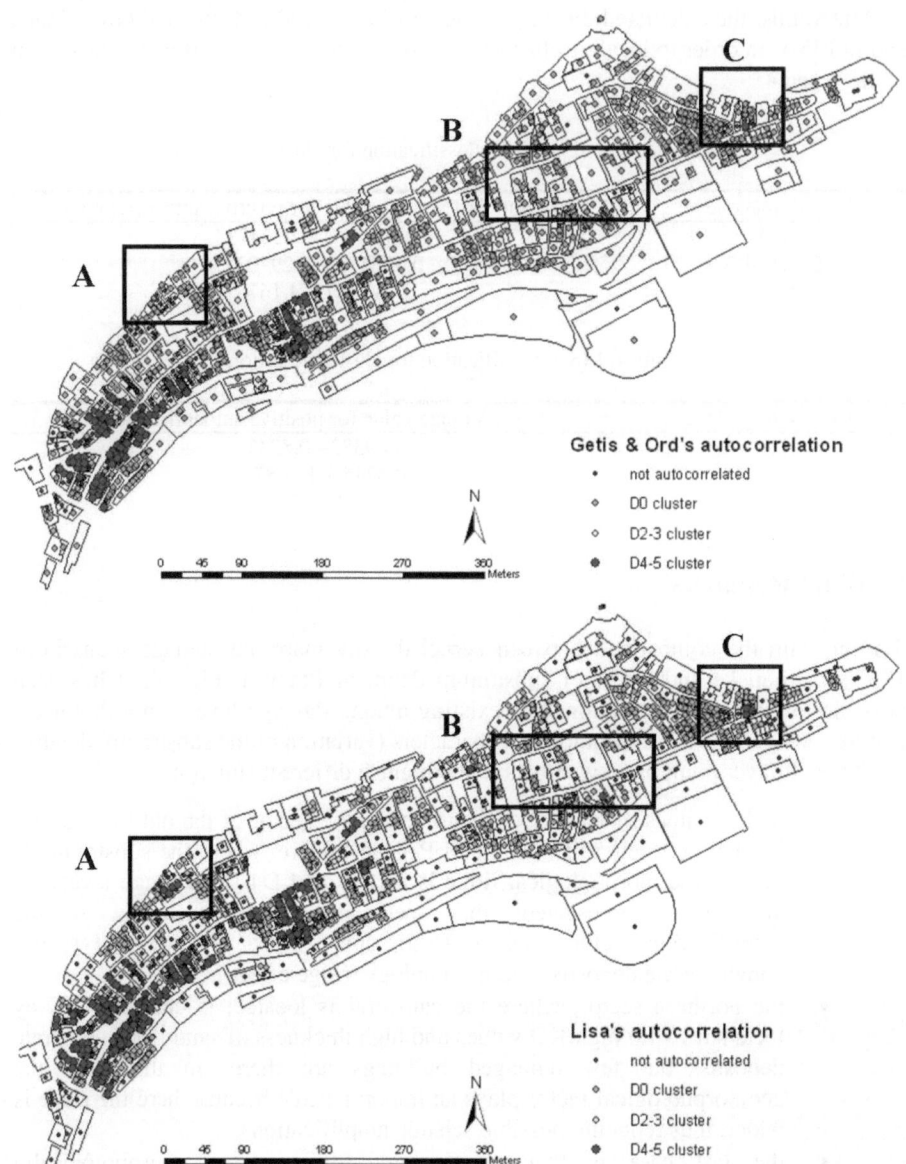

Fig. 4. Clusters obtained by means of Getis and Ord's index and LISA

In particular, figure 4 shows the results of two methods used to analyze spatial distribution of data and seem to be very similar between them. Besides, some differences are also shown in LISA's autocorrelation, whereas the clusterization of D4 and D5 damage values (sectors A and C) seem to be emphasized respect to Getis and Ord's index. Moreover the axial sector B of the town is characterized by unclustered values in LISA respect to Getis and Ord's G (Fig. 4).

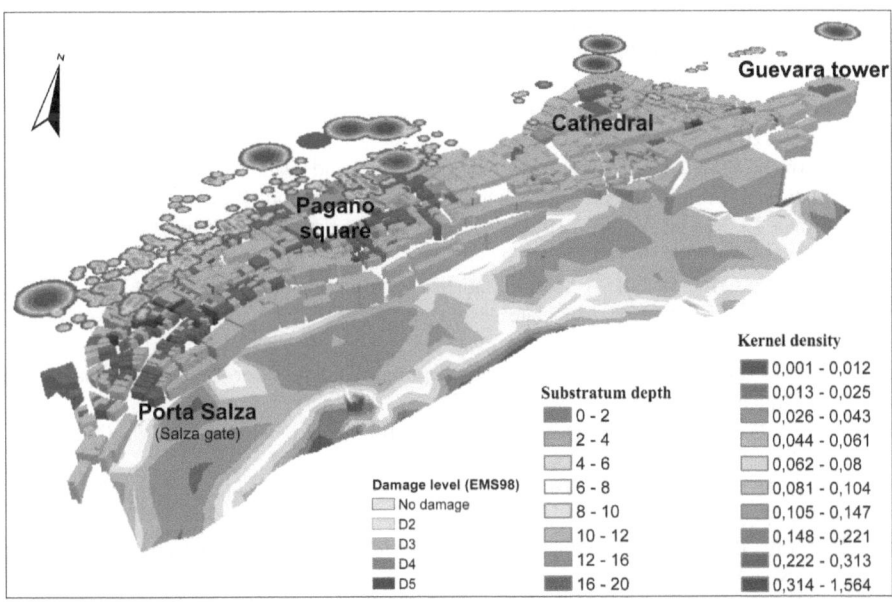

Fig. 5. Overlay between the DEM of substratum depth, 1857 earthquake macro-seismic damage scenario and kernel density final map

The use of geostatistics to process historical macro-seismic data is a new field of application of these techniques and represents a new approach to territorial analysis. Particularly, our application allows to define urban areas, historically most exposed to seismic risk, achieving useful knowledge bases for emergency planning in case of earthquakes. This work could be also a good basis for Civil Defence Plan re-examination concerning the definition of waiting and refuge areas and strategic points of entrance to old town centre.

References

1. Tobler, W.R.: A Computer Model Simulating Urban Growth in the Detroit Region. Economic Geography 46, 234–240 (1970)
2. Gatrell, A.C., Bailey, T.C., Diggle, P.J., Rowlingson, B.S.: Spatial Point Pattern Analysis and Its Application in Geographical Epidemiology. Transaction of Institute of British Geographer 21, 256–271 (1996)
3. Bailey, T.C., Gatrell, A.C.: Interactive Spatial Data Analysis. Longman Higher Education, Harlow (1995)
4. O'Sullivan, D., Wong, D.W.S.: A Surface-Based Approach to Measuring Spatial Segregation. Geographical Analysis, 147–168 (2007)
5. Danese, M., Lazzari, M., Murgante, B.: Kernel Density Estimation Methods for a Geostatistical Approach in Seismic Risk Analysis: the Case Study of Potenza Hilltop Town (southern Italy). In: Gervasi, O., Murgante, B., Laganà, A., Taniar, D., Mun, Y., Gavrilova, M.L. (eds.) ICCSA 2008, Part I. LNCS, vol. 5072, pp. 415–429. Springer, Heidelberg (2008)

6. Danese, M., Lazzari, M., Murgante, B.: Integrated Geological, Geomorphological and Geostatistical analysis to study macroseismic effects of 1980 Irpinian earthquake in urban areas (southern Italy). In: Gervasi, O., Murgante, B., Laganà, A., Taniar, D., Mun, Y., Gavrilova, M. (eds.) ICCSA 2009. LNCS, vol. 5592, pp. 302–9743. Springer, Heidelberg (in press, 2009)
7. Silverman, B.W.: Density Estimation for Statistics and Data Analysis. Monographs on Statistics and Applied Probability. Chapman and Hall, London (1986)
8. Breiman, L., Meisel, W., Purcell, E.: Variable Kernel Estimates of Multivariate Densities. Technometrics 19, 135–144 (1977)
9. Levine, N.: CrimeStat III: A Spatial Statistics Program for the Analysis of Crime Incident Locations. Ned Levine & Associates, Houston, TX, and the National Institute of Justice, Washington, DC (2004)
10. Sacks, J., Ylvisaker, D.: Asymptotically Optimum Kernels for Density Estimates at a Point. Annals of Statistics 9, 334–346 (1981)
11. Borruso, G.: Network Density and the Delimitation of Urban Areas. Transaction in GIS, 7 2, 177–191 (2003)
12. Miller, H.: A Measurement Theory for Time Geography. Geographical Analysis 37, 17–45 (2005)
13. Burt, J.E., Barber, G.M.: Elementary Statistics for Geographers, 2nd edn. The Guilford Press, New York (1996)
14. Bailey, T.C., Gatrell, A.C.: Interactive Spatial Data Analysis. Longman Higher Education, Harlow (1995)
15. Epanechnikov, V.A.: Nonparametric Estimation of a Multivariate Probability Density. Theory of Probability and Its Applications 14, 153–158 (1969)
16. Downs, J.A., Horner, M.W.: Characterising Linear Point Patterns. In: Proceedings of the GIS Research UK Annual Conference (GISRUK 2007), Maynooth, Ireland (2007)
17. Jones, M.C., Marron, J.S., Sheather, S.J.: A Brief Survey of Bandwidth Selection for Density Estimation. Journal of the American Statistical Association 91, 401–407 (1996)
18. Sheather, S.J.: An Improved Data-Based Algorithm for Choosing the Window Width When Estimating the Density at a Point. Computational Statistics and Data Analysis 4, 61–65 (1986)
19. Hall, P., Sheather, S.J., Jones, M.C., Marron, J.S.: On Optimal Data-Based Bandwidth Selection in Kernel Density Estimation. Biometrika 78, 263–269 (1991)
20. Hazelton, M.: Optimal Rates for Local Bandwidth Selection. J. Nonparametr. Statist. 7, 57–66 (1996)
21. Sillverman, B.W., Jones, M.C., Fix, E., Hodges, J.L.: An important contribution to nonparametric discriminant analysis and density estimation. International Statistical Review 57(3), 233–247 (1951)
22. Loftsgaarden, D.O., Quesenberry, C.P.: A Nonparametric Estimate of a Multivariate Density Function. Ann. Math. Statist. 36, 1049–1051 (1965)
23. Clark, P.J., Evans, F.C.: Distance to Nearest Neighbour as a Measure of Spatial Relationships in Populations. Ecology 35, 445–453 (1994)
24. Cao, R., Cuevas, A., González-Manteiga, W.: A Comparative study of several smoothing methods in density estimation. Computational Statistic and Data Analysis 17, 153–176 (1994)
25. Wand, M., Jones, M.C.: Kernel Smoothing. Chapman & Hall, London (1995)
26. Simonoff, J.: Smoothing Methods in Statistics. Springer, New York (1996)
27. Chiu, S.T.: A comparative Review of Bandwidth Selection for Kernel Density Estimation. Statistica Sinica 6, 129–145 (1996)

28. Devroye, L., Lugosi, T.: Variable Kernel Estimates: on the Impossibility of Tuning the Parameters. In: Giné, E., Mason, D. (eds.) High-Dimensional Probability. Springer, New York (1994)

29. Murgante, B., Las Casas, G., Danese, M.: Where are the slums? New approaches to urban regeneration. In: Liu, H., Salerno, J., Young, M. (eds.) Social Computing, Behavioral Modeling and Prediction, pp. 176–187. Springer US, Heidelberg (2008)

30. Brunsdon, C.: Estimating Probability Surfaces for Geographical Points Data: An Adaptive Kernel Algorithm. Computers and Geosciences, 21 7, 877–894 (1995)

31. Abramson, L.S.: On Bandwidth Variation in Kernel Estimates - A Square Root Law. Ann. Stat. 10, 1217–1223 (1982)

32. Breiman, L., Meisel, W., Purcell, E.: Variable Kernel Estimates of Multivariate Densities. Technometrics 19, 135–144 (1977)

33. Hall, P., Marron, J.S.: On the Amount of Noise Inherent in Band-Width Selection for a Kernel Density Estimator. The Annals of Statistics 15, 163–181 (1987)

34. Hall, P., Hu, T.C., Marron, J.S.: Improved Variable Window Kernel Estimates of Probability Densities. Ann. Statist. 23, 1–10 (1994)

35. Sain, S.R., Scott, D.W.: On Locally Adaptive Density Estimation. J. Amer. Statist. Assoc. 91, 1525–1534 (1996)

36. Wu, C.O.: A Cross-Validation Bandwidth Choice for Kernel Density Estimates with Selection Biased Data. Journal of Multivariate Analysis 61, 38–60 (1999)

37. Härdle, W.: Smoothing Techniques with Implementation in S. Springer, New York (1991)

38. Bowman, A.W., Azzalini, A.: Applied Smoothing Techniques for Data Analysis: The Kernel Approach with S - Plus Illustrations. Oxford Science Publications, Oxford University Press, Oxford (1997)

39. Murgante, B., Las Casas, G., Danese, M.: The periurban city: Geo-statistical methods for its definition. In: Coors, Rumor, Fendel, Zlatanova (eds.) Urban and Regional Data Management, pp. 473–485. Taylor & Francis Group, London (2008)

40. Chainey, S., Reid, S., Stuart, N.: When is a Hotspot a Hotspot? A Procedure for Creating Statistically Robust Hotspot Maps of Crime. In: Kidner, D., Higgs, G., White, S. (eds.) Innovations in GIS 9: Socio-Economic Applications of Geographic Information Science, pp. 21–36. Taylor and Francis, Abington (2002)

41. Moran, P.: The interpretation of statistical maps. Journal of the Royal Statistical Society 10, 243–251 (1948)

42. Anselin, L.: Local Indicators of Spatial Association – LISA. Geographical Analysis 27, 93–115 (1995)

43. Getis, A., Ord, J.K.: The analysis of spatial association by use of distance statistics. Geographical Analysis 24, 189–206 (1992)

44. Gizzi, F.T., Lazzari, M., Masini, N., Zotta, C., Danese, M.: Geological-Geophysical and Historical-Macroseismic Data Implemented in a Geodatabase: a GIS Integrated Approach for Seismic Microzonation. The Case-Study of Potenza Urban Area (Southern Italy). Geophysical Research Abstracts 9(09522), SRef-ID: 1607-7962/gra/EGU2007-A-09522 (2007)

45. Grünthal, G.G.: European Macroseismic Scale 1998. In: Conseil de l'Europe Cahiers du Centre Européen de Géodynamique et de Séisomologie, Luxembourg, vol. 15 (1998)

A Discrete Approach to Multiresolution Curves and Surfaces

Luke Olsen and Faramarz Samavati

University of Calgary

Abstract. Subdivision surfaces have been widely adopted in modeling in part because they introduce a separation between the surface and the underlying basis functions. This separation allows for simple general-topology subdivision schemes. Multiresolution representations based on subdivision, however, incongruently return to continuous functional spaces in their construction and analysis. In this paper, we propose a discrete multiresolution framework applicable to many subdivision schemes and based only on the subdivision rules. Noting that a compact representation can only afford to store a subset of the detail information, our construction enforces a constraint between adjacent detail terms. In this way, all detail information is recoverable for reconstruction, and a decomposition approach is implied by the constraint. Our framework is demonstrated with case studies in Dyn-Levin-Gregory curves and Catmull-Clark surfaces, each of which our method produces results on par with earlier methods. It is further shown that our construction can be interpreted as biorthogonal wavelet systems.

1 Introduction

Multiresolution (MR) methods have become a standard paradigm for curve and surface editing, allowing a transition between different resolutions while maintaining geometric details between edits. Subdivision curves and surfaces are supported by many popular geometric modeling programs, fitting naturally into an iterative, multi-scale design process. Since subdivision techniques naturally create a hierarchy of different resolutions, many MR approaches require the connectivity structure imposed by subdivision.

An important factor in the adoption of subdivision over parametric approaches is the simplicity that comes from the absence of basis functions: high-quality surfaces can be modeled without any direct functional evaluations. Clean, simple, affine operations are the only necessary background to learn and implement subdivision. However, the underlying theory of subdivision is based on the refinement of scaling functions [1]. An important recent movement is focused on analyzing subdivision surfaces based only on discrete operations, without explicit use of continuous functions [2].

Unlike discrete subdivision formulations, most MR schemes are defined based on wavelet theory that is heavily grounded in a functional representation. MR

M.L. Gavrilova and C.J.K. Tan (Eds.): Trans. on Comput. Sci. VI, LNCS 5730, pp. 342–361, 2009.

settings based on wavelets have been proposed for many curve and surface subdivision types – B-spline [3], Doo-Sabin [4], Loop [5,6,7], and Catmull-Clark [8,9] surfaces. So while wavelet-based approaches can indeed be used to construct MR systems, several related questions have not been addressed. Is it possible to construct the full MR representation of a subdivision scheme only from its discrete description? Is it necessary to use continuous functions (wavelets and scalings) to create a discrete MR system? Like many problems in computer graphics, it is perhaps more natural to employ a discrete approach rather than delving into the "labyrinth of the continuum" [10].

In this paper, we present a discrete method for constructing compact and efficient multiresolution settings for a variety of subdivision schemes directly from the subdivision rules, without any direct use of basis functions (scaling or wavelets). Our construction, described in Sec. 2, is demonstrated with case studies in Dyn-Levin-Gregory curves (Sec. 3) and Catmull-Clark surfaces (Sec. 4), which are shown to offer competitively low approximation error.

1.1   Related Work

Much of the work in multiresolution is based on the theory of wavelets, although it should be noted that the two are not the same. Wavelets are a tool for "hierarchically decomposing functions" [11] into complementary bases; one basis, the coarse scaling functions, encode an approximation of the function, while the wavelet basis encodes the missing details. They are often used in geometric modeling because of their natural fit to a multi-scaling functional representation. That is, wavelet analysis undoes scaling function refinement, and can be used to derive a hierarchical multiresolution representation for curves and surfaces [12,13]. Since our paper seeks to diverge from the wavelet approach, we'll refer the curious reader to the book of Stollnitz et al. [1] for further details.

Samavati and Bartels [14,4] investigated an alternative to wavelets for subdivision curves and Doo surfaces, constructing multiresolution systems by reversing subdivision rules. The methodology is based on discrete least-squares and can be considered to be wavelet-free, but the construction is not easily extended to general topology surfaces.

More recent multiresolution approaches for general-topology surfaces follow Sweldens' lifting method [15], a two-stage process for constructing so-called second-generation wavelet systems. First an initial "lazy" wavelet with poor fitting properties is selected. It is then improved with respect to some criteria (eg. fitting quality or support) by one or more lifting stages. Instances of the lifting method include work on Loop [5,6] and Catmull-Clark surfaces [8,9].

Each of these methods requires a semi-regular input surface, i.e. one with subdivision connectivity. This is not a particularly restrictive condition in modeling, save for surfaces extracted from range or point-set data. Several remeshing methods have been proposed [12,16,17,18] to make semi-regular meshes from irregular ones. Remeshing comes at the expense of precision and complexity, as a remeshed surface often has more vertices and contains error relative to the original. If the original surface must be maintained, Valette and Prost [19] proposed a multiresolution method for irregular meshes, but the lack of stationary subdivision rules hampers the run-time efficiency.

Some common applications of multiresolution systems are surface editing [20] and compression [21,8]. Many other representations have been considered for surface editing, such as displacement volumes [22], Laplacian encodings [23], mean value coordinates [24], and coupled prisms [25].

1.2 Notation

Let $V = \{v_1, \ldots, v_n\}$ denote a set of vertices defining an object. For a curve, V represents a piecewise-linear approximation of a (usually smooth) curve. A polygon mesh M can be defined as $M = \{V, F\}$, where V defines the geometry, and F is a set of faces defining the topology. The valence n of a vertex is the number of incident edges, while the valence n^f of a face is the number of sides.

To represent relative positioning in a subdivision sequence or multiresolution hierarchy, a superscript k is used. Thus, subdivision is a process for computing a set of *fine* vertices V^{k+1} from *coarse* vertices V^k. The topological information F^{k+1} must be constructed according to F^k and the subdivision rules.

A key property of subdivision is that the subdivided position of a vertex results from a *linear* combination of a *local* neighborhood of vertices, implying that subdivision can be implemented in linear time with discrete operations. Typically a matrix $\mathbf{P}$ is used to capture all vertex interactions during subdivision, such that $V^{k+1} = \mathbf{P}V^k$.

Decomposition is a process for converting fine data V^{k+1} to a coarse approximation V^k plus some details D^k. This should also be a linear process – let $\mathbf{A}$ and $\mathbf{B}$ be the associated matrices for which $V^k = \mathbf{A}V^{k+1}$ and $D^k = \mathbf{B}V^{k+1}$. Together V^k and D^k should be enough to reconstruct V^{k+1}, i.e. for some matrix $\mathbf{Q}$, $V^{k+1} = \mathbf{P}V^k + \mathbf{Q}D^k$. A biorthogonal MR system is one that satisfies

$$\begin{bmatrix} \mathbf{A} \\ \mathbf{B} \end{bmatrix} [\mathbf{P}|\mathbf{Q}] = \mathbf{I} \, .$$

In other words, decomposition and reconstruction are inverse processes. There are other orthogonality classifications [1], but anything stronger than biorthogonality is hard to satisfy in a mesh multiresolution.

Though both reconstruction and decomposition can be expressed with matrices, each matrix is sparse (due to the locality of each operation) and exhibits regularity within the columns or rows (because the same linear weights are applied to each vertex neighborhood). Thus the notion of *filters* (for curves) or *masks* (for surfaces) is useful for expressing the regular entries in a matrix. The $\mathbf{P}$ and $\mathbf{Q}$ matrices contain regular columns, each shifted downward by two elements from the previous column. Let $p = \{p_0, p_{\pm 1}, p_{\pm 2}, \ldots\}$ and $q = \{q_0, q_{\pm 1}, q_{\pm 2}, \ldots\}$ represent the non-zero entries of a regular column in $\mathbf{P}$ and $\mathbf{Q}$, respectively; call these *reconstruction filters*. Similarly, $\mathbf{A}$ and $\mathbf{B}$ are characterized by regular entries across the rows, denoted by the *decomposition filters*, $a = \{a_0, a_{\pm 1}, \ldots\}$ and $b = \{b_0, b_{\pm 1}, \ldots\}$.

2 Method Overview

Our approach to constructing MR systems is based on the observation that all subdivisions impose a particular structure. New vertices are created, old vertices are (possibly) displaced, and in the mesh case, the face structure is changed (eg. one triangle split into four). Common terminology – descended from the indexing of 1D subdivision curves – refers to new vertices as *odd* and old vertices as *even*, a notation which we will use. (For vertex-splitting schemes such as Chaikin [26] or Doo-Sabin [27], one split vertex can be classified as even, the remaining as odd.)

This inherent structure recommends a decomposition strategy: even vertices are a natural choice for coarse vertices, and then, to have a compact representation, only odd vertices can be replaced with details. Notice that in addition to compactness, this approach naturally fits to the structure of subdivision.

An important question is, how can we fully reconstruct a surface if only some of the details are stored? Our idea is to place a constraint on the details. Specifically, the coarse surface is chosen such that the associated details satisfy a particular relationship, implicitly defining the missing details. Consider Fig. 1: if v^k is chosen such that its associated detail d^k is a linear combination of adjacent details – eg. $d^k = \alpha(d_l^k + d_r^k)$ – then d^k need not be stored explicitly.

In the surface case, this idea can be generalized to: choose the coarse position of even vertices such that the associated detail vector is a linear combination of the details from adjacent odd vertices. For uniform subdivisions, it is sensible for adjacent details of the same type to contribute equally to an even-vertex detail, i.e. have the same weight. Therefore, the detail constraint has a general form of

$$d^e = \alpha \sum d_j^o \ , \tag{1}$$

where d^e and d^o denote even and odd details respectively, and j enumerates the odd details adjacent to d^e.

Our method is as follows. First, we define a detail constraint that expresses even details in terms of adjacent odd details, and then use the subdivision rules to find an initial (or *trial*) decomposition mask that satisfies this constraint. To improve the fitting behavior, we then compute a refinement of the coarse surface by a local optimization step.

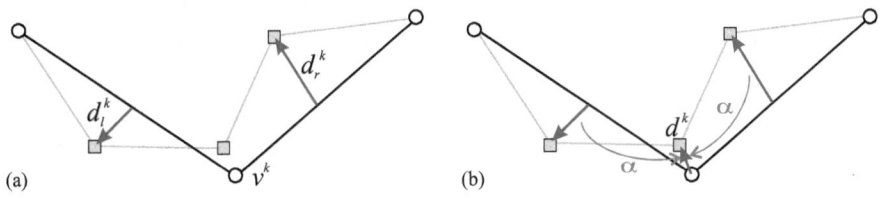

Fig. 1. (a) When a surface is decomposed, some vertices are coarsened and others are replaced with detail vectors; (b) If the details at even vertices can be computed from adjacent odd details, then a compact MR is possible

The initial masks that arise from satisfying the detail constraint represent the equivalent of a lazy wavelet decomposition. They typically have narrow support, which can lead to poor fitting performance. To increase the fitting quality, we consider a refinement of each coarse vertex v_i^k, in the form of a per-vertex displacement vector δ_i: $v_i^k \leftarrow v_i^k + \delta_i$.

The optimal value of δ_i depends on details adjacent to v_i^k, which in turn depend on their local neighborhood of original data, and so the refinement process is equivalent to widening the support of the decomposition mask. If Δ is a vector of all such refinements, then $\Delta = \mathbf{R}D^k$ for some sparse matrix $\mathbf{R}$. It was noted earlier that refinement is equivalent to widening the mask support. In fact, it is equivalent to lifting, because $V^k = \mathbf{A}V^{k+1} + \mathbf{R}D^k = (\mathbf{A} + \mathbf{R}\mathbf{B})V^{k+1}$, exactly as in lifting.

Since the initial decomposition mask reverses the subdivision rules, and the size of V^k and D^k together equals the size of V^{k+1}, the initial masks are biorthogonal. According to the lifting theory, any matrix $\mathbf{R}$ preserves biorthogonality. Therefore, the MR systems produced by our construction are biorthogonal.

This construction is best illustrated with an example. First we consider a simple curve subdivision proposed by Dyn et al. (Sec. 3). We then present a more complex surface subdivision example in Sec. 4.

3   Dyn-Levin-Gregory Curve Subdivision

Dyn et al. [28] describe an interpolating subdivision scheme based on a 4-point filter for odd points (even points are not moved), which will be referred to as *DLG* subdivision. Their construction contains a parameter w that relates to the smoothness of the limit curve. We use the typical value of $w = \frac{1}{16}$, which results in the following filters (see Fig. 2):

$$v_{2i}^{k+1} = v_i^k \text{ , and,} \tag{2}$$

$$v_{2i+1}^{k+1} = \frac{9}{16}(v_i^k + v_{i+1}^k) - \frac{1}{16}(v_{i-1}^k + v_{i+2}^k) \ . \tag{3}$$

3.1   Trial Filter

For an interpolating scheme, there is an obvious choice for the initial filter. Since even vertices are not displaced by subdivision, a multiresolution setting can be attained by simply retaining the even vertices in V^{k+1} to form V^k, and replacing the odd vertices with details. That is, even vertices do not move:

$$\tilde{v}_i^k = v_{2i}^{k+1} \ . \tag{4}$$

Odd vertices can be replaced with detail vectors that capture the difference between V^{k+1} and $\mathbf{P}V^k$:

$$
\begin{aligned}
d_{i+1}^k &= v_{2i+1}^{k+1} - \left(\frac{9}{16}\left(\tilde{v}_i^k + \tilde{v}_{i+1}^k\right) - \frac{1}{16}\left(\tilde{v}_{i-1}^k + \tilde{v}_{i+2}^k\right) \right) \\
&= v_{2i+1}^{k+1} - \frac{9}{16}\left(v_{2i}^{k+1} + v_{2i+2}^{k+1}\right) + \frac{1}{16}\left(v_{2i-2}^{k+1} + v_{2i+4}^{k+1}\right) \ .
\end{aligned}
\tag{5}
$$

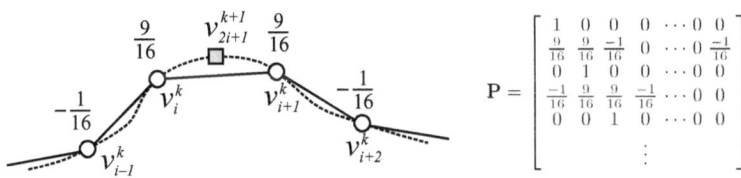

Fig. 2. The DLG subdivision filter for odd vertices, with $w = \frac{1}{16}$; even vertices are unmoved (Eqn. 2). Right: the corresponding subdivision matrix $\mathbf{P}$.

The details at even vertices are $d_i^k = \mathbf{0}$ (i.e. $\alpha = 0$ in Eqn. 1). In this case, the detail constraint is not needed to derive an initial decomposition filter.

In matrix form, the trial filters can be summarized as:

$$\tilde{\mathbf{A}} = \begin{bmatrix} 1\,0\,0\,0\,0\cdots \\ 0\,0\,1\,0\,0\cdots \\ 0\,0\,0\,0\,0\cdots \\ 0\,0\,0\,0\,1\cdots \\ \vdots \end{bmatrix}, \tilde{\mathbf{B}} = \begin{bmatrix} \frac{-9}{16}\,1\,\frac{-9}{16}\,0\,\frac{1}{16}\,0\,0\,0\cdots\,\frac{1}{16}\,0 \\ \frac{1}{16}\,0\,\frac{-9}{16}\,1\,\frac{-9}{16}\,0\,\frac{1}{16}\,0\cdots\,0\,0 \\ \vdots \end{bmatrix}, \tilde{\mathbf{Q}} = \begin{bmatrix} 0\,0\,0\,0\cdots \\ 1\,0\,0\,0\cdots \\ 0\,0\,0\,0\cdots \\ 0\,1\,0\,0\cdots \\ \vdots \end{bmatrix}.$$

3.2   Refinement

As Bartels and Samavati note [29], in the general case the fine data V^{k+1} will have non-zero details. In such cases, we may be able to achieve lower approximation error by choosing a non-trivial decomposition filter.

Our approach is to improve the trial filter by a refinement step, with the goal of reducing the local error, call it E, at each vertex $v_i^k \in V^k$. Based on a three-point neighborhood around v_i^k, the local error is given by the magnitude of the details. In particular, recalling that $d_i^k = \mathbf{0}$, we have

$$E = \|d_{i-1}^k\|^2 + \|d_{i+1}^k\|^2 .$$

If V^k is a poor approximation of V^{k+1}, the details (and therefore the error) will be large.

After refinement, $\tilde{v}_i^k$ will be replaced by $v_i^k = \tilde{v}_i^k + \delta_i$. This has an impact on the local error of v_i^k and neighboring vertices: d_i^k becomes $-\delta_i$, and d_{i+1}^k becomes $d_{i+1}^k - \frac{9}{16}\delta_i$ by Eqn. 5. Thus, after refinement the local error becomes

$$E(\delta) = \| - \delta_i\|^2 + \|d_{i-1}^k - \frac{9}{16}\delta_i\|^2 + \|d_{i+1}^k - \frac{9}{16}\delta_i\|^2 .$$

We should choose δ_i to minimize this error.

After a bit of algebraic manipulation, $E(\delta)$ becomes

$$E(\delta) = \frac{209}{128}\|\delta_i\|^2 - \left(\frac{9}{8}(d_{i-1}^k + d_{i+1}^k)\right)\cdot\delta_i + E$$
$$= a\|\delta_i\|^2 - \mathbf{g}\cdot\delta_i + E .$$

A minimal solution occurs where the derivate equals 0, or

$$2a\delta_i - \mathbf{g} = 0 \rightarrow \delta_i = \frac{a^{-1}}{2}\mathbf{g} .$$

Therefore, an optimal displacement for vertex v_i^k is

$$\delta_i = \frac{72}{209}(d_{i-1}^k + d_{i+1}^k) . \tag{6}$$

The refinement step can be incorporated into a closed-form filter by expressing d^k in terms of elements from V^{k+1}.

$$v_i^k = \tilde{v}_i^k + \delta_i$$
$$= v_{2i}^{k+1} + \frac{72}{209}(d_{i-1}^k + d_{i+1}^k)$$
$$= \frac{128}{209}v_{2i}^{k+1} + \frac{72}{209}v_{2i\pm1}^{k+1} - \frac{36}{209}v_{2i\pm2}^{k+1} + \frac{9}{418}v_{2i\pm4}^{k+1} .$$

Our MR setting for DLG subdivision can be summarized by the following filters:

$$a = \left\{\frac{128}{209}, \frac{72}{209}, -\frac{36}{209}, 0, \frac{9}{418}\right\} ,$$
$$b = \left\{1, -\frac{9}{16}, 0, \frac{1}{16}\right\} , \text{and}$$
$$q = \left\{\frac{128}{209}, -\frac{72}{209}, -\frac{36}{209}, 0, \frac{9}{418}\right\} .$$

The q filter given above results from refining the trivial q filter ($q = \{1, 0, \ldots, 0\}$).

That is, $\mathbf{Q} \leftarrow \mathbf{Q} - \mathbf{PR}$ where $\mathbf{R}$ is the refinement matrix defined by Eqn. 6. This set of filters allows for perfect reconstruction of a decomposed curve.

3.3   Results

To evaluate our MR construction, the work of Samavati and Bartels [29] is used as a comparison. Their local least-squares approach yields an optimal decomposition filter $a_{opt} = \{\frac{107}{161}, \frac{48}{161}, -\frac{24}{161}, 0, \frac{3}{161}\}$. Based on the width of the filter and the relative magnitude of the weights, we anticipate that our DLG filter will perform close to optimally.

$$
\mathbf{A} = \tfrac{1}{418}
\begin{bmatrix}
256 & 144 & -72 & 0 & 9 & 0 & 9 & 0 & -72 & 144 \\
-72 & 144 & 256 & 144 & -72 & 0 & 9 & 0 & 9 & 0 \\
9 & 0 & -72 & 144 & 256 & 144 & -72 & 0 & 9 & 0 \\
9 & 0 & 9 & 0 & -72 & 144 & 256 & 144 & -72 & 0 \\
-72 & 0 & 9 & 0 & 9 & 0 & -72 & 144 & 256 & 144
\end{bmatrix},
$$

$$
\mathbf{Q} = \tfrac{1}{418}
\begin{bmatrix}
-144 & 0 & 0 & 0 & -144 \\
256 & -72 & 9 & 9 & -72 \\
-144 & -144 & 0 & 0 & 0 \\
-72 & 256 & -72 & 9 & 9 \\
0 & -144 & -144 & 0 & 0 \\
9 & -72 & 256 & -72 & 9 \\
0 & 0 & -144 & -144 & 0 \\
9 & 9 & -72 & 256 & -72 \\
0 & 0 & 0 & -144 & -144 \\
-72 & 9 & 9 & -72 & 256
\end{bmatrix}
$$

Results from a number of curves (shown in Fig. 3 and 4) are listed in Table 1. For each curve V^k, the decomposition filter is applied k times to get a coarse approximation V^0 of the original curve; the difference between the approximation $P^k V^0$ and the original gives an error measure $L = \|V^k - P^k V^0\|^2$ for comparing different filters. In each case, our refined filter is a substantial improvement over the trial and virtually identical to the optimal result.

To illustrate the better fitting behavior of the refined filter relative to the trial filter, consider the coastline data depicted in Fig. 4. At the southwest tip of the island, for instance, the refined filter (c) is able to approximate the original data more closely than the trial filter (b). This makes the refined filter more suitable for compression applications.

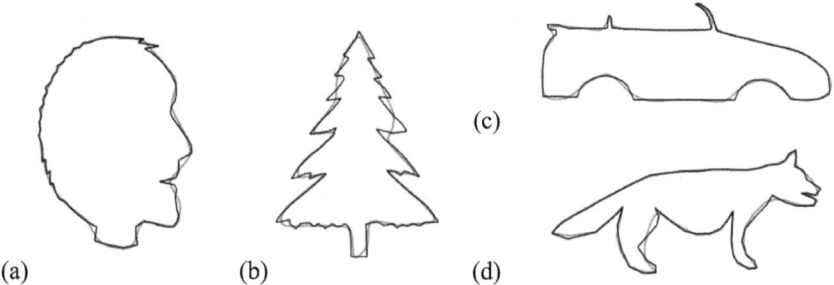

(a)   (b)   (c)   (d)

Fig. 3. Curves used to evaluate the DLG filters: (a) face; (b) car; (c) tree; (d) wolf. The original curve is shown in black, and the curve approximated by our refined filter in red. Numeric results are listed in Table 1.

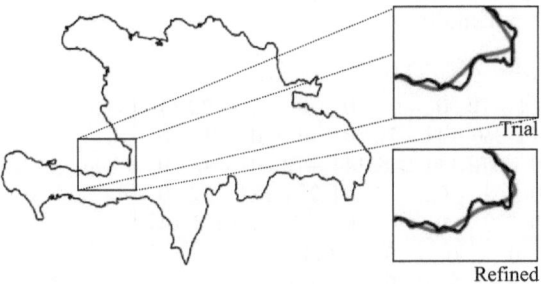

Fig. 4. Comparison of our 4-point filters on real coastline data (from [30]). The original data (4096 points) is decomposed six times to obtain a 64-point approximation, which is then subdivided to the original resolution. The refined filter is able to better approximate the original data (numeric results given in Table 1).

Table 1. L_2 error relative to the original curve for the trial, refined, and optimal DLG filters (normalized against the trial error value). Shorter bars are better.

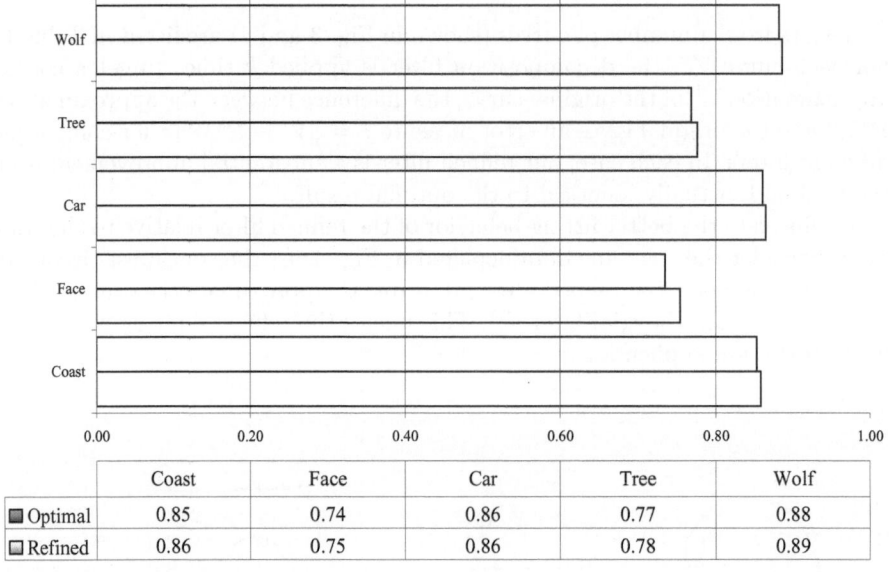

	Coast	Face	Car	Tree	Wolf
▣ Optimal	0.85	0.74	0.86	0.77	0.88
▢ Refined	0.86	0.75	0.86	0.78	0.89

4   Catmull-Clark Surface Subdivision

Catmull-Clark subdivision [31,32] is a popular scheme [33] for manifold surfaces of arbitrary topology. It produces a quadrilateral mesh from any base mesh by adding vertices at each edge and face, then inserting edges from each edge vertex to adjacent face vertices; see Fig. 5. The limit surface is C^2-continuous at regular (valence-4) vertices, C^1 elsewhere. This subdivision scheme is a popular

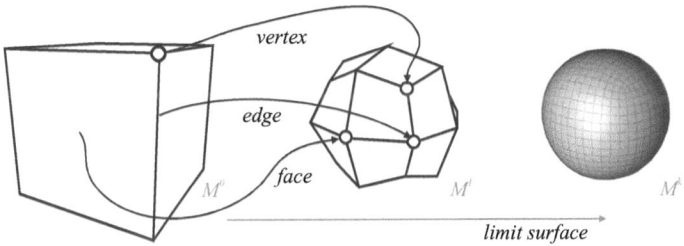

Fig. 5. An iteration of Catmull-Clark subdivision creates new vertices at each face (f) and edge (e), while displacing old vertices (v), to create a smooth surface in the limit

alternative to triangle-mesh schemes such as Loop, providing more pleasing results when the input mesh contains mostly quadrilateral faces.

The subdivision masks for Catmull-Clark subdivision are given by:

$$f_i^{k+1} = \frac{1}{n_i^f} \left(v^k + e_i^k + e_{i+1}^k + \sum_j f_{i,j}^k \right) , \tag{7}$$

$$e_i^{k+1} = \frac{1}{4} \left(v^k + e_i^k + f_{i-1}^{k+1} + f_i^{k+1} \right) , \tag{8}$$

$$v^{k+1} = \frac{n-2}{n} v^k + \frac{1}{n^2} \sum_i e_i^k + \frac{1}{n^2} \sum_i f_i^{k+1} , \tag{9}$$

according to the notation illustrated in Fig. 6 (indices are computed modulo n). Note that these masks, as presented, assume that the face vertices f^{k+1} are computed first.

4.1   Trial Mask

The general form of the detail constraint (Eqn. 1) has only one free parameter. For Catmull-Clark surfaces, the heterogeneity of odd vertices necessitates two free parameters; we denote these weights α_e for edge-vertex details, and α_f for face-vertex details. The detail constraint is then

$$d^v = \alpha_e \sum d_i^e + \alpha_f \sum d_i^f , i = 1, \dots, n , \tag{10}$$

where d^v, d^e, and d^f represent the details of vertex-, edge-, and face-vertices respectively.

The vertex subdivision mask (Eqn. 9) can be rewritten so that the e^k terms are replaced with e^{k+1} terms:

$$v^{k+1} = \frac{n-3}{n} v^k + \frac{4}{n^2} \sum_i e_i^{k+1} - \frac{1}{n^2} \sum_i f_i^{k+1} .$$

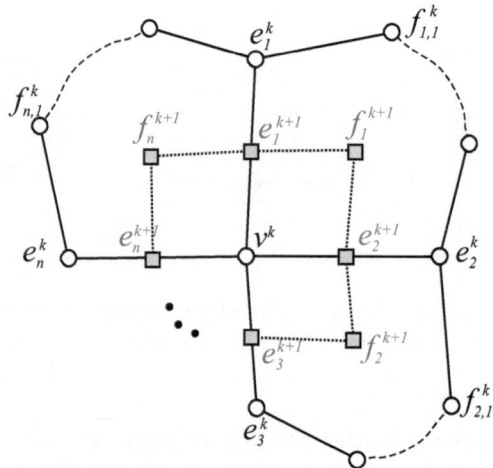

Fig. 6. Notation for the neighborhood of vertex v^k. The 1-ring of v^k consists of edge neighbors e_i^k and face neighbors $f_{i,j}^k$. After subdivision, the 1-ring of v^{k+1} contains edge neighbors e_i^{k+1} and face neighbors f_i^{k+1}.

With this new form, the vertex-, edge- and face-details can be expressed by

$$d^v = v^{k+1} - \left(\frac{n-3}{n} v^k + \frac{4}{n^2} \sum_i \tilde{e}_i^{k+1} - \frac{1}{n^2} \sum_i \tilde{f}_i^{k+1} \right),$$

$$d_i^e = e_i^{k+1} - \tilde{e}_i^{k+1}, \text{ and,}$$

$$d_i^f = f_i^{k+1} - \tilde{f}_i^{k+1},$$

where $\tilde{e}_i^{k+1}$ and $\tilde{f}_i^{k+1}$ represent the approximations of e_i^{k+1} and f_i^{k+1} (i.e. subdivision of the coarse approximation).

To determine α_e and α_f, we must compare the left and right sides of Eqn. 10: $\sum_i \tilde{e}_i^{k+1}$ and $\sum_i \tilde{f}_i^{k+1}$ appear on both sides, with respective weights of $\frac{4}{n^2}$ and $-\frac{1}{n^2}$ on the left, and α_e and α_f on the right side. By setting $\alpha_e = \frac{4}{n^2}$ and $\alpha_f = -\frac{1}{n^2}$, those terms are eliminated, leaving

$$v^{k+1} - \frac{n-3}{n} v^k = \frac{4}{n^2} \sum_i e_i^{k+1} - \frac{1}{n^2} \sum_i f_i^{k+1}.$$

Thus, the detail constraint is satisfied if this relationship between coarse and fine vertices holds, or equivalently, if

$$v^k = \frac{1}{n-3} \left(n v^{k+1} - \frac{4}{n} \sum_i e_i^{k+1} + \frac{1}{n} \sum_i f_i^{k+1} \right). \tag{11}$$

Using this equation to decompose all even vertices v^{k+1} to coarse vertices v^k will ensure that the detail constraint is satisfied for $\alpha_e = \frac{4}{n^2}$ and $\alpha_f = -\frac{1}{n^2}$.

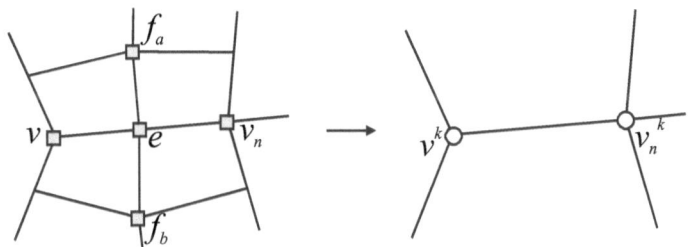

Fig. 7. Valence-3 vertices must be handled with special-case masks

Note that this represents an inversion of the vertex subdivision mask (Eqn. 9). That is, if a surface is the product of subdivision, then decomposition with this mask will produce a zero-error coarse approximation.

Valence-3 Vertices. The decomposition given by Eqn. 11 is undefined for valence-3 vertices, due to the $\frac{1}{n-3}$ term. Fortunately, a decomposition mask that works for most valence-3 vertices can be found by an alternate method.

Consider a valence-3 vertex v with at least one non-valence-3 coarse neighbor v_n (see Fig. 7). Furthermore, let e be the edge vertex between v and v_n, and f_a and f_b be the face-vertices adjacent to e. According to Eqn. 8, these vertices are related by $e \approx \frac{1}{4}\left(v^k + v_n^k + f_a + f_b\right)$.

Since v_n is not valence-3, it can be decomposed via Eqn. 11 to v_n^k, leaving v^k as the only unknown. Therefore,

$$v^k \approx 4e - v_n^k - f_a - f_b \ . \tag{12}$$

Although it was assumed that v_n is not valence-3, Eqn. 12 allows v to be decomposed as long as v_n has already been decomposed, regardless of v_n's valence. Therefore, isolated valence-3 vertices can be decomposed by cascading inwards from non-valence-3 vertices. In practice there is rarely a need for this approach, though: because all edge vertices are valence-4, two valence-3 vertices can only be adjacent in the base mesh.

One downside to this approach is that the detail constraint is not satisfied, meaning that a detail term must be explicitly stored for each of these vertices. Another drawback is that the position of v^k does not depend on v^{k+1}; in cases where v^{k+1} deviates largely from a smooth position, Eqn. 12 will produce a poor approximation. To alleviate this issue somewhat, each already-decomposed neighbor of v^{k+1} (a maximum of 3) can nominate a "candidate" position for v^k based on Eqn. 12. The final position of v^k is then taken to be the average of all candidates.

4.2   Refinement

The initial decomposition mask has small support, considering that a coarse vertex depends on only one level of vertex neighbors. Multiresolution systems

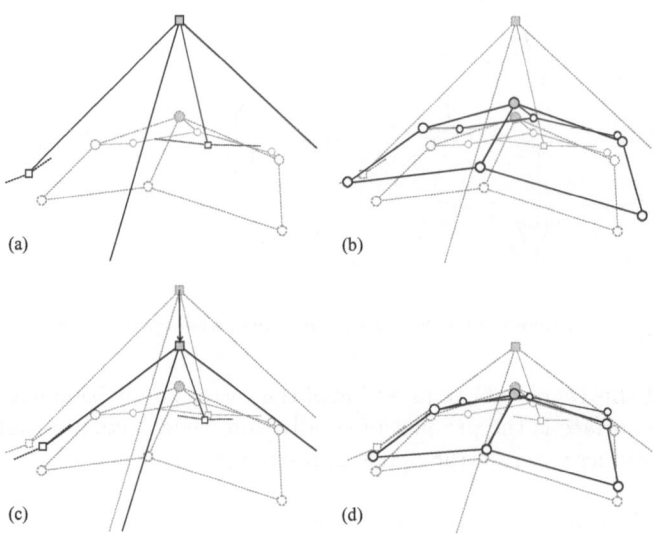

Fig. 8. Refinement reduces the error in coarse data: (a) initial coarse approximation; (b) subdivision of (a); (c) coarse surface after refinement; (d) subdivision of (c). The surface in (d) is "closer" to the original surface than (b).

generally exhibit greater stability and better fitting properties when the decomposition mask has wider support [29].

We address the stability and fitting properties from another direction, recalling the geometric interpretation of detail vectors. A detail vector captures the difference between a surface and the subdivision of its coarse approximation, i.e the *error* in the coarse surface. The larger the error, the larger the magnitude of the detail vectors.

We can look at error on local and global scales. Since efficiency is a primary concern in our construction, we restrict ourselves to a local examination. If the local error can be reduced everywhere in a mesh, then the global error is also reduced. The local error of a vertex is represented by the magnitude of a k-ring of details. Here we consider the 1-ring, in which case the local error of v^k is

$$E = \|d^v\|^2 + \sum \|d_i^e\|^2 + \sum \|d_i^f\|^2 .$$

Figure 8a-b depicts the error introduced by decomposition.

If the error is non-zero, there should be a vector δ for which $v^k + \delta$ is a more optimal position than v^k. To find δ – recalling that the error is a measure of how far a subdivided surface is from the original – we must determine how the displacement of v^k impacts $\tilde{v}^{k+1}$, $\tilde{e}_i^{k+1}$, and $\tilde{f}_i^{k+1}$. If chosen correctly, we should see a reduction in the error (Fig. 8c-d).

From the subdivision rules, it can be seen that the weight applied to v^k is r for $\tilde{v}^{k+1}$, s_i for $\tilde{e}_i^{k+1}$, and t_i for $\tilde{f}_i^{k+1}$, where

$$r = \frac{n-2}{n} + \frac{1}{n^2} \sum_i \frac{1}{n_i^f} \ ,$$

$$s_i = \frac{1}{4} \left(1 + \frac{1}{n_i^f} + \frac{1}{n_{i-1}^f} \right), \quad \text{and}$$

$$t_i = \frac{1}{n_i^f} \ .$$

When v^k is displaced, the error changes. For instance, $\tilde{v}^{k+1}$ becomes $\tilde{v}^{k+1} + r\delta$, and therefore $d^v = v^{k+1} - \tilde{v}^{k+1}$ becomes $v^{k+1} - (\tilde{v}^{k+1} + r\delta) = d^v - r\delta$; similarly, d_i^e and d_i^f are impacted according to s_i and t_i. Thus, after displacement the local error becomes

$$E(\delta) = \|d^v - r\delta\|^2 + \sum \|d_i^e - s_i\delta\|^2 - \sum \|d_i^f - t_i\delta\|^2 \ .$$

This expression expands to

$$E(\delta) = a\|\delta\|^2 - \mathbf{g} \cdot \delta + E \ , \tag{13}$$

where

$$a = r^2 + \sum_i s_i^2 + \sum_i t_i^2 \ , \quad \text{and}$$

$$\mathbf{g} = 2 \left(rd^v + \sum_i s_i d_i^e + \sum_i t_i d_i^f \right) \ .$$

Equation 13 is quadratic in δ, so the minimizing value can be found analytically by differentiating and finding a zero-crossing.

$$E'(\delta) = 2a\delta - \mathbf{g} = 0 \quad \rightarrow \quad \delta = \frac{\mathbf{g}}{2a} \ .$$

This is a minimum, because $E''(\delta) = 2a > 0$. Therefore

$$\delta = \frac{\mathbf{g}}{2a} = \frac{\sum(r\alpha_e + s_i)d_i^e + \sum(r\alpha_f + t_i)d_i^f}{r^2 + \sum s_i^2 + \sum t_i^2} \ ,$$

where the detail constraint (Eqn. 10) is used to replace the d^v term. (Because the detail constraint does not apply to valence-3 vertices, the d^v term should be retained when refining them.)

Due to the interrelationships between vertices (eg. coarse vertices sharing an edge each contribute to the edge vertex), this refinement is only optimal when applied to a single vertex. That is, if a neighbor of v^k has been displaced by refinement, the details of any shared face- and edge-vertices have already been reduced, leading to an over-refinement of v^k.

We have observed [7] that scaling the displacements by the central weight of the subdivision mask adequately accounts for these interrelationships. Here, the central weight (i.e. the contribution of v^k to v^{k+1}) is r. Scaling δ by r yields

$$\delta = \frac{\sum r(r\alpha_e + s_i)d_i^e + \sum r(r\alpha_f + t_i)d_i^f}{r^2 + \sum s_i^2 + \sum t_i^2} . \tag{14}$$

4.3 Boundaries and Creases

Because Catmull-Clark surfaces are a generalization of cubic B-spline curves, curve subdivision is typically used along boundary vertices and edges [33]. Sharp features can also be accommodated with boundary masks, increasing the versatility of the representation. Similarly, a multiresolution system can employ a B-spline curve multiresolution (such as [3,14,7]) when decomposing a mesh with boundaries and sharp features. As illustrated in Fig. 9, using boundary masks can provide much better results after decomposition, introducing less error and also making the mesh easier to edit.

4.4 Results

To evaluate the fitting quality of our construction, we use a similar approach as the previous section. Each of the models in Figs. 9–12 was decomposed k times, then subdivided without details back to the original resolution. For comparison purposes, the Catmull-Clark MR system described by Bertram et al. [8] was implemented.

Table 2 summarizes the L_2-norm error of the approximation relative to the original. The approximation error introduced by our method is comparable to Bertram's method; in some instances ours outperforms theirs and vice versa, depending on the characteristics of the particular model. Based on visual inspection and the numeric results, our method provides good fitting behavior suitable for use in compression and mesh editing applications.

A benefit of subdivision-based multiresolution systems is their run-time efficiency; decomposition and reconstruction are linear, if designed well. For the

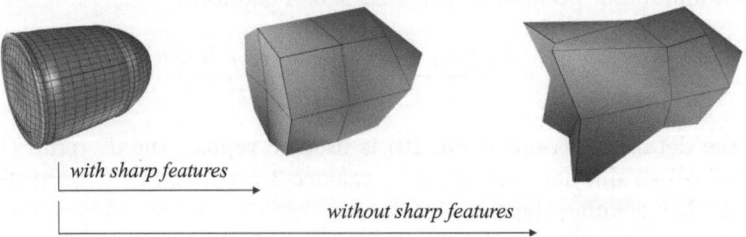

with sharp features

without sharp features

Fig. 9. Sharp features can be handled with B-spline curve masks, providing a better coarse approximation of the original data. In this example, the boundary between the silver and blue sections was automatically tagged as a sharp feature.

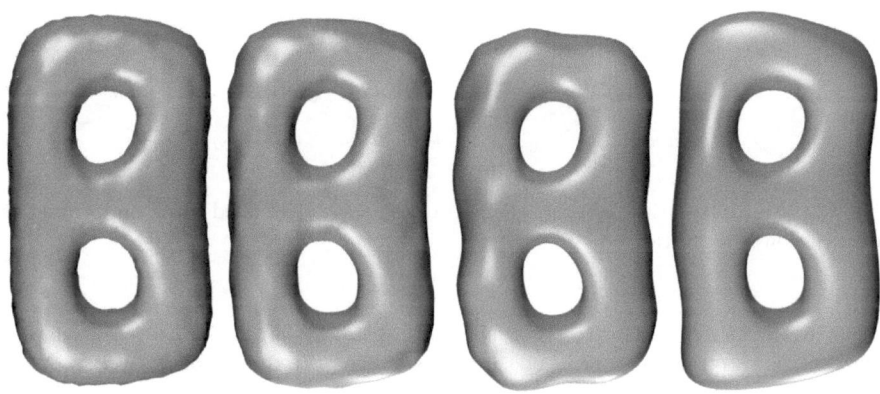

Fig. 10. De-noising of a double-torus with our refined masks, from level 3 (left) to level 0 (right)

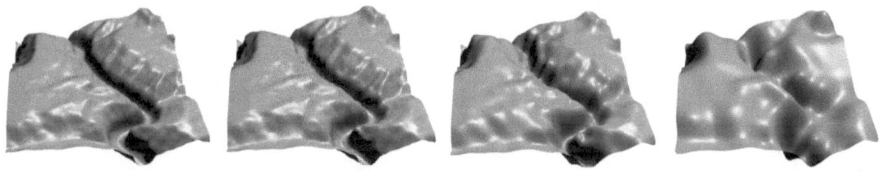

Fig. 11. Decomposition of a terrain section, from level 3 (left) to level 0 (right). (Data from USGS[34].)

Table 2. L_2 error relative to the original surface (normalized against the trial error). Shorter bars are better.

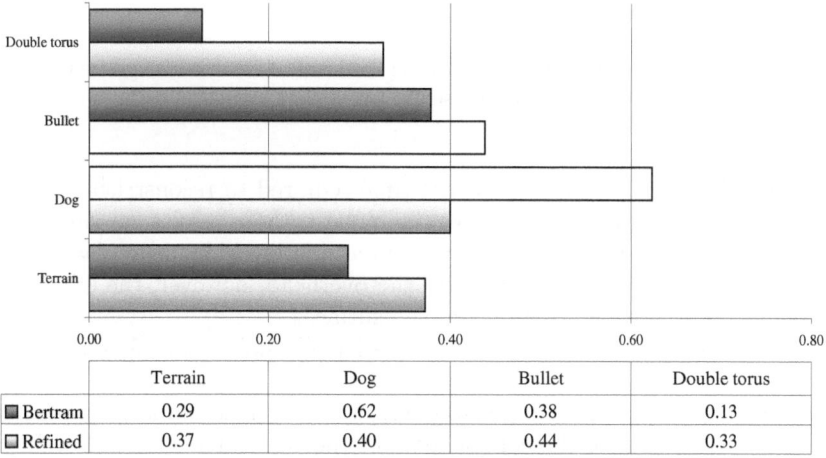

	Terrain	Dog	Bullet	Double torus
■ Bertram	0.29	0.62	0.38	0.13
▢ Refined	0.37	0.40	0.44	0.33

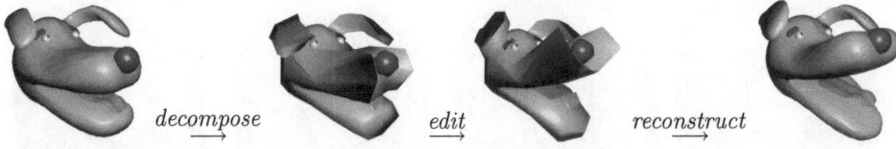

$$\xrightarrow{decompose} \qquad \xrightarrow{edit} \qquad \xrightarrow{reconstruct}$$

Fig. 12. Multiresolution editing: geometric details are retained after performing an edit at a lower resolution

Table 3. Time required to reconstruct a surface from the previous level, i.e. level 1 refers to the reconstruction of level 1 from level 0 (reported in seconds)

Model	Reconstruction time (s)			
	M^1	M^2	M^3	M^4
Double torus	0.62	0.63	1.47	3.50
Terrain	0.15	0.31	0.36	1.21
Bullet	0.10	0.15	0.17	–
Dog	0.17	0.53	1.22	–

108K faces   27K faces   6800 faces

Fig. 13. Decomposition of a complex surface: with only 1/16th of the geometric detail, the rightmost surface is a faithful representation of the original

models in Figs. 9–12, we measured the time required to reconstruct the model from the lowest resolution (level 0) to the highest; these results are summarized in Table 3. The system was implemented in a high-level language (C#), so performance is less than optimal. Despite the overhead, however, the efficiency is suitable for interactive editing of complex surfaces.

For a visual analysis of the fitting quality, consider Fig. 13. One level of decomposition reduces the geometric complexity by a factor of 4, yet has very little effect on the visual quality of the object. Further decomposition results in a surface with 1/16th of the complexity, yet the coarse model is still a faithful representation of the original.

5 Conclusion

We have proposed a wavelet-free multiresolution construction that is applicable to a variety of subdivision surfaces. By constraining the details and then choosing the coarse surface to satisfy a detail constraint, it is possible to store only a subset of the detail terms and compute the rest. To have high fitting quality, it is necessary to perform a local optimization on the coarse surface.

Though our construction can be viewed through the lens of wavelet analysis, and in fact is an instance of the lifting method, no knowledge of wavelets or scaling functions is required to understand or apply the method. Thus, the simplicity of subdivision is carried over to multiresolution systems, in both description and implementation.

Our construction was illustrated by constructing MR systems for Dyn-Levin-Gregory subdivision curves and Catmull-Clark surfaces, including a consideration of boundary masks. The resulting systems are stable and provide fitting behavior that is comparable to other methods, showing that our constructions are suitable for editing and compression applications.

Acknowledgments

This research was supported in part by the National Science and Engineering Research Council of Canada.

References

1. Stollnitz, E., DeRose, T., Salesin, D.: Wavelets for Computer Graphics: Theory and Applications, pp. 152–159. Morgan Kaufmann Publishers, Inc., San Francisco (1996)
2. Warren, J., Weimer, H.: Subdivision Methods for Geometric Design: A Constructive Approach. Morgan Kaufmann, San Francisco (2001)
3. Finkelstein, A., Salesin, D.: Multiresolution curves. In: Proc. of SIGGRAPH 1994, pp. 261–268. ACM Press, New York (1994)
4. Samavati, F., Mahdavi-Amiri, N., Bartels, R.: Multiresolution surfaces having arbitrary topologies by a reverse doo subdivision method. Computer Graphics Forum 21(2), 121–136 (2002)
5. Bertram, M.: Biorthogonal Loop-Subdivision Wavelets. Computing 72(1-2), 29–39 (2004)
6. Li, D., Qin, K., Sun, H.: Unlifted loop subdivision wavelets. In: 12th Pacific Conference on Computer Graphics and Applications (2004)
7. Olsen, L., Samavati, F., Bartels, R.: Multiresolution for curves and surfaces based on constraining wavelets. Computers and Graphics (2007)
8. Bertram, M., Duchaineau, M., Hamann, B., Joy, K.: Generalized b-spline subdivision-surface wavelets for geometry compression. IEEE Transactions on Visualization and Computer Graphics 10, 326–338 (2004)
9. Wang, H., Qin, K.H., Tang, K.: Efficient wavelet construction with catmull-clark subdivision. The Visual Computer 22, 874–884 (2006)

10. Leibniz, G.: The Labyrinth of the Continuum: Writings on the Continuum Problem, pp. 1672–1686. Yale University Press, New Haven (2001)

11. Stollnitz, E., DeRose, T., Salesin, D.: Wavelets for computer graphics: A primer, part 1. IEEE Computer Graphics and Applications 15(3), 76–84 (1995)

12. Eck, M., DeRose, T., Duchamp, T., Hoppe, H., Lounsbery, M., Stuetzle, W.: Multiresolution analysis of arbitrary meshes. In: Proc. of SIGGRAPH 1995, pp. 173–182 (1995)

13. Lounsbery, M., DeRose, T.D., Warren, J.: Multiresolution analysis for surfaces of arbitrary topological type. ACM Trans. Graph. 16(1), 34–73 (1994)

14. Samavati, F., Bartels, R.: Multiresolution curve and surface representation by reversing subdivision rules. Computer Graphics Forum 18(2), 97–120 (1999)

15. Sweldens, W.: The lifting scheme: A construction of second generation wavelets. SIAM J. Math. Anal. 29(2), 511–546 (1997)

16. Lee, A., Sweldens, W., Schröder, P., Cowsar, L., Dobkin, D.: Maps: multiresolution adaptive parameterization of surfaces. In: Proc. of SIGGRAPH 1998, pp. 95–104. ACM Press, New York (1998)

17. Litke, N., Levin, A., Schröder, P.: Fitting subdivision surfaces. In: IEEE Visualization 2001, pp. 319–324 (2001)

18. Boier-Martin, I., Rushmeier, H., Jin, J.: Parameterization of triangle meshes over quadrilateral domains. In: Proc. of ACM Symposium on Geometry Processing (SGP 2004), pp. 193–203 (2004)

19. Valette, S., Prost, R.: Wavelet-based multiresolution analysis of irregular surface meshes. IEEE Transactions on Visualization and Computer Graphics 10(2), 113–122 (2004)

20. Zorin, D., Schröder, P., Sweldens, W.: Interactive multiresolution mesh editing. In: Proc. of SIGGRAPH 1997, pp. 259–268. ACM Press, New York (1997)

21. Gioia, P.: Reducing the number of wavelet coefficients by geometric partitioning. Computational Geometry: Theory and applications 14(1-3), 25–48 (1999)

22. Botsch, M., Kobbelt, L.: Multiresolution surface representation based on displacement volumes. Computer Graphics Forum 22, 483–491 (2003)

23. Sorkine, O., Cohen-Or, D., Lipman, Y., Alexa, M., Rössl, C., Seidel, H.P.: Laplacian surface editing. In: Proc. of ACM Symposium on Geometry Processing (SGP 2004), pp. 175–184 (2004)

24. Ju, T., Schaefer, S., Warren, J.: Mean value coordinates for closed triangular meshes. In: Proc. of SIGGRAPH 2005, pp. 561–566 (2005)

25. Botsch, M., Pauly, M., Gross, M., Kobbelt, L.: Primo: Coupled prisms for intuitive surface modeling. In: Proc. of ACM Symposium on Geometry Processing (SGP 2006), pp. 11–20 (2006)

26. Chaikin, G.: An algorithm for high speed curve generation. Computer Graphics and Image Processing 3(4), 346–349 (1974)

27. Doo, D., Sabin, M.: Behaviour of recursive subdivision surfaces near extraordinary points. Computer-Aided Design 10(6), 356–260 (1978)

28. Dyn, N., Levine, D., Gregory, J.: A 4-point interpolatory subdivision scheme for curve design. CAGD 4, 257–268 (1987)

29. Bartels, R., Samavati, F.: Reversing subdivision rules: Local linear conditions and observations on inner products. Journal of Computational and Applied Mathematics 119, 29–67 (2000)

30. National Geophysical Data Center: Coastline extractor (2007), http://rimmer.ngdc.noaa.gov/mgg/coast/getcoast.html

31. Catmull, E., Clark, J.: Recursively generated b-spline surfaces on arbitrary topological surfaces. Computer-Aided Design 10(6), 350–355 (1978)
32. Zorin, D., Schröder, P.: Subdivision for modeling and animation. In: SIGGRAPH 2000 Course Notes. ACM Press, New York (2000)
33. DeRose, T., Kass, M., Truong, T.: Subdivision surfaces in character animation. In: Proc. of SIGGRAPH 1998, pp. 85–94 (1998)
34. United States Geological Survey: Seamless data distribution system (2004), http://seamless.usgs.gov/website/seamless

On the Structure of Straight Skeletons

Kira Vyatkina

Saint Petersburg State University, Department of Mathematics and Mechanics,
28 Universitetsky pr., Stary Peterhof, Saint Petersburg 198504, Russia
kira@meta.math.spbu.ru
http://meta.math.spbu.ru/~kira

Abstract. For a planar straight line graph G, its straight skeleton $S(G)$ can be partitioned into two subgraphs $S^c(G)$ and $S^r(G)$ traced out by the convex and by the reflex vertices of the linear wavefront, respectively. By further splitting $S^c(G)$ at the nodes, at which the reflex wavefront vertices vanish, we obtain a set of connected subgraphs $M_1, \ldots, M_k$ of $S^c(G)$. We show that each M_i is a pruned medial axis for a certain convex polygon Q_i closely related to G, and give an optimal algorithm for computation of all those polygons, for $1 \le i \le k$. Here "pruned" means that M_i can be obtained from the medial axis $M(Q_i)$ for Q_i by appropriately trimming some (if any) edges of $M(Q_i)$ incident to the leaves of the latter.

1 Introduction

The straight skeleton was first introduced for simple polygons by Aichholzer et al. [2], and soon generalized to the case of planar straight line graphs by Aichholzer and Aurenhammer [1]. It has promtly found a number of applications in such areas as surface reconstruction [3], computational origami [6], and many others; besides, it has served as a basis for another type of skeleton called linear axis [9,10,12]. Its advantage over the only previously known skeleton – the medial axis – resides in the fact that all its edges are straight line segments, while the medial axis for a non-convex polygonal domain necessarily contains parabolic edges as well.

Both the straight skeleton and the medial axis can be defined through wavefront propagation; we shall illustrate this for the case of polygons. Initially, the wavefront coincides with the given polygon. To obtain the straight skeleton, we let the wavefront edges move inside the polygon at equal speed, thereby remaining parallel to themselves, and keep track of the movement of its vertices. The underlying process is referred to as a *linear wavefront propagation*. To obtain the medial axis, we apply a *uniform wavefront propagation*, during which all the wavefront points move inside at constant speed. Thus, at time $t > 0$, the uniform wavefront consists of the interior points of the polygon at the distance t from its boundary. In the process, the wavefront vertices trace out the edges of the medial axis.

However, the straight skeleton is computationally more expensive than the medial axis: for an n-gon with r reflex vertices, the fastest known deterministic

M.L. Gavrilova and C.J.K. Tan (Eds.): Trans. on Comput. Sci. VI, LNCS 5730, pp. 362–379, 2009.

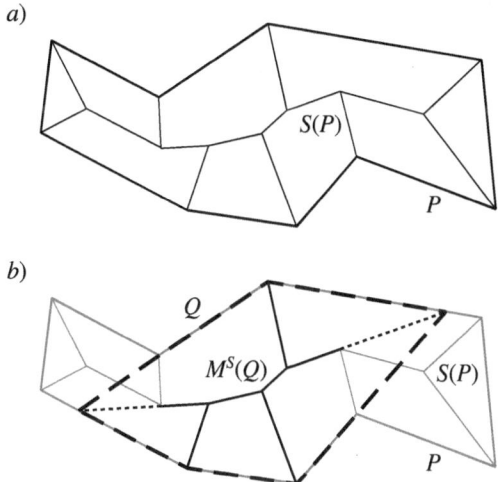

Fig. 1. a) A simple polygon P and its straight skeleton $S(P)$. b) The subtree $M^S(Q)$ of $S(P)$ is a pruned medial axis for the convex polygon Q.

algorithm for its construction, proposed by Eppstein and Erickson [7], requires $O(n^{1+\varepsilon} + n^{8/11+\varepsilon} r^{9/11+\varepsilon})$ time, where ε is an arbitrarily small positive constant; the same bounds hold for planar straight line graphs. The best existing randomized algorithm by Cheng and Vigneron [4] computes the straight skeleton for a non-degenerate simple polygon in $O(n \log^2 n + r\sqrt{r} \log r)$ expected time; for a degenerate one, the expected time bound amounts to $O(n \log^2 n + r^{17/11+\varepsilon})$. For a non-degenerate polygon with h holes, their algorithm takes $O(n\sqrt{h+1} \log^2 n + r\sqrt{r} \log r)$ expected time. But the medial axis for a simple polygon can be obtained in linear time [5], and for a planar straight line graph – in $O(n \log n)$ time [13]. It is a common belief that the straight skeleton can be computed in a more efficient way than it is possible nowadays. Yet development of such methods is likely to require investigation of additional properties of the straight skeleton. In this work, we take one step further in that direction.

Our main observation is that during the linear wavefront propagation for a planar straight line graph G, the pieces of the wavefront locally interact exactly in the same way as if they originated from the boundary of a (bounded or unbounded) convex polygon, the sides of which lie on the lines through certain edges of G. Since for a convex polygon, the two kinds of propagation proceed identically, this implies that some pieces of the medial axes for such polygons are embedded in the straight skeleton $S(G)$ for G. (Figure 1 illustrates this observation for a simple polygon P and its straight skeleton.) We formalize our ideas by indicating those pieces in $S(G)$, providing an efficient algorithm for computation of the corresponding convex polygons, and pointing out that for each such polygon Q, the piece $M^S(Q)$ of its medial axis $M(Q)$ present in $S(G)$ can be obtained by appropriately trimming the edges of $M(Q)$ incident to the vertices of Q not being those of G, and possibly, the unbounded ones, if any exist. Consequently, we say that $M^S(Q)$ is a *pruned* medial axis for Q.

A preliminary version of this paper, in which the case of simple polygons only is addressed, has appeared as [11].

In the next section, we specify classification of events that occur during the linear wavefront propagation. Section 3 analyzes the structure of the straight skeleton for a simple polygon. In Section 4, the obtained results are extended to polygons of holes, and then to planar straight line graphs. We conclude by indicating a potential direction for future research.

2   Event Classification

For clarity of exposition, we shall first develop our reasoning for the case of simple polygons, and in Section 4, a passage to general cases will be performed.

Let P be a simple polygon. Consider the process of constructing the straight skeleton $S(P)$ for P through the linear wavefront propagation. During the propagation, the wavefront structure changes as certain events occur. The very first event taxonomy [2,1] distinguishes between *edge events* and *split events*. More elaborated variations [7,4,12] additionally recognize *vertex events*. The most refined version can be found in [8].

For our purposes, we find it convenient to follow the classification we propose below.

1. *Edge event*: an edge incident to two convex vertices shrinks to zero.
2. *Sticking event*: a reflex vertex runs into an edge, thereby giving rise to precisely one (convex) vertex in the wavefront.
3. *Split event*: a reflex vertex collides into an edge, thereby splitting a wavefront component into two, and giving rise to a convex vertex in either part.
4. *Vertex event*: two reflex vertices collide together, thereby giving rise to a new reflex vertex in the wavefront.

A sticking event is attended either with annihilation of a wavefront edge incident to the reflex vertex and adjacent to the edge involved in the event, or with an edge collision. In the latter case, two parallel wavefront edges meet, thereby producing an edge of the straight skeleton, and the nodes incident to the former are brought forth either by two sticking events, or by a sticking event and an edge event. If the second event is a sticking one as well, we shall be left with two components of the wavefront instead of one (the found skeleton edge is no longer part of the wavefront).

At a vertex event, either two reflex vertices and nothing else meet at the same point, or the vertex collision is attended with simultaneous annihilation of two wavefront edges, which form a chain connecting the two vertices, and are adjacent at a convex vertex of the wavefront. Vertex events often either require special handling [7], or are ruled out by non-degeneracy assumptions [4]. We do not have to exclude vertex events from our consideration. All types of events are illustrated in Fig. 2.

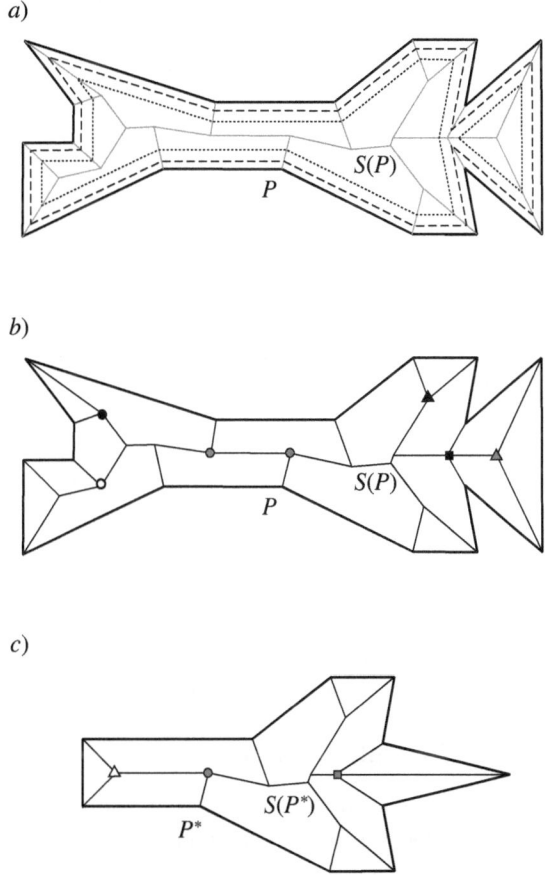

Fig. 2. a) A polygon P and its straight skeleton $S(P)$. The linear wavefront at two times is shown: soon after the propagation starts (dashed), and soon after the vertex event occurs (dotted). b) For some inner nodes of $S(P)$, the type of the corresponding event is indicated. Black triangle: an edge event. Gray triangle: three simultaneous edge events, which lead to annihilation of a wavefront component. Black circle: a sticking event. Two gray circles: two sticking events that occur simultaneously; at the same moment, the edge of $S(P)$ between the two nodes is generated entirely. White circle: a split event. Black box: a vertex event, at which two reflex vertices and nothing else meet at the same place. c) A polygon P^* and its straight skeleton $S(P^*)$. For three nodes of $S(P^*)$, the type of the underlying event is indicated. The gray circle and the white triangle denote a sticking event and an edge event, respectively; those two events occur simultaneously, and at the same moment, the edge of $S(P^*)$ between the corresponding nodes is generated entirely. The gray box denotes a vertex event attended with annihilation of the two wavefront edges forming a chain between the two colliding vertices.

A vertex event may lead to appearance of a degenerate vertex in the wavefront, with an internal angle of π. Such vertices should be handled as reflex ones.

Any inner node u of $S(P)$, which emerged not from a vertex event, but has degree $d \geq 4$, is produced by $(d-2)$ edge and/or split and/or sticking events that simultaneously occur at the same location. Those events can be handled one at a time, with any two consecutive ones being separated by a zero time interval. Therefore, any such node of $S(P)$ can be interpreted as $(d-2)$ coinciding nodes of degree three connected by $(d-3)$ edges of zero length in such a way that the subgraph induced by those nodes is a tree.

On the other hand, at any point, at most one vertex event can occur. Thus, if a node u of degree $d \geq 5$ has an associated vertex event, then it is produced by that vertex event and $(d-4)$ edge and/or split and/or sticking events that simultaneously occur at the same location. All those events can be handled as in the above case, and u admits an analogous interpretation.

Consequently, we may further suppose that any node of $S(P)$ resulting from a vertex event has degree four, and any other its inner node has degree three. A wavefront component annihilation then corresponds to three simultaneous edge events at the same point.

3   Separation of the Pruned Medial Axes

In this section, we outline step by step both extraction of the pruned medial axes from the straight skeleton and reconstruction of the respective convex polygons.

3.1   Partition of the Straight Skeleton

Let P be a simple polygon with n vertices, r of those being reflex; assume that $r \geq 1$. First, let us decompose the straight skeleton $S(P)$ for P into two subgraphs $S^c(P)$ and $S^r(P)$, being parts of $S(P)$ traced out by the convex and by the reflex wavefront vertices, respectively. By construction, $S^r(P)$ is a forest. To make it more precise, if no vertex events occur during the propagation, $S^r(P)$ consists of r edges of $S(P)$ incident to the reflex vertices of P; otherwise, at least one tree from $S^r(P)$ has three or more edges.

Next, let us split $S^c(P)$ at the nodes, at which the reflex wavefront vertices vanish (Fig. 3a,b). We shall be left with a decomposition of $S^c(P)$ into $k \leq r+1$ connected subgraphs $M_1, M_2, \ldots, M_k$. We claim that for any i, $1 \leq i \leq k$, M_i is a part of the medial axis for a convex polygon. Strictly speaking, there are infinitely many such polygons; of course, we would like to retrieve one with the least computational effort. Below we shall formalize our intent.

3.2   Extraction of the Boundary Chains

For any edge e of P, we define its corresponding *cell* $C(e)$ to be the face of the partition of P induced by $S(P)$, which is adjacent to e. Equivalently, $C(e)$ is the region swept in the propagation by the portion of the linear wavefront originating from e.

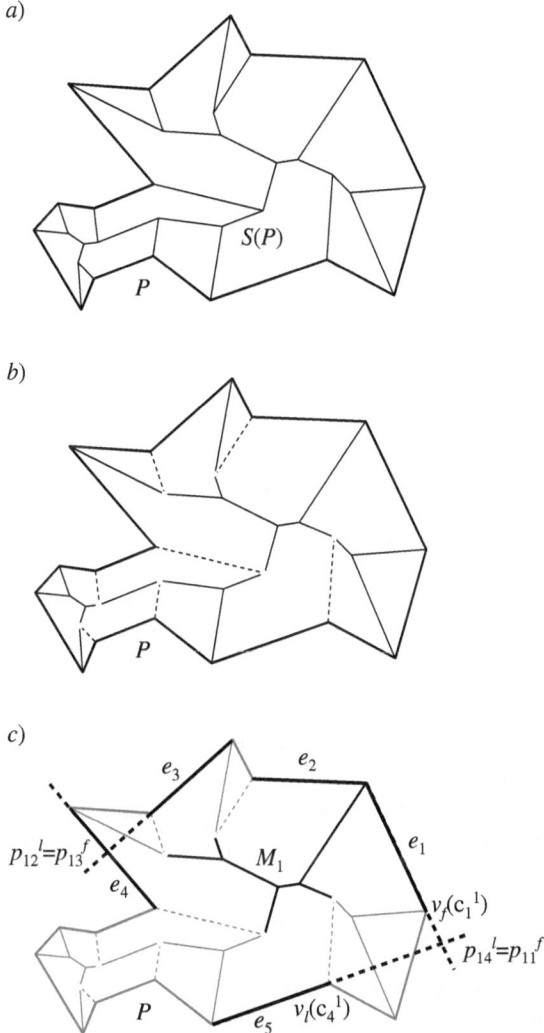

Fig. 3. a) A simple polygon P and its straight skeleton $S(P)$. b) Decomposition of the part $S^c(P)$ of the straight skeleton traced out by the convex vertices. The edges of $S(P)$ traced out by the reflex vertices are shown dashed. c) For the fragment M_1, $E_1 = \{e_1, e_2, e_3, e_4, e_5\}$; $C_1 = \{c_1^1, c_2^1, c_3^1, c_4^1\}$, where the chain c_1^1 is formed of e_1 and e_2, and each of c_2^1, c_3^1, and c_4^1 consists of a single edge – of e_3, e_4, and e_5, respectively. By prolonging the edges as shown in dotted lines, we obtain the rays $\bar{f}(c_1^1)$, $\bar{l}(c_2^1)$, $\bar{f}(c_3^1)$, and $\bar{l}(c_4^1)$, respectively; here $p_{14}^l = p_{11}^f = \bar{l}(c_4^1) \cap \bar{f}(c_1^1)$, and $p_{12}^l = p_{13}^f = \bar{l}(c_2^1) \cap \bar{f}(c_3^1)$.

Consider any M_i. Since $S(P)$ is a tree, M_i is a tree as well. The embedding of M_i in the plane induces a cyclic order of its leaves. For any consecutive pair of leaves, when walking from one of them to the other along the edges of M_i, we follow the boundary of some cell. Moreover, in can be easily verified that for any two such pairs of leaves, the corresponding cells must be different. These cells are also cyclically ordered, in compliance with the ordering of the leaves.

Now retrieve all the edges of P, such that the boundaries of their cells contribute to M_i; denote the resulting set by E_i (Fig. 3c). From the above discussion, it follows that $|E_i|$ equals the number of the leaves in M_i. Let the edges in E_i inherit the cyclic order of the cells. For any two consecutive edges $e, e' \in E_i$, their cells $C(e)$ and $C(e')$ share an edge of M_i incident to a leaf. If the leaf corresponds to a convex vertex of P, then e and e' share this vertex. Otherwise, the leaf corresponds to an inner node u of $S(P)$ adjacent to a reflex vertex of P. In this case, e and e' can be (but not necessarily are) adjacent only if E_i consists solely of e and e'. To see this, suppose e and e' are adjacent at vertex v of P. Then $C(e)$ and $C(e')$ must share the edge of $S(P)$ incident to v. On the other side, any two cells can share at most one edge of $S(P)$. Therefore, the edge of M_i shared by $C(e)$ and $C(e')$ must be uv. It follows immediately that v is convex, M_i consists of a single edge uv, and e and e' are the only two edges in E_i.

Thus, we conclude that the edges from E_i together compose one or a few disjoint convex chains cut out of the boundary of P. Let $\mathcal{C}_i = \{c_1^i, \ldots, c_{m_i}^i\}$ denote the set of those chains; observe that m_i equals the number of the leaves of M_i that correspond to the inner nodes of $S(P)$. Denote by $f(c_j^i)$ and $l(c_j^i)$ the first and the last edge of the chain c_j^i, respectively; assume that c_j^i is traversed from $f(c_j^i)$ to $l(c_j^i)$ when walking counterclockwise along the boundary of P, where $1 \leq j \leq m_i$. If c_j^i consists of a single edge, then $f(c_j^i)=l(c_j^i)$. Let $v_f(c_j^i)$ and $v_l(c_j^i)$ denote the first and the last vertex of c_j^i, respectively. Without loss of generality, suppose that the chains in $\mathcal{C}_i$ are enumerated in such a way that $f(c_{d+1}^i)$ follows $l(c_d^i)$ in the cyclic order of the edges from E_i, where $1 \leq d < m_i$. To unify the notation, let $c_{m_i+1}^i = c_1^i$, and let $c_0^i = c_{m_i}^i$.

Lemma 1. *For any j, $1 \leq j \leq m_i$, at least one of $v_l(c_j^i)$ and $v_f(c_{j+1}^i)$ is reflex.*

Proof. Consider the edge (u, x) of M_i shared by $C(l(c_j^i))$ and $C(f(c_{j+1}^i))$. Assume that when walking along (u, x) from u to x, we follow counterclockwise the boundary of $C(l(c_j^i))$; then u is necessarily a leaf of M_i, which corresponds to an inner node u' of $S(P)$.

If u' appeared as a result of a vertex event, then for any of the four cells incident to u', its generative edge is incident to a reflex vertex of P, which is adjacent with u' in $S(P)$. In particular, this holds for the edges $l(c_j^i)$ and $f(c_{j+1}^i)$. Otherwise, u' has degree three; therefore, precisely three cells meet at u'. Two of those are $C(l(c_j^i))$ and $C(f(c_{j+1}^i))$. On the other hand, u' is adjacent to a reflex vertex r of P, and the cells of the both edges of P incident to r are incident to u'. This implies that r is incident either to $l(c_j^i)$ or to $f(c_{j+1}^i)$.

To prove our claim, it remains to demonstrate that u' can be adjacent neither to the first vertex of $l(c_j^i)$, nor to the last vertex of $f(c_{j+1}^i)$. By symmetry, it suffices to show the first statement. To this end, recall that $C(l(c_j^i))$ is a simple polygon. Let v denote the first vertex of $l(c_j^i)$. Note that $v_l(c_j^i)$ is encountered immediately after v, and x – immediately after u', when walking counterclockwise along the boundary of $C(l(c_j^i))$. Consequently, u' and v are non-adjacent vertices of $C(l(c_j^i))$, and thus, they cannot be adjacent in $S(P)$.

Corollary 1. *For any j, $1 \leq j \leq m_i$, one of the following three possibilities occurs in the propagation:*

- *$v_l(c_j^i)$ is a reflex vertex that runs into $f(c_{j+1}^i)$;*
- *$v_f(c_{j+1}^i)$ is a reflex vertex that runs into $l(c_j^i)$;*
- *$v_l(c_j^i)$ and $v_f(c_{j+1}^i)$ are both reflex and collide.*

3.3   Edge Delineation

Consider any tree M_i, where $1 \leq i \leq k$. We shall analyze how the edges of M_i could have been traced out.

Let us first assume that M_i has precisely one edge g. Then E_i consists of two edges e_1 and e_2, which can represent either a single or two separate chains comprising C_i. In the former case, g is incident to the convex vertex of P shared by e_1 and e_2, and its delineation is terminated by a sticking event, at which a reflex vertex incident to a wavefront edge emanating from one of e_1 and e_2, runs into an edge originating from the other one.

In the latter case, g could either appear entirely at once, or be traced out from one endpoint to another. The first possibility could be realized only if g emerged as a result of collision between two wavefront edges originating from e_1 and e_2, respectively, attended by two simultaneous sticking events (see Section 2). Then, in particular, e_1 and e_2 must be two disjoint parallel edges lying not on the same line. Otherwise, g is traced out properly. It means that during the propagation, two wavefront edges emanating from e_1 and e_2, respectively, become adjacent – through an event involving a reflex vertex – at a convex wavefront vertex w, which then traces out g. The delineation of g ends as in the above case.

If M_i has more than one edge, then it must have at least three edges, and any its edge incident to a leaf is also incident to an inner node.

Lemma 2. *Among the edges of M_i incident to a leaf, at most one was not traced out starting from the leaf.*

Proof. If M_i has a single edge, the claim trivially holds. Now assume that M_i has at least three edges.

Let z be any inner node of M_i. Observe that z must have appeared as a result of an edge event, and is incident to three edges of M_i. Two of those were delineated by the convex wavefront vertices incident to the vanishing edge; therefore, they were traced out towards z. The third one might have been traced out towards z as well, or delineated starting from z, or generated entirely at once (Fig. 4).

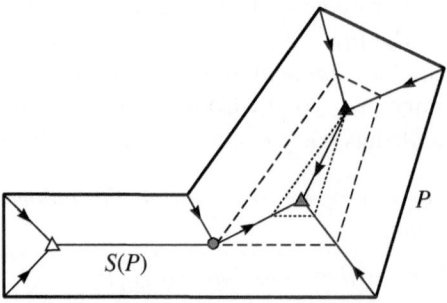

Fig. 4. A polygon P and its straight skeleton $S(P)$. Directions, in which the edges of $S(P)$ are traced out, are shown with arrows. Edge events of the three possible kinds are indicated with triangles. The horizontal edge of $S(P)$ is generated entirely at once; at the same moment, an edge event (white triangle) and a sticking event (gray circle) occur. The linear wavefront thereby becomes as depicted dash. Next, an edge event marked by the black triangle occurs, and the wavefront becomes as shown dotted. Finally, an edge event marked by the gray triangle occurs, and the wavefront vanishes.

First, suppose that M_i contains an edge (x, y), which emerged from an edge collision and is incident to two inner nodes. Having removed this edge from M_i, we shall be left with two trees M_i^x and M_i^y containing the nodes x and y, respectively. Let us consider M_i^x. From the above observation, it follows that the two edges (x', x) and (x'', x) of M_i incident to x must have been traced out towards x. Having rooted M_i^x at x and traversed it in a breadth first (or depth first) order, at every step applying the same argument, we shall conclude that any edge of M_i^x incident to a leaf is delineated starting from that leaf. The same holds for M_i^y, and hence, for M_i.

In case M_i has an edge (u, x), which emerged from an edge collision and is incident to a leaf u, a similar reasoning implies that any other edge of M_i incident to a leaf was traced out starting from the leaf.

Finally, assuming that all the edges of M_i have been delineated properly, let us suppose for contradiction that at least two edges (u, x) and (u', x') of M_i were traced out towards the leaves u and u', respectively. Consider the path $(x_0 = x, x_1, \ldots, x_h = x')$ between x and x' in M_i, where $h \geq 1$. Any node x_l is an inner node of M_i, where $1 \leq l \leq h$. Since the edge (u, x) was traced out from x to u, this implies that each edge (x_l, x_{l-1}) must have been traced out from x_l to x_{l-1}, for $1 \leq l \leq h$, and finally, the edge (u', x') must have been traced out from u' to x', which is a contradiction.

Corollary 2. *At most one edge of M_i could have appeared as a result of an edge collision. If g is such an edge, then the delineation of M_i ended at the moment when g was generated.*

Corollary 3. *Let (u, x) be an edge of M_i incident to a leaf u. If (u, x) had been traced out from x to u, then the delineation of M_i ended at the moment when u was generated.*

For any edge $e \in E_i$, denote by p_e^i the path in M_i consisting of the edges that belong to $\partial C(e)$. Obviously, p_e^i connects two consecutive (with respect to the cyclic order) leaves of M_i. By analyzing the process of edge delineation, we shall prove the following lemma.

Lemma 3. *For any $e \in E_i$, p_e^i is a convex chain strictly monotone with respect to the line through e.*

Proof. Let u and w denote the two consecutive leaves of M_i being the endpoints of p_e^i, u preceding w in the counterclockwise order of the leaves. Without loss of generality, assume that e is horizontal, and $t_w \geq t_u \geq 0$, where t_w and t_u denote the time, at which the nodes w and u appeared, respectively. By definition of M_i, each of u and w either coincides with a convex vertex of P, or was generated at an event involving a reflex vertex, and any inner node of p_e^i must have been generated at an edge event.

Let us first consider a special case when M_i has a single edge g that appeared through an edge collision. Then so does p_e^i; moreover, g is parallel to e, and thus, the claim holds.

Otherwise, the first edge (u, x) of p_e^i must have been traced out starting from u. Consequently, as soon as u comes into existence, it becomes a convex vertex of the wavefront incident to two edges $\hat{e}$ and $\hat{e}_u$ of the latter, the first of which originated from e, and the second – from the edge $e_u \in E_i$, which precedes e in the cyclic order of the edges composing E_i. Observe that $\hat{e}$ lies on the left of $\hat{e}_u$, if the latter is oriented towards u, and the bisector of the interior wavefront angle between $\hat{e}$ and $\hat{e}_u$, on which (u, x) lies, is inclined to the right.

If the node x coincides with w, our claim follows immediately. Otherwise, x is generated at an edge event. First, suppose that $\hat{e}$ thereby neither collapses, nor collides with another edge; then $\hat{e}_u$ must shrink to zero. Note that $\hat{e}_u$ must have been incident to two convex vertices straight before the event. Let $\hat{e}_x$ denote the second wavefront edge adjacent to $\hat{e}_u$. At the event, $\hat{e}$ and $\hat{e}_x$ become adjacent at a convex wavefront vertex v, the interior angle at which is less than was the one between $\hat{e}$ and $\hat{e}_u$; therefore, the next edge (x, x') of p_e^i traced out by v will be inclined to the right as well, and have a smaller slope than that of (u, x). The process continues until the node w is reached, or $\hat{e}$ shrinks to zero, or the generation of the next node involves collision of $\hat{e}$ with another edge.

In case we finally reach w, both convexity and strict x-monotonicity of p_e^i are guaranteed by the above reasoning. Otherwise, if $\hat{e}$ shrinks to zero, then the node x_0 generated thereby is the topmost vertex of $C(e)$, at which the part of p_e^i traced out starting from u meets the one delineated starting from w. Since symmetric arguments apply to the second part of p_e^i, our claim holds in this case.

Finally, if the generation of the next node $\bar{x}$ resulting from an edge event is also attended by a collision of $\hat{e}$ with another edge, then a horizontal edge $(\bar{x}, \bar{y})$ of p_e^i must appear at that moment. Two cases are possible: either $\bar{y} = w$, or $\bar{y}$ appears simultaneously due to another edge event. The former case is similar to the one when we reach w, having started the delineation of p_e^i from u. In the latter case, the part of p_e^i traced out starting from w terminates at $\bar{y}$; the rest is similar to the case when the two parts meet at the topmost vertex of $C(e)$.

3.4 Chain Prolongation

For an edge (u, x) of M_i incident to a leaf u, let $r_{u,x}$ denote the open ray with the endpoint u, collinear to (u, x), such that $r_{u,x} \cap (u, x) = \emptyset$.

Observation 1. *Let (u, x) be the edge of M_i shared by the cells $C(l(c_j^i))$ and $C(f(c_{j+1}^i))$ for some j, $1 \le j \le m_i$; assume that u denotes a leaf. If (u, x) is traced out from u to x, the ray $r_{u,x}$ and the lines through $l(c_j^i)$ and $f(c_{j+1}^i)$ intersect at a common point. Otherwise, neither line intersects $r_{u,x}$.*

For any chain c_j^i, let us take the edge $l(c_j^i)$ and prolong it to infinity, thereby eliminating $v_l(c_j^i)$; denote the resulting ray by $\bar{l}(c_j^i)$. Similarly, let $\bar{f}(c_j^i)$ denote the ray obtained by prolonging $f(c_j^i)$ to infinity beyond the vertex $v_f(c_j^i)$ (see Fig. 3c).

For any j, $1 \le j \le m_i$, consider the edge (u, x) of M_i shared by $C(l(c_j^i))$ and $C(f(c_{j+1}^i))$. Assume that when walking along (u, x) from u to x, we follow counterclockwise the boundary of $C(l(c_j^i))$; then u is necessarily a leaf of M_i corresponding to an inner node of $S(P)$.

If the lines through $l(c_j^i)$ and $f(c_{j+1}^i)$ intersect the ray $r_{u,x}$ at a point p, let $p_{ij}^l = p_{i,j+1}^f = p$; otherwise, let p_{ij}^l and $p_{i,j+1}^f$ be two points at infinity lying on $\bar{l}(c_j^i)$ and on $\bar{f}(c_{j+1}^i)$, respectively.

Lemma 4. *Let c_j^i be a chain formed of at least two edges. Then $p_{ij}^f \in \bar{f}(c_j^i)$, and $p_{ij}^l \in \bar{l}(c_j^i)$.*

Proof. Let us demonstrate that $p_{ij}^l \in \bar{l}(c_j^i)$; the second statement is symmetric. Consider the edge (u, x) of M_i shared by $C(l(c_j^i))$ and $C(f(c_{j+1}^i))$. Assume that when walking along (u, x) from u to x, we follow counterclockwise the boundary of $C(l(c_j^i))$; then u is necessarily a leaf of M_i corresponding to an inner node of $S(P)$. Without loss of generality, suppose that (u, x) is vertical, and x lies above u; then $C(l(c_j^i))$ locally lies to the left of (u, x).

If (u, x) was traced out from x to u or generated entirely at once, then p_{ij}^l is infinite, and the claim holds by definition of p_{ij}^l. Otherwise, the lines through $l(c_j^i)$, $f(c_{j+1}^i)$, and (u, x) intersect at a common point below u (see Observation 1), and the polygon locally lies above the edges $l(c_j^i)$ and $f(c_{j+1}^i)$.

Let z denote the vertex of P incident to $l(c_j^i)$ other than $v_l(c_j^i)$. If z lies on the left of the vertical line l through (u, v), the claim holds. Otherwise, z must be a reflex vertex: if z was convex, the horizontal projection of its speed would be positive, and the wavefront edge originating from $l(c_j^i)$ would never cross l. Thus, $C(l(c_j^i))$ would lie entirely to the right of l, contradicting the fact that $C(l(c_j^i))$ locally lies to the left of (u, x). But if z is reflex, then c_j^i must consist of a single edge, which is a contradiction.

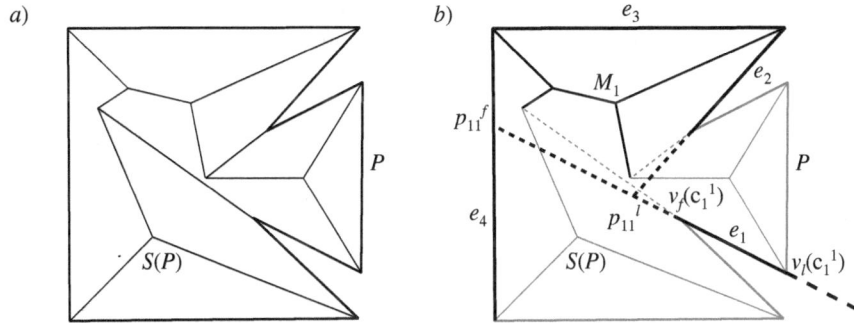

Fig. 5. a) A simple polygon P and its straight skeleton $S(P)$. b) The polygon P and its straight skeleton $S(P)$ are depicted gray; the edges of $S(P)$ incident to the reflex vertices of P are marked dotted. For the fragment M_1 of $S(P)$, $E_1 = \{e_1, e_2, e_3, e_4\}$; $\mathcal{C}_1 = \{c_1^1, c_2^1\}$, where the chain c_1^1 consists solely of e_1, and c_2^1 is formed of e_2, e_3, and e_4. By prolonging e_1 beyond $v_l(c_1^1)$ as shown dashed, we obtain the ray $\bar{l}(c_1^1)$. Here $p_{1,1}^l$ lies outside $\bar{l}(c_1^1)$; the chain $\bar{c}_1^1$ is the segment between $p_{1,1}^f$ and $p_{1,1}^l$, and $\bar{c}_1^1 \cap c_1^1 = \emptyset$.

If c_j^i is formed of a single edge, an analogous statement is not necessarily true (Fig. 5).

However, the following property will hold.

Lemma 5. *Let c_j^i be a chain formed of a single edge $(v_f(c_j^i), v_l(c_j^i))$. Then the vectors $\overrightarrow{p_{ij}^f p_{ij}^l}$ and $\overrightarrow{v_f(c_j^i) v_l(c_j^i)}$ have the same direction.*

Proof. If at least one of p_{ij}^l and p_{ij}^f is infinite, the claim obviously holds. Otherwise, it is implied by Lemma 3.

3.5   Formation of the Convex Polygons

For each j, $1 \le j \le m_i$, consider the chain c_j^i. If c_j^i is formed of at least two edges, construct a chain $\bar{c}_j^i$ from c_j^i by adjusting the first and the last edge of the latter, so that they will terminate at p_{ij}^f and p_{ij}^l, respectively, if those are finite, or become unbounded, if the corresponding point is infinite. Otherwise, let $\bar{c}_j^i$ be the segment between p_{ij}^f and p_{ij}^l (see Fig. 5); if one or both of p_{ij}^f and p_{ij}^l are infinite, $\bar{c}_j^i$ will become a ray or a line, respectively. Let $\bar{c}^i = \cup_j \bar{c}_j^i$.

Lemma 6. *$\bar{c}^i$ bounds a convex region in the plane.*

Proof. Applying Observation 1 and Lemma 2, we derive that $\bar{c}^i$ can be either composed of two parallel lines or a polygonal chain (closed or open). In the former case, $\bar{c}^i$ bounds an infinite strip in the plane, and the claim trivially holds.

Let us take M_i and prolong to infinity each its edge incident to a leaf, thereby eliminating all the leaves. Lemma 3 assures that no two neighbor unbounded edges will intersect; consequently, the obtained tree-like structure $\mathcal{M}_i$ induces a partition $\mathcal{P}_i$ of the plane into $|E_i|$ unbounded convex regions.

Suppose that $\bar{c}^i$ is a closed polygonal chain. It is easy to verify that the unbounded edges of $\mathcal{M}_i$ pass precisely through the vertices of $\bar{c}^i$. Any edge $\bar{e}_h$ of $\bar{c}^i$ cuts away a convex polygon $\overline{P}_h$ from a separate unbounded region of $\mathcal{P}_i$, where $1 \leq h \leq |E_i|$; in particular, no two such polygons overlap. Having glued all the polygons $\overline{P}_h$, for $1 \leq h \leq |E_i|$, along the corresponding bounded edges or parts of the unbounded edges of $\mathcal{M}_i$, we shall obtain a plane simply connected domain D^i bounded by $\bar{c}^i$. Our construction implies that any interior angle of D^i is less than π; therefore, D^i is a convex polygon.

Now let $\bar{c}^i$ be an open polygonal chain. If M_i has a single edge, then $\bar{c}^i$ consists of two half-infinite edges and bounds a wedge with an apex angle less than π, which is an unbounded convex polygon.

Otherwise, for some j, $1 \leq j \leq m_i$, the points p_{ij}^l and $p_{i,j+1}^f$ are both infinite. Let g denote the edge of M_i shared by $C(l(c_j^i))$ and $C(f(c_{j+1}^i))$; observe that g is incident to a leaf. Consider the unbounded edge $\bar{g}$ of $\mathcal{M}_i$ obtained from g. Either half-infinite edge of $\bar{c}^i$ cuts away an unbounded convex polygon from a separate region of $\mathcal{P}_i$ incident to $\bar{g}$; either such polygon will retain $\bar{g}$ in its boundary. Any finite edge of $\bar{c}^i$ cuts away a bounded convex polygon from some region of $\mathcal{P}_i$, as in the previous case. Following a similar reasoning as above, we conclude that $\bar{c}^i$ bounds an unbounded convex polygon.

Let Q_i denote the convex region bounded by $\bar{c}^i$ (see Fig. 6). We shall refer to Q_i as to a convex polygon, either bounded or unbounded.

Now construct a tree-like structure $\overline{M}_i$ from the tree M_i as follows. For each j, $1 \leq j \leq m_i$, consider the edge $g_j = (u_j, x_j)$ of M_i shared by $C(l(c_j^i))$ and $C(f(c_{j+1}^i))$; assume that it is traversed from u_j to x_j when walking counter-clockwise along the boundary of $C(l(c_j^i))$. Consequently, u_j is necessarily a leaf of M_i, which corresponds to an inner node of $S(P)$. Prolong g_j beyond u_j until the point p_{ij}^l, if p_{ij}^l is finite, and to infinity, otherwise. The reasoning carried out above implies correctness of the proposed construction.

Lemma 7. $\overline{M}_i$ *is the medial axis for* Q_i.

Proof. Since Q_i is a convex polygon, its medial axis coincides with its Voronoi diagram. Therefore, it is sufficient to demonstrate that $\overline{M}_i$ partitions the interior of Q_i into the Voronoi cells of its edges.

By construction, $\overline{M}_i$ partitions Q_i into convex polygonal faces, each adjacent to a separate edge of Q_i. For an edge e of Q_i, let $F(e)$ denote its adjacent face of the partition.

Suppose for contradiction that for some edge e of Q_i, the face $F(e)$ contains a point x, for which the closest edge of Q_i is e' other than e. Let us drop a perpendicular from x onto e'; note that its foot a must fall inside e'. Since x lies outside $F(e')$, the segment xa intersects the boundary $\partial F(e')$ of $F(e')$ at some point z belonging to $\overline{M}_i$. Such point is unique by convexity of $F(e')$.

Recall that $\partial F(e')$ is composed of e' and of pieces of bisectors between e' and some other edges of Q_i. If z is interior to some edge g of $\overline{M}_i$, let e'' denote the

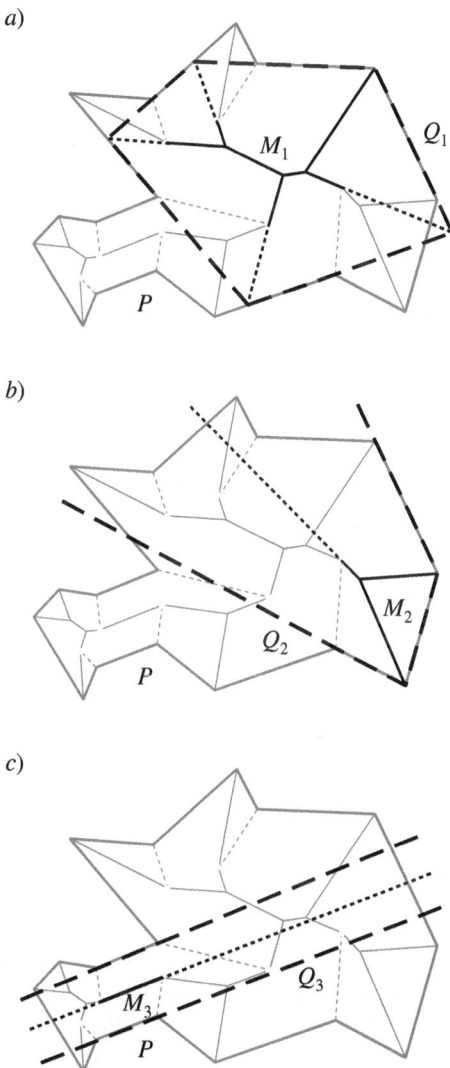

a)

b)

c)

Fig. 6. A simple polygon P (bold gray) and its straight skeleton (gray); the edges of the latter incident to the reflex vertices of the former are depicted dotted gray. The convex region Q_i bounded by $\bar{c}^i$ (dashed) can be of one of the three types: a) a convex polygon; b) an infinite convex region bounded by a chain, the first and the last edges of which are infinite; c) an infinite strip bounded by two parallel lines. For any Q_i, the corresponding subtree M_i of $S(P)$ is shown bold, and the parts of the edges of $M(Q_i)$ not belonging to M_i are shown dotted, where $1 \leq i \leq 3$.

edge of Q_i, such that the bisector of e' and e'' contains g. In particular, e'' and e may be the same edge. In case z coincides with a node u of $\overline{M}_i$, any edge other than e', such that its adjacent face is incident to u, can be chosen as e''.

Next, let us drop a perpendicular from z onto e''. Lemma 3 together with our construction imply that its foot b will fall inside e''. But since, by triangle inequality, $|xb| < |xz + zb| = |xz + za| = |xa|$, it follows that x is closer to e'' than to e', which contradicts our assumption.

Corollary 4. *M_i is part of $M(Q_i)$, and can be obtained from the latter by appropriately trimming its edges incident to the vertices of Q_i not being those of P, and the unbounded ones, if any exist.*

It is easy to see that each M_i is a *maximal* fragment of a medial axis, in a sense that it cannot be extended along the edges of $S(P)$ while remaining a part of the medial axis for any polygon.

Given P and $S(P)$, and assuming that the representation of the latter provides information on the partition of P induced by $S(P)$, it is straightforward to decompose $S^c(P)$ into the subtrees $M_1, \ldots, M_k$, and to retrieve the corresponding sets of chains $\mathcal{C}_1, \ldots, \mathcal{C}_k$. For any i, $1 \leq i \leq k$, the convex polygon Q_i can then be constructed from $\mathcal{C}_i$ following the procedure described above.

We summarize our results in the next theorem.

Theorem 1. *Let P be a simple polygon. The subgraph $S^c(P)$ of the straight skeleton $S(P)$ for P, traced out by the convex vertices of the linear wavefront, can be uniquely partitioned into a set of maximal fragments of medial axes. Each of those fragments represents a pruned medial axis for a certain convex polygon. Both the partition and the corresponding set of convex polygons can be computed from P and $S(P)$ in linear time.*

4   General Cases

In this section, we generalize our results to the case of polygons with holes, and further to the case of planar straight line graphs. To this end, we demonstrate that in either case, by extracting from the straight skeleton its subgraph traced out by the convex vertices of the wavefront, and splitting it at the nodes, at which the reflex wavefront vertices vanish, we again obtain a forest. Subsequently, to any connected component of the forest, which is a tree, the previously developed reasoning fully applies. Special attention is paid to the unbounded edges of the straight skeleton for a planar straight line graph.

The only remark to be made is that in general cases, a split event causes either a break of a linear wavefront component into two (if the edge and the reflex vertex involved in the event belong to the same connected component of the wavefront), or a merge of two wavefront components (if the respective edge and vertex belong to different components). Similarly, a vertex event can result in merging two wavefront components. Though in the case of simple polygons, different wavefront components can never merge, this distinction does not affect the above discussion.

4.1 Polygons with Holes

Let P be a polygon with holes; let $S(P)$ denote the straight skeleton of P. Consider the subgraph $S^c(P)$ of $S(P)$ traced out by the convex vertices of the linear wavefront. Having split $S^c(P)$ at the inner nodes, at which the reflex wavefront vertices vanish, we obtain a partition of $S^c(P)$ into a number of connected components $M_1, \ldots, M_k$.

Lemma 8. *For any* i, $1 \leq i \leq k$, M_i *is a tree.*

Proof. Suppose for contradiction that for some i, $1 \leq i \leq k$, M_i contains cycles. Let c denote any cycle of M_i; note that c bounds a simple polygon P_c formed as a union of the cells and of the holes of P that lie inside c. In particular, the generative edge of any cell being part of P_c lies inside c. Thus, at least one connected component of ∂P falls inside c.

Any vertex of c is incident to a cell lying inside c, and to a cell lying outside c. It follows that the first generated vertex u of c must have appeared as a result of interaction between two different connected components of the linear wavefront, one emanating from inside c, and the other – from outside c. But such interaction must have involved a reflex wavefront vertex. Therefore, c must have been split at u, which is a contradiction.

4.2 Planar Straight Line Graphs

Let G be a planar straight line graph; let $S(G)$ denote the straight skeleton for G. It follows from the definition of $S(G)$ that for any face F of G, the restriction of $S(G)$ to the interior of F is the straight skeleton $S(F)$ for F. Consequently, $S(G)$ can be viewed as a union of the straight skeletons for all the faces of G.

Any bounded face of G is a polygon, either simple or with holes. Thus, for each bounded face, its straight skeleton can be processed by means of the technique developed above. Let F_∞ denote the unbounded face of G. Note that $S(F_\infty)$ contains unbounded edges, any of which is incident to one finite and one infinite node. Consider the subgraph $S^c(F_\infty)$ of $S(F_\infty)$ traced out by the convex vertices of the linear wavefront, and split $S^c(F_\infty)$ at the inner nodes, at which the reflex wavefront vertices vanish. As a result, we obtain a number of connected subgraphs $M_1, \ldots, M_k$ of $S^c(F_\infty)$. Absence of cycles in any M_i, where $1 \leq i \leq k$, can be demonstrated in the same way as in the case of polygons with holes; therefore, each M_i can be handled as described in Section 3, unless it contains an unbounded edge.

Lemma 9. M_i *contains at most one unbounded edge.*

Proof. Observe that an unbounded edge of M_i must be traced out towards the infinite node, and apply a similar reasoning as in the last part of the proof of Lemma 2.

However, the case when M_i contains an unbounded edge is fully similar to that when all the edges of M_i are finite, and one of its edges incident to a leaf is traced out towards the leaf. The only difference is that at the very last step, there is no need to clip the infinite edge of $M(Q_i)$ in order to obtain M_i.

Thus, we can generalize Theorem 1 as follows.

Theorem 2. *Let G be a planar straight line graph. The subgraph $S^c(G)$ of the straight skeleton $S(G)$ for G, traced out by the convex vertices of the linear wavefront, can be uniquely partitioned into a set of maximal fragments of medial axes. Each of those fragments represents a pruned medial axis for a certain convex polygon. Both the partition and the corresponding set of convex polygons can be computed from G and $S(G)$ in linear time.*

5   Conclusion

The principal objective of this research was to enhance understanding of the geometry of the straight skeleton. We restricted our attention to its subgraph traced out by the convex vertices of the linear wavefront, and claimed that by splitting it at the nodes, at which the reflex vertices of the wavefront vanish, we would obtain a set of pruned medial axis for certain convex polygons. Moreover, any subgraph we thus get is a maximal fragment of a medial axis embedded in the straight skeleton, in a sense that it cannot be extended along the straight skeleton edges, while remaining a piece of the medial axis for any polygon. Finally, we pointed out that, given a polygon or, more generally, a planar straight line graph, and its straight skeleton, we can easily retrieve this set of pruned medial axes along with the respective convex polygons in total linear time.

An interesting development of our work would be to speed up the computation of the straight skeleton by exploiting structural properties of the latter described and analyzed in this paper.

Acknowledgement

This research was supported by Russian Foundation for Basic Research (grant 07-07-00268-a).

References

1. Aichholzer, O., Aurenhammer, F.: Straight skeletons for general polygonal figures. In: Cai, J.-Y., Wong, C.K. (eds.) COCOON 1996. LNCS, vol. 1090, pp. 117–126. Springer, Heidelberg (1996)
2. Aichholzer, O., Aurenhammer, F., Alberts, D., Gärtner, B.: A novel type of skeleton for polygons. J. Univ. Comp. Sci. 1, 752–761 (1995)
3. Barequet, G., Goodrich, M.T., Levi-Steiner, A., Steiner, D.: Contour interpolation by straight skeletons. Graphical Models (GM) 66(4), 245–260 (2004)
4. Cheng, S.W., Vigneron, A.: Motorcycle graphs and straight skeletons. Algorithmica 47(2), 159–182 (2007)
5. Chin, F., Snoeyink, J., Wang, C.: Finding the medial axis of a simple polygon in linear time. Discr. Comp. Geom. 21(3), 405–420 (1999)

6. Demaine, E.D., Demaine, M.L., Lubiw, A.: Folding and cutting paper. In: Akiyama, J., Kano, M., Urabe, M. (eds.) JCDCG 1998. LNCS, vol. 1763, pp. 104–118. Springer, Heidelberg (2000)
7. Eppstein, D., Erickson, J.: Raising roofs, crashing cycles, and playing pool: applications of a data structure for finding pairwise interactions. Discr. Comp. Geom. 22(4), 569–592 (1999)
8. Tănase, M.: Shape decomposition and retrieval. Ph.D. Thesis, Utrecht Univ. (2005)
9. Tănase, M., Veltkamp, R.C.: Straight skeleton approximating the medial axis. In: Proc. 12th Annu. European Symp. on Algorithms, pp. 809–821 (2004)
10. Trofimov, V., Vyatkina, K.: Linear axis for general polygons: properties and computation. In: Gervasi, O., Gavrilova, M.L. (eds.) ICCSA 2007, Part I. LNCS, vol. 4705, pp. 122–135. Springer, Heidelberg (2007)
11. Vyatkina, K.: A Search for Medial Axes in Straight Skeletons. In: Proc. 24th European Workshop on Comp. Geom., Nancy, France, pp. 157–160 (2008)
12. Vyatkina, K.: Linear Axis for Planar Straight Line Graphs. In: Proc. 15th Computing: Australasian Theory Symp., CRPIT 1994, pp. 137–150 (2009)
13. Yap, C.K.: An $O(n \log n)$ algorithm for the Voronoi diagram of a set of simple curve segments. Discr. Comp. Geom. 2, 365–393 (1987)

Author Index